Yearbook on Space Policy

More information about this series at http://www.springer.com/series/8166

Yearbook on Space Policy

Edited by the
European Space Policy Institute

Director: Jean-Jacques Tortora

Cenan Al-Ekabi • Blandina Baranes •
Peter Hulsroj • Arne Lahcen
Editors

Yearbook on Space Policy 2015

Access to Space and the Evolution of Space Activities

Springer

Editors
Cenan Al-Ekabi
European Space Policy Institute
Vienna, Austria

Blandina Baranes
European Space Policy Institute
Vienna, Austria

Peter Hulsroj
European Space Policy Institute
Vienna, Austria

Arne Lahcen
European Space Policy Institute
Vienna, Austria

ISSN 1866-8305 ISSN 2197-9405 (electronic)
Yearbook on Space Policy
ISBN 978-3-7091-4859-4 ISBN 978-3-7091-4860-0 (eBook)
DOI 10.1007/978-3-7091-4860-0

Cover illustration: Artist's impression of Herschel, ESA - C. Carreau

Printed on acid-free paper

This Springer imprint is published by Springer Nature
The registered company is Springer-Verlag GmbH Austria
The registered company address is: Prinz-Eugen-Strasse 8-10, 1040 Wien, Austria

Preface

The market for launchers and launch services has developed rapidly over the last decade. More competitive markets, sectoral reforms, changing policies, and institutional changes are all affecting the playing field that determines our access to space. What is more, recent innovation dynamics, increasing private involvement, and a continuing trend of globalization will be impacting the launch sector and what we launch even more in the future. In the mid-term, it is expected that these changes will lead to significant cost reductions in bringing payloads to the different Earth orbits and even beyond. In order to clarify and shed additional light on these expected impacts of these trends, ESPI has decided to focus on the topic of access to space and the evolution of space activities from a wide variety of angles in its Yearbook on Space Policy 2015.

Traditionally, the first part of the Yearbook sets out a comprehensive overview of the economic, political, technological, and institutional trends that affected space activities in 2015. It is prepared in-house in ESPI and while its perspective is European, it also provides a comparative analysis of space developments around the world.

The second part of the ESPI Yearbook approaches the overall theme from an analytical perspective. This year it includes ten external contributions that bring together the views of eminent professionals and experts coming from different branches of the space sector. The first contributions in this part of the Yearbook offer a panoramic perspective on the elements that ultimately define how other space activities are affected in the long run. This starts with a contribution by CNES Director of Launchers Jean-Marc Astorg, who presents an overview of the state-of-the-art capabilities of the European launching sector and their outlook. Following this, Cristina Chaplain, Director in the U.S. Government Accountability Office, explains how the commercial space sector in the United States is currently affecting governmental space programs. Subsequently, the future potential of new space activities is explored in a contribution by Richard DalBello, Vice President Business Development and Government Affairs at Virgin Galactic. Subsequent contributions take the assumption of declining launch cost in the future a step further, as they assess the impact of falling launch costs on various aspects of space utilization. In this respect, Eurisy Secretary General Stefaan de Mey explores the future potential of space applications. This is followed by a contribution written by Professor Shuang-Nan Zhang of the Chinese Academy of Sciences on the potential impacts of what he coined "efficient access to space" on space astronomy and scientific progress. Leopold Summerer, Head of the ESA Advanced Concepts Team, offers an outlook on the evolution of other enabling and game-changing space technologies that might fundamentally impact the status quo of human activities in outer space. The final contributions in Part II of the Yearbook address the evolving needs in terms of governance, strategies, and approaches that will arise in the playing field of space anno 2030. Young researchers Nina Witjes (Scientific Researcher at the Munich Center for Technology in Society), Philipp Olbrich (PhD Student at the Rijksuniversiteit Groningen), and Isabella Rebasso (Trainee at the Austrian Institute for International Affairs) write about how future remote sensing capabilities might impact transparency building and create a new landscape for various societal activities. Professors James Schwartz (Department of Philosophy, Wichita State University, USA) and Tony Milligan (Department of Theology and Religious Studies, King's College London, UK) then present their views on the main ethical constraints on near-Earth resource exploitation in a wide sense. The European options to deal with the prospect of falling cost of utilizing space are explored by Jesse Phaler, Head of the Industrial Return Management Office at ESA. Finally, Staffordshire University professor and Head of Research at the British Royal Aeronautical Society Keith Hayward sheds light on the transition to a new business model in the launcher markets and its economic implications.

The third part of the Yearbook continues the character of the Yearbook as an archive of space activities. Again prepared in-house by ESPI, a bibliography, chronology, and data about institutions are provided where readers of the now nine volumes of the Yearbook can identify statistical developments and evolutions.

In closing, we would like to thank the contributors of the articles in Part Two for their engagement in this publication. Moreover, we are very grateful to Frances Brown, former editor-in-chief of Space Policy and current member of the ESPI Advisory Council, for her support and inspiration as we prepared the ESPI Autumn Conference 2015. The contributions in Part II of the Yearbook reflect the presentations made by the authors at the Autumn Conference.

Vienna, Austria

Cenan Al-Ekabi
Blandina Baranes
Peter Hulsroj
Arne Lahcen

Contents

Part III Facts and Figures

List of Acronyms: Acronym Explanation

#

21AT	Twenty-First Century Aerospace Technology Co.

A

A3R	Arkyd 3 Reflight
AAD	Advanced Air Defence
ABS	Asia Broadcast Satellite
ACS	Alcântara Cyclone Space
ADF	Aerospace Defence Forces
ADM	Atmospheric Dynamics Mission
AG	Aktiengesellschaft
AI	Artificial intelligence
AIA	Atmospheric Imaging Assembly
Airbus D&S	Airbus Defence and Space
AIS	Automatic Identification Satellites
AMS	Alpha Magnetic Spectrometer
ARM	Asteroid Redirect Mission
ASAT	Anti-Satellite
ASI	Agenzia Spaziale Italiana (Italian Space Agency)
ASL	Airbus Safran Launchers
ASNARO	Advanced Satellite with New System Architecture for Observation
AST	Office of Commercial Space Transportation
ATK	Alliant Techsystems Inc.
ATV	Automated Transfer Vehicle
AU	Astronomical unit

B

BAES	BAE SYSTEMS
BMD	Ballistic Missile Defence

C

CAA	Civil Aviation Authority
CAGR	Compound Annual Growth Rate
CALET	CALorimetric Electron Telescope
CALT	China Academy of Launch Vehicle Technology
CASC	China Aerospace Science and Technology Corp.
CAST	China Academy of Space Technology
CATS	Cheap Access To Space
CBERS	China-Brazil Earth Resources Satellite
CCME	Central Command for Maritime Emergencies
CD	Conference on Disarmament
CERN	Conseil Européen pour la Recherche Nucléaire (European Organization for Nuclear Research)
CGWIC	China Great Wall Industry Corporation
CHF	Swiss franc
CIRS	Composite Infrared Spectrometer
CLA	Centro de Lançamento de Alcântara (Alcântara Launch Center)
CME	Coronal Mass Ejections
CMSA	China Manned Space Agency
CNES	Centre National d'Études Spatiales (French Space Agency)
COP	Conference of the Parties
COPUOS	Committee on the Peaceful Uses of Outer Space
CPM	Conference Preparatory Meeting
CRS	Commercial Resupply Services
CSA	Canadian Space Agency
CSLCA	Commercial Space Launch Competitiveness Act

D

DAMPE	Dark Matter Particle Explorer
DARPA	Defense Advanced Research Projects Agency
DARS	Digital Audio Radio Service
DBS	Direct Broadcast Services
DG	DigitalGlobe
DHN	Digital Humanitarian Network
DLR	Deutsches Zentrum für Luft- und Raumfahrt (German Aerospace Center)
DoD	Department of Defence
DPAC	Data Processing and Analysis Consortium
DRDO	Defence Research and Development Organisation
DSI	Deep Space Industries
DTH	Direct To Home

E

EARSC	European Association of Remote Sensing Companies
EATS	Efficient Access to Space
EBITDA	Earnings Before Interest, Taxes, Depreciation and Amortization
EC	European Commission
ECA	Evolution Cryotechnique type A
ECB	European Central Bank
ECMWF	European Centre for Medium-Range Weather Forecasts
EDA	European Defence Agency
EDC	Export Development Canada
EDM	Entry, Descent and Landing Demonstrator Module
EDRS	European Data Relay Satellite System
EELV	Evolved Expendable Launch Vehicle
EFS	Equator-Facing Slopes
EGNOS	European Geostationary Navigation Overlay Service
EIP	European Innovation Partnership
ELIRG	Extremely luminous infrared galaxies
EMG	Environment Management Group
EMSA	European Maritime Safety Agency
EO	Earth Observation
EPS-SG	European Polar System Second Generation
ESA	European Space Agency
ESA DG	ESA Director General
ESOA	European Satellite Operators Association
EU	European Union
EUMETSAT	The European Organisation for the Exploitation of Meteorological Satellites
EUSC	European Union Satellite Centre
EUTELSAT	European Telecommunications Satellite Organisation
EVE	EUV Variability Experiment
Ex-Im Bank	U.S. Export-Import Bank

F

FAA	Federal Aviation Administration
FSS	Fixed Satellite Services
FWF	Austrian Science Fund
FY	Fiscal Year

G

GAO	Government Accountability Office
GCR	Galactic Cosmic Rays
GDP	Gross Domestic Product
GEO	Geostationary Earth Orbit

GEOSS	Group on Earth Observation System-of-Systems
GERD	Gross Domestic Expenditure on Research and Development
GLONASS	Globalnaya Navigatsionnaya Sputnikovaya Sistemya (Russian GNSS Constellation)
GmbH	Gesellschaft mit beschränkter Haftung
GNSS	Global Navigation Satellite Systems
GPS	Global Positioning System
GSLV	Geosynchronous Satellite Launch Vehicle
GSSAP	Geosynchronous Space Situational Awareness Program
GTO	Geosynchronous Transfer Orbit

H

He3	Helium-3
HMI	Helioseismic and Magnetic Imager
HST	Hubble Space Telescope
HTV	H-2 Transfer Vehicle

I

IAC	International Astronautical Congress
IAEA	International Atomic Energy Agency
IAP	Integrated Applications Promotion
ICC	International Criminal Court
ICG	International Committee on Global Navigation Satellite Systems
ICoC	International Space Code of Conduct for Outer Space Activities
ICS	Information and Communication Systems
ICT	Information and Communications technology
IGS	International GNSS Service
ILS	International Launch Services
IMF	International Monetary Fund
IMT	International Mobile Telecommunication
IMU	Inertial Measurement Unit
INDC	Intended Nationally Determined Contribution
IOV	In-Orbit Validation
IRIS	Interface Region Imaging Spectrograph
IRNSS	India Regional Navigation Satellite System
ISA	International Seabed Authority
ISIS	Islamic State of Iraq and Syria
ISL	International Launch Services
ISRO	Indian Space Research Organization
ISS	International Space Station
IT	Information Technology
ITAR	International Traffic in Arms Regulations
ITU	International Telecommunication Union

IUVS	Imaging Ultraviolet Spectrograph
IXV	Intermediate Experimental Vehicle

J

JAXA	Japan Aerospace Exploration Agency
JCPOA	Joint Comprehensive Plan of Action
JPS	Joint Polar System
JUICE	JUpiter ICy moon Explorer

K

K2	Kepler 2
KSA	Kazakh Space Agency

L

L2	Earth-Moon Lagrange point on the opposite side of the Moon
LEND	Lunar Exploration Neutron Detector
LEO	Low Earth Orbit
LHC	Large Hadron Collider
LOLA	Lunar Orbiter Laser Altimeter
LPW	Langmuir Probe and Waves
LRO	Lunar Reconnaissance Orbiter
LTS	Large Technical Systems

M

MATS	Mesospheric Airglow/Aerosol Tomography and Spectroscopy
MAVEN	Mars Atmosphere and Volatile EvolutioN (NASA)
MDA	MacDonald, Dettwiler and Associates Ltd.
Melco	Mitsubishi Electric Co.
MENCA	Mars Exospheric Neutral Composition Analyser
MEO	Medium Earth Orbit
MESSENGER	MErcury Surface, Space ENvironment, GEochemistry and Ranging
Metop	Meteorological Operational Satellite
Metop-SG	Metop Second Generation
MFF	Multiannual Financial Framework
MFG	Meteosat First Generation
MMO	Mercury Magnetospheric Orbiter
MOD	Ministry of National Defense
MOKV	Multi-object kill vehicle
MOM	Mars Orbiter Mission
MPCV	Multi-Purpose Crew Vehicle
MPO	Mercury Planetary Orbiter
MRO	Mars Reconnaissance Orbiter

MSG	Meteosat Second Generation
MSL	Mars Science Laboratory
MSM	Methane Sensor for Mars
MSS	Mobile Satellite Services
MTG	Meteosat Third Generation

N

NASA	National Aeronautics and Space Administration
NATO	North Atlantic Treaty Organisation
NDAA	National Defense Authorization Act
NEA	Near-Earth Asteroids
NEC	Nippon Electric Company
NEO	Near-Earth Object
NEOWISE	Reboot of the Wide-field Infrared Survey Explorer mission to find NEOs
NER	Near-Earth Resources
NGA	National Geospatial-Intelligence Agency
NOAA	National Oceanic and Atmospheric Administration
NRO	National Reconnaissance Office
NTER	Thermal and Electric Nuclear Power

O

OCX	GPS Next Generation Operational Control System
OECD	Organisation for Economic Co-operation and Development
OGLE	Optical Gravitational Lensing Experiment
OHB	Orbitale Hochtechnologie Bremen
OSM	OpenStreetMap
OSTP	Office of Science and Technology Policy
OPEC	Organization of the Petroleum Exporting Countries

P

PAD	Prithvi Air Defense
PCP	Pre-Commercial Procurement
PFS	Pole-Facing Slopes
PGM	Platinum-Group Metals
PHA	Potentially hazardous asteroids
PLA	People's Liberation Army
PND	Portable Navigation Devices
PPI	Public Procurement of Innovative solutions
PPP	Public-Private Partnership
PPWT	Treaty on Prevention of the Placement of Weapons in Outer Space and of the Threat or Use of Force Against Outer Space Objects
PROBA	PRoject for OnBoard Autonomy

PRS	Public Regulated Service
PSLV	Polar Satellite Launch Vehicle

Q

QZSS	Quasi-Zenith Satellite System

R

R&D	Research and Development
REID	Risk of Exposure Induced Death
RHESSI	Ramaty High Energy Solar Spectroscopic Imager
RKV	Redesigned kill vehicle
RLV	Reusable Launch Vehicle
Roscosmos	Russian Federal Space Agency
ROSINA	Rosetta's Orbiter Spectrometer for Ion and Neutral Analysis
RP15	Resource Protector
RRS	Regional Radiocommunication Seminar
rSLV	reusable Suborbital Launch Vehicles

S

SAFER	Simplified Aid For EVA Rescue
SAM	Sample Analysis at Mars instrument
SAR	Synthetic Aperture Radar
SBTF	Standby Volunteer Task Force
SDO	Solar Dynamics Observatory
SES	Société Européenne des Satellites
SIA	Satellite Industry Association
SLA	Service Level Agreement
SLI	Sustainable Land Imaging
SLS	Space Launch System
SM	Standard Missile
SME	Small and Medium-sized Enterprises
SNSB	Swedish National Space Board
SOHO	SOlar and Heliospheric Observatory
SPACE Act	Spurring Private Aerospace Competitiveness and Entrepreneurship Act
SpaceX	Space Exploration Technologies
SPE	Solar Particle Events
SS2	SpaceShipTwo
SS/L	Space Systems/Loral
SSO	Sun-synchronous orbit
SSTL	Surrey Satellite Technology Ltd.

STEM	Science, Technology, Engineering and Mathematics
STEREO	Solar TErrestrial RElations Observatory
SWF	Secure World Foundation

T

TCBM	Transparency and Confidence Building Measures
TGO	Trace Gas Orbiter
THEMIS	Thermal Emission Imaging System
TPP	Trans-Pacific Partnership
TRL	Technology Readiness Level

U

UAV	Unmanned Aerial Vehicle
UCAR	University Corporation for Atmospheric Research
UK	United Kingdom
ULA	United Launch Alliance
UN	United Nations
UNCTAD	United Nations Conference on Trade and Development
UNDSS	United Nations Department of Safety and Security
UNFCCC	United Nations Framework Convention on Climate Change
UNGA	United Nations General Assembly
UNGIWG	United Nations Graphic Information Working Group
UNIDIR	United Nations Institute for Disarmament Research
UNOOSA	United Nations Office for Outer Space Affairs
UNPSA	United Nations Programme on Space Applications
UNSC	United Nations Security Council
UNSDI	United Nations Spatial Data Infrastructure
UN-SPIDER	United Nations Platform for Space-based Information for Disaster Management and Emergency Response
URSC	United Rocket and Space Corporation
US	United States of America
USAT	Ultra Small Aperture Terminals
USMDA	Missile Defense Agency

V

VKO	Russian Aerospace Defence Forces
VMC	Venus Monitoring Camera
VSAT	Very Small Aperture Terminals

W

WGP	World Gross Product
WISE	Wide-field Infrared Survey Explorer
WRC	World Radiocommunication Conference

Part I
The Year in Space 2015

Chapter 1
European Space Activities in the Global Context

Cenan Al-Ekabi

1.1 Global Political and Economic Trends

1.1.1 Global Economic Outlook

The United Nations Annual Report "World Economic Situation and Prospects" reported a slight stumble in global growth by the end of 2015, as the persistent effects of the financial crisis in 2007 continued to discourage investment and global growth. In mid-2015, the growth of World Gross Product (WGP) was estimated to be 2.8 %, yet weak aggregate demand, falling commodity prices and increasing financial market volatility in major economies resulted in a year-end revision of WGP growth to 2.4 %.[1]

Developed economies contributed more to WGP growth, reaching a 1.9 % increase in global output in 2015; they will likely continue to pick up momentum in 2016, surpassing 2 %, which has not been seen since 2010. In the eurozone, new EU Members showed the most growth at 3.2 %, while Western European economies continued to be the main drivers of growth in the region with 1.8 % for 2015; and the European Union as a whole reached 1.9 % for 2015. U.S. growth in global output reached 2.4 % in 2015, in line with the 2.4 % in 2014, and is expected to contribute more in 2016. Japan's global output also increased by 0.5 % in 2015, a notable change from the 0.1 % decrease in 2014; Japan's GDP growth should reach 1.3 % in 2016.[2]

[1] "World Economic Situation and Prospects 2016." 30 Nov. 2015. United Nations 20 June 2016 <http://www.un.org/en/development/desa/policy/wesp/wesp_current/2016wesp_full_en.pdf>.

[2] Ibid. at 2.

C. Al-Ekabi (✉)
European Space Policy Institute, Vienna, Austria
e-mail: cenan.al-ekabi@espi.or.at

C. Al-Ekabi et al. (eds.), *Yearbook on Space Policy 2015*, Yearbook on Space Policy, DOI 10.1007/978-3-7091-4860-0_1

Growth slowed in developing and transition economies to 3.8 % and −2.8 % respectively in 2015, due to the earlier-mentioned headwinds (i.e. sharply lower commodity prices, large capital outflows, and increased financial market volatility). While China may have contributed to a reduction of growth in East and South Asia, the region will likely remain the fasted growing as many of the region's economies will benefit from importing low-cost oil, metals, and food commodities. In contrast, Russia and Brazil were mired in severe economic downturns, accompanied by elevated inflation, reaching −3.8 % and −2.8 % in 2015 respectively.[3]

Generally less restrictive fiscal and still accommodative monetary stances worldwide are likely to support WGP growth by 2.9 % in 2016 and 3.2 % in 2017. That increased growth is also predicated on the easing of downward pressures on commodity prices and on the anticipated pace of normalization of the United States monetary policy stance that should help to reduce policy uncertainties. Yet, amid the moderate pace of global growth and in an environment of weak investment, employment figures continue to fall short of closing the gap in the employment rate that had opened up during the global financial crisis. In developed economies, particularly in the euro area, the pattern of work has been shifting towards more part-time employment, which raises concerns about job security, working poverty, and low long-term earnings. Moreover, given the sharp economic slowdown and declining labour force participation in several economies in the developing world, masked by large informal sectors in these regions, employment opportunities in the developing world for 2015 are likely to have deteriorated. The persistence of these factors may undermine the United Nations 2030 Agenda for Sustainable Development, which focuses on promoting "inclusive and sustainable economic growth, employment and decent work for all".[4]

1.1.2 Political Developments

1.1.2.1 Geopolitics

A number of significant world events in 2015 remained unresolved by year's end.

While the United States sought to unseat Islamic State (ISIS) terrorists from occupied territory in Syria throughout 2015, Russia's surprise airstrike intervention starting in September 2015 worked to muddle the operation, as the U.S. targeted ISIS forces while Russian counterparts aimed at Syrian rebel groups seeking to oust Syria's President and long-time Russian ally, Bashar al-Assad. Russia's military operations were not coordinated with the U.S. and its allies which raised concerns of unintended confrontations, such as on 24 November when Turkish F-16s shot

---

[3]Ibid.

[4]"World Economic Situation and Prospects 2016." 30 Nov. 2015. United Nations 20 June 2016: 1–8 <http://www.un.org/en/development/desa/policy/wesp/wesp_current/2016wesp_full_en.pdf>.

down a Russian Su-24 fighter jet following repeated warnings not to fly over Turkish airspace.[5] As the campaign went on throughout the year, ISIS terrorists also managed to carry out attacks in France, Turkey, and in the United States.[6] France experienced its first shock at the beginning of 2015, when gunmen attacked its satirical magazine Charlie Hebdo offices and a Jewish supermarket in Paris on 7 January 2015.[7] On 13 November 2015, three suicide bombings took place outside the Stade de France stadium during a football match while other ISIS fighters attacked young concertgoers, resulting in a death toll of 130 people, with another 351 injured.[8] In Turkey, a suicide-bomber with reported links to ISIS killed a group of 32 youth activists on 22 July 2015,[9] while in the United States, a self-radicalized couple killed 14 people and wounded another 21 in California on 4 December 2015.[10]

In Europe, Greece's debt crisis came back into the spotlight at the beginning of 2015 with the election of Syriza party leader Alexis Tsipras, who pledged to renegotiate the terms of the 240 billion euros ($268 billion) bailout Greece received in 2010 in the wake of the financial crisis.[11] His anti-austerity position worried investors and euro-zone supporters, and following a series of negotiations with Europe's so-called 'Troika' (EU, IMF, and ECB) and a 4-month extension of the bailout to 30 June 2015, it became apparent that Greece might default on its June payment if it did not receive an infusion of cash from the final bailout instalment. The deadline was not met, and with Greek banks closed along with the threat of financial collapse and complete exit from the euro zone looming, Tsipras agreed to the Troika's conditions on measures, including taxes and pensions, along with intensive international oversight in order to qualify for the aid.[12] Yet even with its course correction, Greece's struggles with its bailout looked to be headed toward

[5]"Turkey's downing of Russian warplane – what we know." 1 Dec. 2015. BBC News 23 June 2016 <http://www.bbc.com/news/world-middle-east-34912581>.

[6]Lindsay, James M. "Top Ten Most Significant World Events in 2015." 15 Dec. 2015. Council on Foreign Relations 22 June 2016 <http://blogs.cfr.org/lindsay/2015/12/15/ten-most-significant-world-events-in-2015/>.

[7]"Charlie Hebdo attack: Three days of terror." 14 Jan. 2015. BBC News 23 June 2016 <http://www.bbc.com/news/world-europe-30708237>.

[8]Fuller, Jaime. "Paris Attacks Death Toll Rises to 130." 20 Nov. 2015. New York Magazine 23 June 2016 <http://nymag.com/daily/intelligencer/2015/11/paris-attacks-death-toll-rises-to-130.html>.

[9]"Suruc massacre: 'Turkish student' was suicide bomber." 22 July 2015. BBC News 23 June 2016 <http://www.bbc.com/news/world-europe-33619043>.

[10]Schmidt, Michael S., and Richard Pérez-Peña. "F.B.I. Treating San Bernardino Attack as Terrorism Case." 4 Dec. 2015. The New York Times 23 July 2016 <http://www.nytimes.com/2015/12/05/us/tashfeen-malik-islamic-state.html>.

[11]"Greece election: Syriza leader Tsipras vows to end austerity 'pain'." 26 Jan. 2015. BBC 22 June 2016 <http://www.bbc.com/news/world-europe-30978052>.

[12]Daley, Suzanne and Liz Alderman. "Premier of Greece, Alexis Tsipras, Accepts Creditors' Austerity Deal." 13 July 2015. The New York Times 22 June 2016 <http://www.nytimes.com/2015/07/14/world/europe/greece-debt-plan.html>.

another 'Grexit' scenario from the eurozone near the end of the year,[13] just as Europe was faced with a wave of refugees escaping the calamity in Syria and nearby regions.[14]

Early in 2015, an unprecedented influx of migrants began crossing the Mediterranean and travelling through the Balkans to enter Europe—sometimes at their peril.[15] In addition to violence in Iraq and Afghanistan, and abuses in Eritrea, the ongoing conflict in Syria became the biggest driver of migration in the year, amounting to a total of more than one million migrants coming by sea and land; by comparison, 280,000 migrants had entered Europe by sea and land in 2014.[16] More than 800,000 people crossed by sea from Turkey to Bulgaria and Greece—half of whom were Syrian asylum seekers—other migrants travelled by water to Spain and Italy from Morocco, Tunisia, Libya, and Egypt, while just 34,000 migrants crossed by land to Bulgaria or Greece travelling from Turkey.[17] European countries struggled with the influx, sparking divisions in the EU on how best to deal with resettling people; Germany alone received more than 476,000 asylum applications in 2015. As tensions rose in the EU due to the disproportionate burden placed on some of its member states, ministers had to find a way to relocate and integrate refugees throughout Europe; of the 292,540 asylum applications approved in 2015, 48.2 % were granted by Germany, 11.0 % Sweden, 10.1 % Italy, 7.1 % France, 5.6 % Netherlands, 4.8 % UK, with the remaining 13.3 % granted by other EU member states.[18]

Iran and the five permanent members of the United Nations Security Council[19] (UNSC) plus Germany (P5+1) reached a long awaited agreement on scaling back Iran's Nuclear Programme on 14 July 2015. In development since 2003, the agreement, known as the Joint Comprehensive Plan of Action (JCPOA), was endorsed by the UNSC by 20 July 2015.[20] Yet, concerns over the JCPOA's viability existed while the agreement underwent a 60-day Congressional Review Period in

[13]Khan, Robert. "Greece's Bailout Dead End." 9 Nov. 2015. Council on Foreign Relations 23 June 2016 <http://blogs.cfr.org/kahn/2015/11/09/greeces-bailout-dead-end/>.

[14]Lindsay, James M. "Top Ten Most Significant World Events in 2015." 15 Dec. 2015. Council on Foreign Relations 22 June 2016 <http://blogs.cfr.org/lindsay/2015/12/15/ten-most-significant-world-events-in-2015/>.

[15]"European leaders fret over porous borders." 12 Mar. 2015. CBS News 23 June 2016 <http://www.cbsnews.com/news/eu-borders-isis-migrants-refugees-smuggling-human-trafficking-libya/>.

[16]"Migrant crisis: Migration to Europe explained in seven charts." 4 Mar. 2016. BBC News 23 June 2016 <http://www.bbc.com/news/world-europe-34131911>.

[17]"Migrant crisis: One million enter Europe in 2015." 22 Dec. 2015. BBC News 23 June 2016 <http://www.bbc.com/news/world-europe-35158769>.

[18]"Migrant crisis: Migration to Europe explained in seven charts." 4 Mar. 2016. BBC News 23 June 2016 <http://www.bbc.com/news/world-europe-34131911>.

[19]Namely China, France, Russia, the United Kingdom, and the United States.

[20]Williams, Jennifer R. "A comprehensive timeline of the Iran nuclear deal." 21 July 2015. The Brookings Institution 22 June 2016 <http://www.brookings.edu/blogs/markaz/posts/2015/07/21-comprehensive-timeline-iran-nuclear-deal>.

the U.S., during which Republicans in both the House and Senate sought unsuccessfully to block the agreement through various motions.[21] In return for sanctions relief, Iran has agreed to give up 97 % of its stockpile of enriched uranium, cut its number of centrifuges by two-thirds, shut down a heavy water reactor, and allow onsite inspections by the International Atomic Energy Agency (IAEA), with some provisions lasting into 2040.[22]

China continued to rapidly form seven new islands, piling sand dredged from the ocean floor on what were previously coral reef beds in the Spratly Island region of the South China Sea. Those new islands are among a number of other nearby islands claimed by the Philippines, Malaysia, Vietnam, Brunei and Taiwan, and appear to be meant more for asserting China's territorial claims in the region, as demonstrated by sustained Chinese air and sea patrols of the area, than to support large military units.[23] China also claimed the 12 nautical miles surrounding each island as its exclusive territorial waters; a position China's neighbours contested, and the U.S. considers has no basis in international law.[24] At the heart of their concern is whether China will try to use the islands to choke off freedom of navigation in that strategic area where more than $5 trillion in trade passes through each year, while also containing rich fisheries and the potential for vast oil and mineral deposits.[25] That prospect seems particularly unsettling for the U.S. which has about $1.2 trillion in trade travelling through the South China Sea each year, and which also in October finally reached an agreement on the Trans-Pacific Partnership (TPP) trade deal with 11 other Pacific Rim nations (including Canada, Mexico, Peru, Chile, Japan, Vietnam, Malaysia, Brunei, Singapore, Australia, and New Zealand).[26] If approved by the U.S. Congress, the TPP would set trade rules that govern roughly 40 % of the global economy.[27]

---

[21]Demirjian, Karoun. "Senate rejects attempt to derail Iran deal in victory for Obama." 10 Sept. 2015. The Washington Post 22 June 2016 <https://www.washingtonpost.com/news/powerpost/wp/2015/09/10/senate-set-to-vote-on-iran-nuclear-deal/>.

[22]Lindsay, James M. "Top Ten Most Significant World Events in 2015." 15 Dec. 2015. Council on Foreign Relations 22 June 2016 <http://blogs.cfr.org/lindsay/2015/12/15/ten-most-significant-world-events-in-2015/>.

[23]Watkins, Derek. "What China Has Been Building in the South China Sea." 27 Oct. 2015. The New York Times 23 June 2016 <http://www.nytimes.com/interactive/2015/07/30/world/asia/what-china-has-been-building-in-the-south-china-sea.html>.

[24]Lindsay, James M. "Top Ten Most Significant World Events in 2015." 15 Dec. 2015. Council on Foreign Relations 22 June 2016 <http://blogs.cfr.org/lindsay/2015/12/15/ten-most-significant-world-events-in-2015/>.

[25]Glaser, Bonnie S. "Conflict in the South China Sea." 7 Apr. 2015. Council on Foreign Relations 23 June 2016 <http://www.cfr.org/asia-and-pacific/conflict-south-china-sea/p36377>.

[26]Calmes, Jackie. "Trans-Pacific Partnership Is Reached, but Faces Scrutiny in Congress." 5 Oct. 2015. The New York Times 23 June 2016 <http://www.nytimes.com/2015/10/06/business/trans-pacific-partnership-trade-deal-is-reached.html>.

[27]Lindsay, James M. "Top Ten Most Significant World Events in 2015." 15 Dec. 2015. Council on Foreign Relations 22 June 2016 <http://blogs.cfr.org/lindsay/2015/12/15/ten-most-significant-world-events-in-2015/>.

1.1.2.2 Environment

The 21st UN Framework Convention on Climate Change Conference of Parties (UN FCCC/COP), which took place in Paris, France from 30 November to 12 December 2015, reached a landmark agreement among 195 Parties on 12 December 2015.[28] The Paris Agreement aims to keep global average temperature increases to below 2 °C above pre-industrial levels, and to make more ambitious efforts to limit the temperature increases even further to 1.5 °C and eliminate the increase of greenhouse gas emissions in the second half of the century.[29] Around 188 countries contributed intended nationally determined contributions (INDCs), essentially national climate action plans that that will be updated and enhanced every 5 years—the next occurring in 2020. The members will also continue to address mitigation and adaptation opportunities, in addition to developing a clear roadmap for obtaining $100 billion in climate funding by 2020, while also setting another $100 billion floor before 2025.[30] Yet for now, the requirements of the Paris Agreement are non-binding until 30 days after the date on which at least 55 Parties accounting in total for at least an estimated 55 % of total global greenhouse gas emissions have deposited their instruments of ratification, acceptance, approval or accession[31]; reaching that goal will be challenging without having China, the U.S. or the EU on board, which together account for 45 % of the world's greenhouse emissions.[32] And while the regular review and submission of emission reduction targets and the $100 billion fund from developed economies to help emerging and developing nations decarbonise their energy mix will be binding, each country's INDC targets will not be binding.[33]

[28]"OUTCOMES OF THE U.N. CLIMATE CHANGE CONFERENCE IN PARIS | 21st Session of the Conference of the Parties to the United Nations Framework Convention on Climate Change (COP 21) November 30-December 12, 2015." 12 Dec. 2015. Centre for Climate and Energy Solutions 24 June 2016 <http://www.c2es.org/international/negotiations/cop21-paris/summary>.

[29]"The Paris agreement marks an unprecedented political recognition of the risks of climate change." 12 Dec. 2015. The Economist 24 June 2016 <http://www.economist.com/node/21683990/>.

[30]UN Climate Change Newsroom. "Historic Paris Agreement on Climate Change | 195 Nations Set Path to Keep Temperature Rise Well Below 2 Degrees Celsius." 12 Dec. 2015. UNFCCC 24 June 2016 <http://newsroom.unfccc.int/unfccc-newsroom/finale-cop21/>.

[31]United Nations Framework Convention on Climate Change | Conference of the Parties. Adoption of the Paris Agreement, Held in Paris from 30 November to 11 December 2015. UN Doc. FCCC/CP/2015/L.9/Rev.1 of 12 December 2015. United Nations (Annex | Paris Agreement, Article 21.1).

[32]Kinver, Mark. "COP21: What does the Paris climate agreement mean for me?" 14 Dec. 2015. BBC News 24 June 2016 <http://www.bbc.com/news/science-environment-35092127>.

[33]Ibid.

1.1.2.3 Energy

There were increasing signs of change in global energy throughout 2015, following a steep drop in oil prices at the beginning of the year, along with corresponding fluctuations in other fuel prices in many parts of the world. Some signs of stability emerged, as mandatory energy efficiency regulations grew to cover more than 25 % of global consumption, and renewable energy contributed to almost half of the world's new power generation capacity in 2014. Yet energy demand is expected to grow by one-third over the period to 2040, with that increase driven mainly by India, China, and other non-OECD countries in Africa, the Middle East, and Southeast Asia. In contrast, the European Union is expected to decrease its consumption by 15 % over the period to 2040, followed by Japan by 12 %, and the U.S. by 3 %.[34]

India is entering a period of rapid sustained growth, contributing around one quarter of the growth in global energy demand, including oil demand that was higher than any other country in 2015. Meanwhile, China has started to transition to a less energy-intensive phase in its growth, shifting from heavy industry to the service sector; it will be the largest oil-consuming country by the 2030s, and in 2040 its total energy demand will be nearly twice that of the United States. China and the Middle East were the drivers of natural gas demand growth in 2015, surpassing the European Union whose demand for natural gas peaked in 2010, yet the long-term supply of natural gas remains a concern, constrained by efficiency policies and deferred investment in the current low-price environment, which could bring tighter markets in the 2020s. Yet even with low oil prices, and the return of Iran into the hydrocarbon market, an estimated $2.8 trillion of investment will be needed to meet projected energy demand in 2040.[35]

China was both the world's largest producer and consumer of coal in 2015; it has also deployed more renewable power generation capacity than any other country. Moreover, India was the largest source of growth in global coal use in 2015, increasing the demand for coal in power generation and industry to almost half of the global energy mix for the year. India is now the second-largest coal producer in the world, and will overtake Japan, the EU and China as the world's largest coal importer by 2020. Yet renewable energy contributed nearly half of the world's new power generation capacity in 2014, and is projected to have a 50 % a share of generation in the EU, around 30 % in China and Japan, and above 25 % in the U.S. and India by 2040. In order to meet the anticipated one-third growth in world energy demand by 2040, more renewable power generation capacity needs to be added.[36]

---

[34]"International Energy Agency. World Energy Outlook 2015 – Executive Summary" 10 Nov. 2015. IAE 3 Dec. 2015 <http://www.iea.org/publications/freepublications/publication/WEB_WorldEnergyOutlook2015ExecutiveSummaryEnglishFinal.pdf>.

[35]Ibid.

[36]Ibid.

1.1.2.4 Resources

The growth of international trade remained lacklustre continuing at a rate of 2.3 % in 2014, from 2.2 % in 2013; while initial estimates for 2015 expected the rate of growth of international trade to be close to that of global output (2.5 %), this is still well below the 4.0 % growth posted in the years immediately preceding the financial crisis.[37] Developed economies experienced considerable growth in the volume of imports, with the EU and Japan increasing 2.8 %, while the U.S. had an even higher increase of 4.7 % in part due to the increasing value of the dollar. The volume of exports from developed countries also showed signs of acceleration, with the biggest uptick in Japan, whose export growth of 0.6 % seemed to recover from a slump in previous years. Europe's exports to China and the U.S. also showed resiliency at 1.5 % growth, while U.S. exports have been showing signs of deceleration in recent years hovering at 3.1 %. By contrast, import volumes in transition economies dropped by 8.5 % in 2014, and were anticipated to have contracted even further in 2015, due mainly to the economic and financial difficulties involving Russia and Ukraine; and export volumes growth of 0.2 % showed almost no change from 2013. Developing economies also continued to show deceleration, of which only South Asia departed from the downward trend. Overall, developing countries' growth in volume of imports for 2014 slowed to 2.0 % from 6.1 % in 2013; their volume of exports had a similar outcome in 2014 growing by 2.9 % from 4.2 % in 2013.[38]

By the start of 2015, crude oil prices had dropped to $48 per barrel, dropping by 56.7 % from a monthly average of $112 per barrel as at June 2014—lows not seen since 2009. According to the United Nations Conference on Trade and Development (UNCTAD), the plunge in prices was caused by greater global production, particularly by the United States' 15.9 % increase in shale oil production in contrast to 2.3 % increase by the rest of the globe in 2014. The U.S. had increased production by 50.6 % from 2011 until mid-2014, which helped to stabilize global oil prices during production disruptions in other countries, but this led to an oversupply when disruptions became less of a problem. Prices began to fall even quicker following OPEC's decision not to change production quotas in November 2014; that decision was upheld in OPEC's following meeting in June 2015, in effort to undercut higher cost—shale oil, tar sands, and deep-water oil producers. While prices increased to $65 per barrel between the end of April and June 2015, by the end of 2015 the price of a barrel of oil dropped further to $37.28 per barrel.[39]

---

[37]According to the latest data available from UNCTAD.

[38]United Nations Conference on Trade and Development. Trade and Development Report, 2015. Geneva: UNCTAD, 2015. 1–18.

[39]Friedman, Nicole. "U.S. Oil Prices End 2015 Down 30 % for the Year." 31 Dec. 2015. The Wall Street Journal 27 June 2016 <http://www.wsj.com/articles/oil-prices-rise-but-supply-glut-caps-gains-1451560147>.

The decline in oil prices also had an effect on the prices of non-oil commodities, such as in reducing transportation and fertilizer prices, in addition to becoming more competitive with biofuels; however, prices in agriculture markets were still mainly determined by their own supply situation and weather conditions. Favourable weather conditions in 2014 and the first half of 2015 led to ample levels of food commodities, but adverse conditions in mid-2015 and the return of the El-Niño phenomenon raised levels of uncertainty in agricultural markets. The percentage change of non-oil commodity prices continued to contract, decreasing by 6.1 % in 2014, following a 6.7 % reduction in 2013; it also showed a 13.1 % reduction for the first half of 2015.[40] The plentiful supply of agricultural products resulted in production exceeding consumption, with prices driven down due to weak demand and excess stock. The price of metals and minerals also continued to contract, decreasing by 8.5 % in 2014, following a 5.1 % reduction in 2013; it also showed a 15.81 % reduction for the first half of 2015. Zinc, Nickel, and to a lesser extent Aluminium, were the outliers in 2014, showing positive growth in prices with 13.2 %, 12.3 %, and 1.1 % respectively; but had joined the throng of metals and minerals whose prices decreased in 2015.[41] The indexes for gold, silver, and platinum continued their downward trend in 2015, decreasing by 8.3 %, 17.6 %, and 23.9 % respectively in 2015; they are expected to drop even further by 7.3 %, 10.9 %, and 9.8 % in 2016 mainly due to reduced investment demand.[42]

1.1.2.5 Knowledge

By now the advantages of higher education should be seen as worth the effort as employment rates and earnings tend to increase as an adult's level of education and skills increases; moreover, the labour market still regards a diploma or degree as the primary indication of a worker's skills. For Europe, the expansion of its pool of highly skilled and specialised scientists and professionals should be a constant priority if it is to remain a leading actor in the field of space-related scientific and technological R&D. By 2015, 32 % of the European working age population (not including Bulgaria, Cyprus, Malta, Latvia, Lithuania, Romania, and Croatia) held a higher education degree compared to 44 % in the U.S. and 54 % in Canada.[43] Several European countries, including Belgium, Denmark, Estonia, Finland, Ireland, Luxembourg, Norway, Sweden, and Switzerland performed well above the OECD average of 33 % of the working age population (between 25-to-64 years of age) with a higher

[40]United Nations Conference on Trade and Development. Trade and Development Report, 2015. Geneva: UNCTAD, 2015. 12.

[41]Ibid.

[42]World Bank Group. Commodity Markets Outlook. January 2016. Washington, DC: World Bank, 2016: 30–35.

[43]OECD. Education at a Glance 2015: OECD Indicators, OECD Publishing, 2015: 41 <http://download.ei-ie.org/Docs/WebDepot/EaG2015_EN.pdf>.

education degree. However, when looking at the percentage of the working age population (between 25-to-34 years of age) with a higher education degree, other European countries, including the Czech Republic, Germany, Hungary, Italy, Portugal, and Slovakia appeared at risk of falling further behind the that OECD average.[44]

According to the OECD, over 80 % of tertiary-educated adults are employed, compared to over 70 % of people with upper secondary or post-secondary non-tertiary education, and less than 60 % of adults without upper secondary education.[45] According to an earlier study by the European Centre for the Development of Vocational Training, while employment is projected to grow by 2 % over the period 2013–2020, the European skills forecast indicates that the EU will miss its target of reaching 75 % employment by 2020, reaching instead 68.5 % due to weak demand for labour following the financial crisis and the modest recovery and average GDP growth rates. Prior to the financial crisis, between 2000 and 2008, the employment rate increased by 3.7 %. Between 2013 and 2020, average employment growth in Europe is expected to be around 0.35 % per year, wherein the number of high qualification job openings is expected to exceed 55,000 (combining replacement needs and expansion demand), whereas medium qualification openings will be over 40,000 (coming mainly from replacement needs), and low qualification jobs will be lower than 10,000 (due to shrinking demand offsetting replacement needs). In fact, the overall share of the labour force with low qualifications is forecast to fall from 22 % in 2013 to 16.8 % in 2020. The largest relative growth of jobs between 2013 and 2020 will be in Luxembourg, followed by Ireland, Slovakia, Latvia, and France in the top five positions, while only Hungary, Poland, Cyprus, Bulgaria, and Germany are expected to have a decrease in job growth during that period. And while the EU is on course to surpass its education goal of 40 % of 30–34 year-olds having completed higher education by 2020, possibly reaching 45 % by that time, weak high-skilled labour demand could mean that there will be an increased risk of skill mismatch due to the over-qualification of highly skilled labour that has no alternative than to accept lower skilled employment.[46]

1.1.2.6 Mobility

Maritime transport is the most commonly used form of transport for international trade, accounting for about 80 % of global trade by volume (90 % volume in the case of most developing countries) and over 70 % of global trade by value.[47]

[44]OECD. Education at a Glance 2015: OECD Indicators, OECD Publishing, 2015: 41 <http://download.ei-ie.org/Docs/WebDepot/EaG2015_EN.pdf>.

[45]OECD. Education at a Glance 2015: OECD Indicators, OECD Publishing, 2015: 27 <http://download.ei-ie.org/Docs/WebDepot/EaG2015_EN.pdf>.

[46]"Cedefop's latest skill supply and demand forecasts highlight Europe's employment challenge." 1 Apr. 2014. CEDEFOP 7 May 2015 <http://www.cedefop.europa.eu/node/12684>.

[47]United Nations Conference on Trade and Development. Review of Maritime Transport 2015. Geneva: UNCTD, 2015. 48.

Growth in global trade and shipping proceeded at a moderate pace in 2014 and was expected to continue into 2015; however, the risk of continued moderate growth in global demand and merchandise trade, the fragile recovery in Europe, diverging outlooks for net oil consumers and producers, geopolitical tensions, and a potential faster slowdown in developing economies, kept that outlook uncertain. Between 2014 and 2015, the world order book declined for most vessel types except for oil tankers, and stood far below the order peak of 2008–2009. During the 12 months to 1 January 2015, the global fleet of vessels increased by 3.5 %, the lowest annual growth rate in over a decade. In terms of the types of vessels in operation for the year, dry-bulk carriers accounted for 43.5 % of the world fleet capacity, increasing at a rate of 4.4 % from the previous year. Oil tankers accounted for 28.0 % of the share of the world fleet, with only a 1.4 % increase from the previous year. Container vessels accounted for a 13.0 % share, with a substantial increase of 5.5 % in 2015, while general cargo vessels continued to diminish, accounting for a 4.4 % share, down by 1.0 % from 2014. Additionally, the number of ships sold for demolition decreased by 22.9 % to 22,394 in 2014 from 29,052 in 2013. In 2014, dry-bulk carriers accounted for 40.6 % of the total quantity of tonnage sold for demolition, followed by container ships at 21.3 %, and oil tankers at 20.0 %. As the rate of vessel demolitions decreases, a resulting oversupply of vessels may lead to downward pressure on freight and charter rates.[48]

Supply chain security is another challenge for the maritime industry, as there is heightened exposure and vulnerability to piracy, armed robbery, and other crimes. Between 2003 and 2012, around 3436 acts of piracy were reported worldwide.[49] In that period, piracy incidents in East Africa, particularly off the Somalia coast, the Gulf of Aden and the Indian Ocean were particularly alarming, especially since they exceeded those in some of the traditional regions for piracy including Indonesia, Malaysia, Singapore, and the Philippines. Indeed, between 2005 and 2012, piracy off the coast of Somalia accounted for nearly 50 % of all recorded hijackings. This spurred an increased international military presence in this region and together with preventative measures by merchant ships that privately contracted armed personnel, reduced Africa's share of piracy incidents from 50 % in 2008 to 17.3 % in 2012.[50] The downward trend continued into 2014, with just 245 reported piracy incidents during the year. Moreover, the number of Somalia-based piracy incidents dropped from 49 in 2012 to 3 in 2014, although it continued to increase in East Asian waters.[51]

---

[48]United Nations Conference on Trade and Development. Review of Maritime Transport 2015. Geneva: UNCTD, 2015. 29–45.

[49]United Nations Conference on Trade and Development. Maritime Piracy | Part 1: An Overview of Trends, Costs and Trade-related Implications. Geneva: UNCTAD, 2014. 3.

[50]Ibid.

[51]ICC International Maritime Bureau. Piracy and Armed Robbery Against Ships: Report for the Period 1 January–31 December 2014. Jan. 2015. ICC-CSS 11 May 2015 <http://www.hellenicshippingnews.com/wp-content/uploads/2015/01/2014-Annual-IMB-Piracy-Report-ABRIDGED.pdf>.

1.1.3 Main Science and Technology Indicators Relevant for Space Activities

1.1.3.1 Science and Technology Inputs

Science and technology inputs at the European level are constantly measured against the benefits they generate. In 2015, R&D investment continued to push for increased efficiency and effectiveness and stronger R&D integration in broader industrial and macro-economic policies. The result in the reporting period is measured by gross domestic expenditure on R&D (GERD), a statistical tool showing nominal changes in those expenditures.

According to Eurostat, overall GERD as a percentage of GDP spent by the 28 EU countries was provisionally assessed to be 2.03 % in 2014, about level with the 2.03 % recorded in 2013.[52] The EU's GDP grew by 4.9 % to 14.635 trillion euros in 2015 at current prices, up from 13.958 trillion euros in 2014.[53] The EU's performance trails Japan and South Korea whose data is limited to 2013, and the United States, whose data is limited to 2012. Japan's GERD rebounded to 3.47 % in 2013, amounting to 128.645 billion euros (¥16.680 trillion), from a drop to 3.34 % or 154.977 billion euros (¥15.945 trillion) in 2012. In contrast, South Korea's GERD as a percentage of GDP has steadily increased in recent years, reaching 4.15 % or 40.787 billion euros (₩59.300 trillion). Updated U.S. values were not available on Eurostat, although its 2012 GERD as a percentage of GDP was 2.81 %, or around 353.007 billion euros ($453.544 billion).

On the European national level, according to the latest available figures, the top three countries' GERD as a percentage of GDP in 2014 remained Finland (3.17 %), Sweden (3.16 %), and Denmark (3.05 %), followed closely by Austria (2.99 %) and Germany (2.87 %). However, it should be noted that in nominal terms, Germany spent the most on GERD in 2014, reaching 83.636 billion euros, followed by France (48.108 billion euros), and the United Kingdom (38.323 billion euros). Moreover, the number of central and eastern EU member states with expenditure below 1 % dropped to 9 from 10 in 2014.[54]

A further breakdown of statistical data helps to underscore the impact of private R&D investment in Europe, when paired to publicly funded projects. In 2014, innovation investment in countries with higher GERD as a percentage of GDP tended to come from the private enterprise sector, the top five being

---

[52]"Gross domestic expenditure on R&D (GERD) - % of GDP." Eurostat 21 Apr. 2016 <http://ec.europa.eu/eurostat/tgm/table.do?tab=table&init=1&language=en&pcode=t2020_20&plugin=1>.

[53]"Gross domestic product at market prices | At current prices." Eurostat 21 Apr. 2016 <http://ec.europa.eu/eurostat/tgm/refreshTableAction.do?tab=table&plugin=1&pcode=tec00001&language=en>; "Total intramural R&D expenditure (GERD) by sectors of performance – per Million euro." 31 Mar. 2016. Eurostat 21 Apr. 2016 <http://appsso.eurostat.ec.europa.eu/nui/submitViewTableAction.do>.

[54]"Gross domestic expenditure on R&D (GERD) - % of GDP." Eurostat 21 Apr. 2016 <http://ec.europa.eu/eurostat/tgm/table.do?tab=table&init=1&language=en&pcode=t2020_20&plugin=1>.

Finland (2.15 %), Sweden (2.12 %), Austria (2.11 %), Germany (1.95 %), and Denmark (1.95 %). Those countries also led in the amount put toward higher education research. Overall, Europe's private sector funded R&D accounted for 64.0 % of the EU's total GERD in 2014, whereas private sector funded R&D in China, Japan, and South Korea accounted for 76.9 %, 76.1 %, and 78.6 % respectively in 2013; United States private sector funded R&D was at 69.8 % in 2012.[55]

1.1.3.2 Science and Technology Outputs

In contrast to R&D input, the outputs achieved from investments in innovation are harder to measure, due to the complex market dynamics within the private sector. It should be noted that statistics on the number of patents lodged by country, industrial sector, and individual companies are merely indicative of output—i.e. their effect on financial performance is also heavily dependent on external market competition and other factors. As observed in recent years, the recovery of R&D investment levels by companies based in the EU has been slow, whereas U.S. companies appear more willing to return to the high R&D investment levels experienced prior to the financial crisis.

Based on the European Commission's 2015 EU Industrial R&D Investment Scoreboard, in 2014/2015 approximately 2500 companies worldwide each invested more than 18 million euros in R&D with an average growth rate that was significantly higher than the growth rate of their revenues. Taken together, these leading companies' investment in R&D reached 607.2 billion euros in 2014; an increase of 6.8 % from 2013. European companies accounted for 28.1 % (170.9 billion euros) of that figure (an increase of 3.3 % from 2013), while U.S. companies accounted for 38.2 % (232.2 billion euros, up 8.1 %), and Japanese companies accounted for 14.3 % (86.6 billion euros, an increase of 2.6 %). However, companies in the rest of the world showed the highest increases in R&D investment, i.e. companies in China increased their R&D investments by 23.6 % in 2014, followed by Taiwan (12.4 %), and South Korea (10.6 %). Increases in EU performance is largely attributed to German company investment in the Automobiles & Parts sector; Japan also performed well in the Automobiles & Parts sector. R&D investment growth in U.S. companies came from Pharmaceuticals & Biotechnology, Software & Computer Services, and the Technology Hardware & Equipment sector. Moreover, competitors in China also showed outstanding results in the IT-hardware and software sectors.[56]

---

[55]Ibid.

[56]European Commission. "EU R&D Scoreboard | The 2015 EU Industrial R&D Investment Scoreboard." 12 Dec. 2015. EU R&D Scoreboard 22 Apr. 2016 <http://iri.jrc.ec.europa.eu/scoreboard15.html>.

The uneven benefits of corporate innovation on a European scale are further demonstrated in the European Commission's biennial competitiveness assessment conferences, CONCORDi. The fifth CONCORDi meeting, held on 1–2 October 2015, discussed the role of scientific analyses in the design and implementation of an EU industrial policy based on research, innovation and technological advancement. The conference stressed the importance of grounding policy-making in a sound scientific and academic basis, and highlighted challenges in bridging the gap that exists between academic and policy circles and perspectives. The main conclusions of the conference on developing a science-policy interface found that young innovative EU firms and start-ups face entry barriers that still need to be fully identified to ensure an effective shift towards a more knowledge-intensive and innovation-oriented European industry. Another area where further research is needed is in properly identifying the type and intensity of innovation targeted by policy instruments to ensure their effectiveness. And more regular exchanges between science and policy-makers are required to provide better information on the conditions for policy implementation.[57]

1.2 Worldwide Space Policies and Strategies

1.2.1 The United Nations System

Various institutions within or associated with the United Nations are relevant for space policy. In this section, the UN General Assembly, its Committees, and other UN bodies and organs that deal with space activities are discussed.

1.2.1.1 United Nations General Assembly

The United Nations General Assembly (UNGA) passed five Resolutions in its 70th (2015–2016) session relating to the use and exploration of outer space. The first was Resolution A/RES/70/26 adopted on 7 December 2015, on the 'Prevention of an arms race in outer space'.[58] The second space related resolution, adopted on the same day, was Resolution A/RES/70/27 on 'No first placement of weapons in outer

---

[57] Dosso, Mafini, Petros Gkotsis, Fernando Hervás, and Pietro Moncada-Paternò-Castello. "Industrial Research and Innovation: Evidence for Policy." Nov. 2015. European Commission – JRC Policy Brief 22 Apr. 2016 <http://iri.jrc.ec.europa.eu/documents/10180/12238/Industrial%20Research%20and%20Innovation.%20Evidence%20for%20Policy>.

[58] United Nations General Assembly. Resolution adopted by the General Assembly – Prevention of an arms race – Seventieth Session. UN Doc. A/RES/70/26 of 11 Dec. 2015. Vienna: United Nations <http://www.un.org/en/ga/search/view_doc.asp?symbol=A/RES/70/26>.

space'.[59] The third resolution, adopted on 7 December 2015 without a vote, was Resolution A/RES/70/53 on 'Transparency and confidence-building measures in outer space activities'.[60] On 9 December 2015, the UNGA adopted without a vote its annual Resolution A/RES/70/82 on "International cooperation in the peaceful uses of outer space".[61] And on 23 December 2015, the UNGA adopted without a vote its Resolution A/RES/70/230 on 'Matters relating to activities under the United Nations Programme on Space Applications in 2016'.[62]

First, Resolution A/RES/70/26 on the 'Prevention of an arms race in outer space' repeated the positions of the General Assembly in previous years' resolutions. The instrument reaffirmed the importance of transparency and confidence-building measures (TCBM) to avoid an arms race in space and additionally recognised, as stated in the report of the Ad Hoc Committee on the Prevention of an Arms Race in Outer Space, that the legal regime applicable to outer space by itself does not guarantee the prevention of an arms race in outer space, whilst acknowledging that the regime plays a significant role in the prevention of an arms race in that environment, and that there is a need to consolidate and reinforce the regime and enhance its effectiveness. It also recognized the importance of strict compliance with existing bilateral and multilateral agreements. The Conference on Disarmament (CD) was seen as the relevant multilateral disarmament forum for further discussions, and the instrument noted the updated 2014 version of the draft of the Treaty on Prevention of the Placement of Weapons in Outer Space and of the Threat or Use of Force Against Outer Space Objects (PPWT), as introduced by China and the Russian Federation during the CD. In general, it also recognised the complementary nature of multilateral and bilateral efforts in this area.[63]

The second Resolution A/RES/70/27 on 'No first placement of weapons in outer space' expanded on its previous years' Resolution A/RES/69/32, highlighting

---

[59]United Nations General Assembly. Resolution adopted by the General Assembly – No first placement of weapons in outer space – Seventieth Session. UN Doc. A/RES/70/27 of 11 Dec. 2015. Vienna: United Nations <http://www.un.org/en/ga/search/view_doc.asp?symbol=A/RES/70/27>.

[60]United Nations General Assembly. Resolution adopted by the General Assembly – Transparency and confidence-building measures in outer space activities – Seventieth Session. UN Doc. A/RES/70/53 of 11 Dec. 2015. Vienna: United Nations <http://www.un.org/en/ga/search/view_doc.asp?symbol=A/RES/70/53>.

[61]United Nations General Assembly. Resolution adopted by the General Assembly – International cooperation in the peaceful uses of outer space – Seventieth Session. UN Doc. A/RES/70/82 of 15 Dec. 2015. Vienna: United Nations <http://www.un.org/en/ga/search/view_doc.asp?symbol=A/RES/70/82>.

[62]United Nations General Assembly. Resolution adopted by the General Assembly – Matters relating to activities under the United Nations Programme on Space Applications in 2016 – Seventieth Session. UN Doc. A/RES/70/230 of 30 Dec. 2015. Vienna: United Nations <http://www.un.org/en/ga/search/view_doc.asp?symbol=A/RES/70/230>.

[63]United Nations General Assembly. Resolution adopted by the General Assembly – Prevention of an arms race – Seventieth Session. UN Doc. A/RES/70/26 of 11 Dec. 2015. Vienna: United Nations <http://www.un.org/en/ga/search/view_doc.asp?symbol=A/RES/70/26>.

additional concern about the possibility of outer space turning into an arena for military confrontation. It once again welcomed the draft treaty on the prevention of the placement of weapons in outer space and of the threat or use of force against outer space objects introduced by China and the Russian Federation at the CD in 2008, and updated in 2014, and noted the importance of the political statements made by several States (i.e. Argentina, Armenia, Belarus, Brazil, Cuba, Indonesia, Kazakhstan, Kyrgyzstan, the Russian Federation, Sri Lanka, Tajikistan and Venezuela) that they would not be the first to place weapons in outer space. In addition to urging an early start of substantive work based on the updated draft treaty, it encouraged all states, especially space-faring nations, to consider the possibility of making a political commitment not to be the first to place weapons in outer space, and decided to include this item in the provisional agenda of its next session.[64]

Building on the previous years' resolution, Resolution A/RES/70/53 on 'Transparency and confidence-building measures in outer space activities' reaffirmed the right of free exploration and use of outer space by all states in accordance with international law, and that the prevention of an arms race in outer space was in the interest of maintaining international peace and security, in addition to fostering international cooperation in the exploration and use of outer space for peaceful purposes. It went on to note that the UN COPUOS had a fundamental role to play in enhancing transparency and confidence-building among states, as well as in ensuring that outer space was maintained for peaceful purposes. It also noted a request by the Committee for a special report by the Inter-Agency Meeting on Outer Space Activities (UN-Space) for consideration at its 59th session, in 2016, on the implementation of the report of the Group of Governmental Experts on Transparency and Confidence-building Measures in Outer Space Activities. And it welcomed resolution 186 of 7 November 2014 on strengthening the role of the International Telecommunication Union with regard to transparency and confidence-building measures in outer space activities. The resolution also welcomed the joint ad hoc meeting of the Disarmament and International Security Committee (First Committee) and the Special Political and Decolonization Committee (Fourth Committee) on possible challenges to space security and sustainability"; on 22 October 2015, convened in accordance with the report, and the substantive exchange of opinions on various aspects of security in outer space that took place during the meeting. It called upon Member States and the relevant entities and organizations of the United Nations system to support the implementation of the full range of conclusions and recommendations contained in the report.[65]

[64]United Nations General Assembly. Resolution adopted by the General Assembly – No first placement of weapons in outer space – Seventieth Session. UN Doc. A/RES/70/27 of 11 Dec. 2015. Vienna: United Nations <http://www.un.org/en/ga/search/view_doc.asp?symbol=A/RES/70/27>.

[65]United Nations General Assembly. Resolution adopted by the General Assembly – Transparency and confidence-building measures in outer space activities – Seventieth Session. UN Doc. A/RES/70/53 of 11 Dec. 2015. Vienna: United Nations <http://www.un.org/en/ga/search/view_doc.asp?symbol=A/RES/70/53>.

Resolution A/RES/70/82 on 'International cooperation in the peaceful uses of outer space' updated the resolution from 2014. Updates to this year's resolution included it noting with satisfaction the establishment of and work carried out by the International Asteroid Warning Network and the Space Mission Planning Advisory Group to implement the recommendations for an international response to the near-Earth object impact threat. It also noted UNCOPUOS's plans for the 50th anniversary of the United Nations Conference on the Exploration and Peaceful Uses of Outer Space in 2018 (UNISPACE+50), and welcomed the establishment of the steering committee for the preparations for UNISPACE+50. The resolution also welcomed the adoption of the Sendai Framework for Disaster Risk Reduction 2015–2030, which recognizes the value of space-based technology and Earth observation for disaster management and emergency response, and noted with satisfaction the efforts of the United Nations Office for Outer Space Affairs (UNOOSA) and its UN-SPIDER programme towards promoting international cooperation as a way to enhance the use of space-based technologies and related services at the national and local levels.[66]

Resolution A/RES/70/230 on 'Matters relating to activities under the United Nations Programme on Space Applications in 2016' addressed issues that were left unresolved at the end of the 58th session of the UNCOPOUS held in Vienna from 10 to 19 June 2015. Because that meeting did not reach an agreement enabling the implementation of the United Nations Programme on Space Applications (UNPSA) for the remainder of 2015, the UNGA agreed to reinstate the following activities under the UNPSA's ambit to be carried out in 2016 with the same budget as had been allocated to them in 2015: 'United Nations/Costa Rica Workshop on Human Space Technology'; 'United Nations/South Africa Symposium on Basic Space Technology'; 'United Nations/Kenya Workshop on Space Technology and Applications for Wildlife Management and Protecting Biodiversity'; and 'United Nations/Islamic Republic of Iran Workshop on the Use of Space Technology for Dust Storm and Drought Monitoring in the Middle East Region'.[67]

1.2.1.2 UNGA Committees

Several UNGA committees address space policy and associated matters. The main ones are discussed below.

---

[66]United Nations General Assembly. Resolution adopted by the General Assembly – International cooperation in the peaceful uses of outer space – Seventieth Session. UN Doc. A/RES/70/82 of 15 Dec. 2015. Vienna: United Nations <http://www.un.org/en/ga/search/view_doc.asp?symbol=A/RES/70/82>.

[67]United Nations General Assembly. Resolution adopted by the General Assembly – Matters relating to activities under the United Nations Programme on Space Applications in 2016 – Seventieth Session. UN Doc. A/RES/70/230 of 30 Dec. 2015. Vienna: United Nations <http://www.un.org/en/ga/search/view_doc.asp?symbol=A/RES/70/230>.

The Disarmament and International Security Committee

The Disarmament and International Security Committee (First Committee) works in close cooperation with the United Nations Disarmament Commission and the Geneva-based Conference on Disarmament to deal with disarmament, global challenges and threats to peace that affect the international community, and seeks solutions to the challenges in the international security regime, including the prevention of an arms race in outer space.[68] On 6 November 2015, the First Committee forwarded 57 draft resolutions and decisions to UNGA for adoption, with 23 documents focussing on nuclear weapons. Additional thematic areas included other weapons of mass destruction, outer space (disarmament aspects), conventional weapons, other disarmament measures and international security, and disarmament machinery.[69] On 7 December 2015, the UNGA adopted all 57 documents, including a new consideration on the 'ethical imperatives' for nuclear disarmament, which had a recorded vote of 132 in favour to 36 against, with 16 abstentions. However, there still remained some divergence in views on how to move forward, such as with draft II on a convention on the prohibition of the use of nuclear weapons under any circumstances, whose recorded vote was 130 in favour to 48 against, with 8 abstentions; another document, draft XI resolution on a follow-up to nuclear disarmament obligations agreed to at the 1995, 2000 and 2010 Review Conferences of the Parties to the Treaty on the Non-Proliferation of Nuclear Weapons had a recorded vote of 121 in favour to 48 against, with 12 abstentions. There was a 40 % increase in delegate participation in debates this year, with the documents requiring 41 recorded votes in all (including on separate provisions).[70]

The Committee on the Peaceful Uses of Outer Space (COPUOS)

The UN COPUOS's activities included its 58th plenary session on 10–19 June 2015, along with the 52nd session of its scientific and technical subcommittee held from 2 to 13 February 2015, and the 54th session of its legal subcommittee from 13 to 24 April 2015. At the plenary session, the Committee acknowledged the importance attached by the international community to promoting cooperation on protecting the space environment and securing the long-term sustainability of outer space activities. It also received mixed views regarding the proposed draft of the International Code of Conduct for Outer Space Activities (ICoC) prior to the next ICoC meeting that was held in New York from 27 to 31 July 2015. The Committee also endorsed the findings and recommendations of its subcommittees from their

---

[68] "Disarmament and International Security." General Assembly of the United Nations 18 Apr. 2016 <http://www.un.org/en/ga/first/index.shtml>.

[69] "First Committee Sends Last of 57 Drafts to General Assembly at Close of Session After Record Participation in Thematic Debates." 6 Nov. 2015. UN 18 Apr. 2016 <http://www.un.org/press/en/2015/gadis3542.doc.htm>.

[70] "On Recommendation of First Committee, General Assembly Adopts More than 50 Drafts, Including New One on 'Ethical Imperatives' for Nuclear Disarmament." 7 Dec. 2015. UN 18 Apr. 2016 <http://www.un.org/press/en/2015/ga11735.doc.htm>.

preceding sessions, and considered space and sustainable development; the spin-off benefits of space technology (review of current status); space and water, and space and climate change; the use of space technology in the United Nations system; the future role of the committee; and other related matters.[71]

At the 2015 scientific and technical subcommittee meeting, discussions and presentations built on a continuing range of space related issues, including space technology for socioeconomic development in the context of the United Nations Conference on Sustainable Development and the post-2015 development agenda; matters relating to remote sensing of the Earth by satellite (including applications for developing countries and monitoring of the Earth's environment); space debris; space-system-based disaster management; recent developments in global navigation satellite systems; space weather; near-Earth objects; the use of nuclear power sources in outer space; and the long-term sustainability of outer space activities. The subcommittee also examined the physical nature and technical attributes of the geostationary orbit and its utilization and applications, including the field of space communications with particular note of the needs and interests of developing countries.[72]

As in previous years, the proceedings of the 2015 legal subcommittee focussed on exchanging information on the activities of international governmental and non-governmental organizations related to space law, in addition to the status and application of the five United Nations treaties on outer space. Matters relating to the definition and delimitation of outer space were considered, as was the utilization of the geostationary orbit, including consideration of ways and means to ensure the rational and equitable use of the geostationary orbit without prejudice to the role of the International Telecommunication Union. National legislation relevant to the peaceful exploration and use of outer space was another agenda item; in addition to capacity-building in space law; a review and possible revision of the Principles Relevant to the Use of Nuclear Power Sources in Outer Space; a general exchange of information and views on legal mechanisms relating to space debris mitigation measures; a general exchange of information on non-legally binding United Nations instruments on outer space, and a review of international mechanisms for cooperation in the peaceful exploration and use of outer space.[73]

[71]United Nations General Assembly. Report of the Committee on the Peaceful Uses of Outer Space on its Fifty-eighth session, Held in Vienna from 10–19 June 2015. UN Doc. A/70/20 of 26 June 2015. New York: United Nations <http://www.unoosa.org/res/oosadoc/data/documents/2015/a/a7020_0_html/A_70_20E.pdf>.

[72]United Nations General Assembly. Report of the Scientific and Technical Subcommittee on its fifty-second session, held in Vienna from 2 to 13 February 2015. UN Doc. A/AC.105/1088 of 27 February 2015. Vienna: United Nations <http://www.unoosa.org/pdf/reports/ac105/AC105_1088E.pdf>.

[73]United Nations General Assembly. Report of the Legal Subcommittee on its fifty-fourth session, held in Vienna from 13 to 24 April 2015. UN Doc. A/AC.105/1080 of 30 April 2015. Vienna: United Nations <http://www.unoosa.org/res/oosadoc/data/documents/2015/aac_105/aac_1051090_0_html/AC105_1090E.pdf>.

1.2.1.3 Other UN Bodies and Organs Monitoring Outer Space Activities

Beyond the UN General Assembly and its Committees, there are other UN bodies, programmes and organs related to space activities. The following discusses the ITU (a specialised agency of the UN), UN-SPIDER, the UN Programme on Space Applications, the International Committee on Global Navigation Satellite Systems (ICG), the United Nations Spatial Data Infrastructure (UNSDI), the Conference on Disarmament (CD) and UNIDIR.

International Telecommunication Union (ITU)

The International Telecommunication Union (ITU) held a 3-day symposium and workshop on small satellite regulation and communication systems, focussing on the use of radio-frequency spectrum and satellite orbits for small satellite communication systems, in Prague, Czech Republic from 2 to 4 March 2015.[74] It also held a workshop on the Efficient Use of the Orbit/Spectrum Resource on 29 September 2015, followed by the ITU International Satellite Symposium 2015 from 30 September to 1 October 2015, both in Da Nang, Vietnam.[75] The symposium covered updates to satellite international regulations and WRC-15 preparations and key agency items, in addition to developments on a satellite policy and regulatory framework including country practices on satellite communications and broadcasting services; it also covered satellite market and technology trends, as well as industry opportunities.[76]

In the lead-up to the 2015 World Radiocommunication Conference (WRC-15), the ITU conducted its second Conference Preparatory Meeting (CPM15-2) in Geneva, Switzerland from 23 March to 2 April 2015. The CPM15-2 produced a Report to the WRC-15, focussing on technical, operational, and regulatory materials to be considered by ITU Member States in developing their proposals to the Conference.[77] The ITU also conducted a series of Regional Radiocommunication Seminars (RRS15) to help prepare ITU Member States for the WRC-15 conference, covering general issues of spectrum management and international regulations for terrestrial and space services, including the relevant coordination, examination and registration procedures. The first RRS15 meeting for Eastern Europe—CIS took place in Bishkek, Kyrgyzstan from 2 to 6 March 2015; followed by RRS15-Africa

---

[74]"ITU Symposium and Workshop on small satellite regulation and communication systems, Prague, Czech Republic, 2–4 March 2015." 12 Mar. 2015. ITU 19 Apr. 2016 <http://www.itu.int/en/ITU-R/space/workshops/2015-prague-small-sat/Pages/default.aspx>.

[75]"ITU Workshop on the Efficient Use of the Orbit/Spectrum Resource, Danang, Vietnam." 1 Oct. 2015. ITU 19 Apr. 2016 <http://www.itu.int/en/ITU-R/space/workshops/danang-2015/Pages/default.aspx>.

[76]"ITU International Satellite Symposium 2015." 1 Oct. 2015. ITU 19 Apr. 2016 <http://www.itu.int/en/ITU-D/Regional-Presence/AsiaPacific/Pages/Events/2015/October-IISS-2015/home.aspx>.

[77]"Preparatory meeting finalizes Report to 2015 World Radiocommunication Conference." 2 Apr. 2015. ITU 18 Apr. 2016 <http://www.itu.int/net/pressoffice/press_releases/2015/08.aspx>.

in Niamey, Niger from 20 to 24 April 2015, RRS15-Asia-Pacific in Manila, Philippines from 25 to 30 May 2015, and RRS15-Americas in San Salvador, E□l Salvador from 27 to 31 July 2015.[78] The WRC-15 conference was held in Geneva, Switzerland from 2 to 27 November 2015, and addressed over 40 topics related to frequency allocation and frequency sharing for the efficient use of spectrum and orbital resources. Its key outcomes modified spectrum allocations for mobile broadband communications, amateur radio service, emergency communications and disaster relief, search and rescue beacons, Earth observation satellites for environmental monitoring, unmanned aircraft and wireless avionics systems, global flight tracking for civil aviation, enhanced maritime communication systems, automotive radar, mobile receivers of fixed satellite services, and further studies on universal time.[79]

UN-SPIDER

The United Nations Platform for Space-based Information for Disaster Management and Emergency Response (UN-SPIDER) organized several workshops and regional meetings in 2015. Set up by the United Nations General Assembly in 2006, the platform aims to provide universal access to all types of space-based information and services relevant to disaster management support. In this context, the International Charter on Space and Major Disasters was activated many times by the UN Office of Outer Space Affairs (UN OOSA) at the request of other UN entities and member states. Key geopolitical events that activated the Charter included floods in Madagascar, Australia, northern Bolivia, Brazil, northern and central Chile, India, northern Vietnam, Myanmar, Argentina, Bangladesh, and Iraq; hurricanes, tropical storms and torrential rains in Vanuatu, the Northern Mariana Islands, Turkey, the Bahamas, and Yemen; landslides in India, Colombia, Bangladesh, and Guatemala; wildfires in northern Chile, and Russia's Siberian district; along with volcanic activities in Chile, and in the waters of Grenada; and earthquakes in Nepal, Chile, and Afghanistan. The international charter was also activated for: an oil spill off the coast of Vietnam, the crash of Metrojet's Airbus A321 aircraft in Sinai Egypt that killed 224 passengers and crew, and the collapse of the Bento Rodrigues dam in Brazil.[80]

UN Programme on Space Applications (PSA)

The UN Programme on Space Applications (PSA) is concerned with cooperation in space science and technology. In 2015, several activities were carried out under PSA's auspices dealing with topics including workshops on space technology, on the use of global navigation satellite systems, and on integrated space technology

[78]"Radiocommunication Seminars and Workshops." ITU 18 Apr. 2016 <http://www.itu.int/en/ITU-R/seminars/Pages/default.aspx>.

[79]"World Radiocommunication Conference allocates spectrum for future innovation." 27 Nov. 2015. ITU 18 Apr. 2016 <http://www.itu.int/net/pressoffice/press_releases/2015/56.aspx>.

[80]"UN-SPIDER Monthly Updates." 31 Mar. 2016. UNOOSA 20 Apr. 2016 <http://www.un-spider.org/news-and-events/updates>.

applications for climate change. The PSA conducted the United Nations/Japan Workshop on Space Weather "Science and Data Products from ISWI Instruments" in Fukuoka Japan from 2 to 6 March 2015, to assess the status of space weather instruments (in-situ, space-borne), data access, availability, and collection and modelling efforts to advance space weather research and improve space weather forecasting.[81] Its United Nations/Russian Federation Workshop on the Applications of Global Navigation Satellite Systems was held in Krasnoyarsk, Russia from 18 to 22 May 2015, to address the use of Russia's GLONASS in combination with other Global Navigation Satellite Systems (GNSS) for transport and communication, aviation, surveying, the environment and disasters, and high precision mobile applications. The workshop also looked at space weather effects on GNSS and encouraged greater cooperation in developing partnerships and GNSS networks in the framework of the regional reference frames.[82] The PSA also conducted the United Nations/Austria Symposium on Integrated Space Technology Applications for Climate Change in Graz, Austria from 7 to 10 September 2015, to address, *inter alia*, how countries affected by climate change could make better use of space applications to assess their vulnerability; identify potential alternatives in the context of mitigation and adaption to climate change; and improve synergies among space agencies and organizations targeting efforts on climate change.[83]

International Committee on Global Navigation Satellite Systems (ICG)
The aim of the International Committee on Global Navigation Satellite Systems (ICG) is to make positioning, navigation and timing available globally for societal benefits at a lower cost through promoting (1) cooperation in matters of satellite navigation (i.e. by encouraging compatibility in global navigation satellite systems (GNSS)—to be used separately or together without interference among systems); (2) interoperability (as in using satellite navigation services together to provide better capabilities than can be achieved by individual systems alone); and (3) integration of GNSS services into the national infrastructures of industrialized and developing nations. The UNOOSA serves as the Executive Secretariat of the ICG and the associated Providers' Forum. The tenth meeting of the ICG was organized by the United States Department of State and the University Corporation for Atmospheric Research (UCAR) on behalf of the United States government, and took place in Boulder, Colorado, from 1 to 6 November 2015. The meeting brought

[81]"United Nations/Japan Workshop on Space Weather 'Science and Data Products from ISWI Instruments'." UNOOSA 20 Apr. 2016 <http://www.unoosa.org/oosa/events/data/2015/2015_japan_workshop.html>.

[82]"United Nations/Russian Federation Workshop on the Applications of Global Navigation Satellite Systems." UNOOSA 20 Apr. 2016 <http://www.unoosa.org/oosa/events/data/2015/2015_gnss_workshop_russia.html>.

[83]"United Nations/Austria Symposium on Integrated Space Technology Applications for Climate Change." UNOOSA 20 Apr. 2016 <http://www.unoosa.org/oosa/events/data/2015/2015_gnss_workshop_russia.html>.

together stakeholders in industry, government, non-governmental officials and academia to review and discuss developments in global navigation systems.[84]

Throughout the three sessions of its Plenary and a series of working group meetings, presentations were made to update ICG Members, Associate Members, and Observers on various matters of interest. An expert panel on global navigation satellite systems addressed the theme "Global navigation satellite systems: today and preparing for the future", discussing topics ranging from GNSS vulnerabilities to the use of GNSS signals for atmospheric remote sensing, and from the geodetic challenges of GNSS measurement modelling to the design of operational positioning infrastructure (in the form of continuously operating GNSS receivers) to augment GNSS accuracy. The panel also provided industry perspectives on the design of new multi-constellation GNSS chipsets, issues relating to GNSS governance and United Nations initiatives in areas such as geodetic references and geospatial information management.

An expert seminar on global navigation satellite system applications was also held, addressing "Observing Earth processes using GNSS". The following topics were presented: "Development of BeiDou Navigation Satellite System"; "GNSS ionospheric sounding for space weather"; "Review of GNSS as sources of opportunity for Earth observation: radio-occultation and reflectometry"; "Using GNSS signals to measure soil moisture, vegetation water content, snow depth, water levels, permafrost and volcanic plumes"; "GNSS for monitoring regional water resources and cryospheric changes"; "Update on the International Terrestrial Reference Frame and handling deformation caused by large earthquakes"; and "GNSS augmentation to the tsunami early warning system".[85]

The four working groups also met during the Plenary, covering systems, signals and services; enhancement of GNSS performance, new services and capabilities; information dissemination and capacity-building; and reference frames, timing and applications. In addition to discussing cross-cutting issues relating to the international GNSS monitoring assessment, the groups reviewed the progress made in implementing the recommendations made at previous meetings and ways and means of carrying them into the future. The 11th ICG meeting will take place in Sochi, Russia from 6 to 11 November 2016.[86]

---

[84]"International Committee on Global Navigation Satellite Systems (ICG): Annual Meeting." United Nations Office for Outer Space Affairs 21 Apr. 2016 <http://www.unoosa.org/oosa/en/ourwork/icg/meetings/ICG-2015.html>.

[85]United Nations General Assembly. Tenth Meeting of the International Committee on Global Navigation Satellite Systems – Note by the Secretariat. UN Doc. A/AC.105/1104 of 2 Dec. 2015. New York: United Nations <http://www.unoosa.org/oosa/oosadoc/data/documents/2016/aac.105/aac.1051104_0.html>.

[86]"ICG-11: International Committee on GNSS." 14 Jan. 2016. InsideGNSS 21 Apr. 2016 <http://www.insidegnss.com/node/4777>.

United Nations Spatial Data Infrastructure (UNSDI)

The 15th annual United Nations Geographic Information Working Group (UNGIWG) Plenary Meeting was held in New York from 3 to 5 August 2015. The meeting was co-chaired by representatives from the United Nations Department of Safety and Security (UNDSS) and the United Nations Office for Outer Space Affairs (UNOOSA). The UNGIWG functions as a comprehensive, decentralised geospatial information network of UN cartography and geospatial information management science professionals that lay the foundations for the United Nations Spatial Data Initiative (UNSDI), and facilitate decision-making mechanisms between member states, regional organizations and partners that advance social, economic development, environmental and humanitarian interests.[87] Its plenary conclusions confirmed the need to move the voluntary best-effort approach of the group to the next level, and to formalize this geospatial coordination mechanism in the UN system, building on models such as UN-Water, UN-Space or the EMG. It also called for a permanent, dedicated Secretariat for the Working Group, and highlighted that coordination was required both on the policy level and the technical level. It also found that challenges remained in the availability of core datasets and of commercial very high resolution satellite imagery.[88]

Conference on Disarmament (CD)

The Conference on Disarmament (CD) is the key multilateral disarmament and arms control negotiating forum within the international community. The 2015 session of the Conference occurred in three parts: the first part from 19 January to 27 March, the second on 25 May to 10 July, and the third on 3 August to 18 September 2015, during which the Conference held 40 formal plenary meetings, and 33 informal plenary meetings, at which member states as well as non-member states were invited to participate in the discussions outlined their views and recommendations. In the context of preventing an arms race in outer space, development is ongoing with delegations reaffirming their respective positions in 2015. By 14 September 2015, China's representative to the CD and the Chargé d'affaires a.i. of the Russian Federation had submitted a letter to the Secretary-General of the CD transmitting the comments by China and the Russian Federation regarding the U.S. analysis of the 2014 updated Russian and Chinese texts of the draft treaty on prevention of the placement of weapons in outer space and of the threat or use of force against outer space objects (PPWT) submitted in September 2014.[89] Other topics addressed by the CD included the cessation of the nuclear arms

---

[87]"15th UNGIWG Plenary Meeting." UNGIWG 20 Apr. 2016 <http://www.ungiwg.org/meetings/15th-ungiwg-plenary-meeting>.

[88]"United Nations Geographic Information Working Group | Report to the UN Committee of Experts on Global Geospatial Information Management (GGIM)." 25 Aug. 2015. UNGIWG 20 Apr. 2016 <http://ggim.un.org/docs/meetings/GGIM5/UNGIWG%20Introduction.pdf>.

[89]Conference on Disarmament. Report of the Conference on Disarmament to the General Assembly of the United Nations. UN Doc. CD/2046 of 18 September 2015. Geneva: United Nations <https://daccess-ods.un.org/TMP/5747848.15311432.html>.

race and nuclear disarmament, prevention of nuclear war including all related matters, prevention of an arms race in outer space, effective international arrangements to assure non-nuclear-weapon states against the use or threat of use of nuclear weapons, and prohibition of the development and manufacture of new types of weapons of mass destruction and new systems of such weapons—e.g. radiological weapons, transparency in armaments.

Following a meeting on "Multilateral Negotiations on an International Code of Conduct for Outer Space Activities", convened at the initiative of the European Union, with the assistance of the United Nations Office for Disarmament Affairs, on 27–31 July 2015 at the UN headquarters in New York, the European led-initiative to develop a draft International Space Code of Conduct for Outer Space Activities (ICoC) will be pursued within the framework of the UN through a mandate of the General Assembly.[90] The European Union began drafting the ICoC in 2008, in response to UN General Assembly Resolutions 61/75 of 2006 and 62/43 of 2007 which had invited all members to submit concrete proposals on international outer space TCBMs in the interest of maintaining international peace and security and promoting international cooperation and the prevention of an arms race in outer space. Initial draft versions of the ICoC had received lukewarm receptions on the grounds that the code had been developed within the EU system and lacked a parallel diplomatic process involving the considerations of the international community. However, following several revisions and three open-ended multilateral consultations (held in Kiev, Ukraine in May 2013, in Bangkok, Thailand in November 2013, and Luxembourg in May 2014) the ICoC had gained wider acceptance[91] prior to its inclusion within the ambit of the UN framework.

United Nations Institute for Disarmament Research (UNIDIR)
The United Nations Institute for Disarmament Research (UNIDIR) conducts some projects that cover, both directly and indirectly, issues of space security. It seeks to review former proposals and to propose new options for breaking the deadlock in space weaponisation matters at the CD. On 23 July 2015, the UNIDIR submitted the report on the activities of the Institute for the period from January to December 2014 and the programme of work and financial plan for 2015 and 2016. Activities on space-related emerging security issues included a panel, hosted by the Permanent Mission of Kazakhstan to the United Nations, in October 2014, to discuss current United Nations initiatives and the importance of a holistic approach to developing a resilient space security regime. And on 31 March 2015, the UNIDIR also held a small expert round table with the SWF to consider national positions on how existing national commitments on the right to self-defence should be applied in

[90]"Multilateral Negotiations on an International Code of Conduct for Outer Space Activities, New York 27–31 July 2015 | Chair's Summary." 31 July 2015. Ministry of Foreign Affairs of Japan 29 June 2016 <http://www.mofa.go.jp/mofaj/files/000098436.pdf>.

[91]*See further* "Code of Conduct for Outer Space Activities." European Union External Action 24 June 2015 <http://eeas.europa.eu/non-proliferation-and-disarmament/outer-space-activities/index_en.htm>.

the outer space environment. Moreover, the UNIDIR continued as consultant for the Group of Governmental Experts on Developments in the Field of Information and Telecommunications in the Context of International Security, for its work in 2014 and 2015; the final meeting of the Group was held in June 2015.[92] The UNIDIR's 14th annual space security conference was held in Geneva, Switzerland on 24–25 August 2015, on the theme "Underpinning Foundations of Space Security". It focussed both on security in space of space assets, and security on Earth as it is affected by space assets, and provided established and emerging actors a forum to share views and to explore concerns and options for achieving greater understanding and consensus on the topic.[93]

1.2.2 The Group on Earth Observation

The Group on Earth Observation (GEO), with its system-of-systems cooperation platform (GEOSS), is a voluntary partnership of governments and international organisations that facilitate the coordination of current and future Earth observing systems, while also providing structured and sustained data worldwide as a decision support tool for its users. GEO was very active in 2015, conducting around 72 workshops and symposia, covering an assortment of topics ranging from climate change, inland and coastal water quality, urban environmental hazards, and other Earth observation and environmental monitoring topics. Moreover, other diversified and notable activity included conferences such as The Sixth Asia/Oceania Meteorological Satellite Users' Conference; the Earth Observation Open Science 2.0 conference; the Our Common Future Under Climate Change Conference; the third International Conference on Remote Sensing and Geoinformation of Environment; and the third UN World Conference on Disaster Risk Reduction.[94]

1.2.3 Europe

1.2.3.1 European Space Agency

ESA's budget increased by 8 % in 2015, reaching 4.433 billion euros from 4.102 billion euros in 2014. Direct funding from ESA member states decreased by 2.9 %

---

[92] United Nations General Assembly. United Nations Institute for Disarmament Research – Note by the Secretary General, on its Seventieth session. UN Doc. A/70/177 of 23 July 2015. <http://www.unidir.org/files/files/GA_report_2015_Dir_E_A-70-177.pdf> leading up to 2015.

[93] Johnson, Chris, and Victoria Samson. "UNIDIR Space Security Conference 2015: Underpinning Foundations of Space Security." 15 Oct. 2015. UNIDIR 20 Apr. 2016 <http://www.unidir.org/files/publications/pdfs/space-security-2015-en-640.pdf>.

[94] "Past Meetings." GEO 21 Apr. 2016 <http://www.earthobservations.org/meetings.php>.

to 3.241 billion euros, while funding from the EU and Eumetsat grew by 56.2 % to 1.191 billion euros for the year.[95] While much attention leading up to 2015 was on ESA's long-term launch vehicle programme, funding for launchers changed only marginally for 2015—intended for the upkeep of the Ariane 5 launcher and the current Vega launcher. Funding for the new Ariane 6 launcher and an upgraded Vega launcher is expected to increase in the coming years. In contrast, funding for Earth observation increased by 28.3 % to 1254.3 billion euros, while Navigation had a 15.0 % increase to 664.5 million euros with much of that funding coming from the EU's Copernicus and Galileo Flagship programmes. Moreover, in December 2014, the European Commission agreed to procure six new replacement satellites for the Copernicus programme (i.e. Sentinel 1, Sentinel 2, and Sentinel 3). Those satellites will be identical to their predecessors, and will assure data continuity well into the 2030s.[96]

2015 also marked the end of ESA's Automated Transfer Vehicle (ATV) programme, after its fifth and last ATV Georges Lemaître departed the International Space Station (ISS) on 15 February 2015. Funding for the ISS for the period between 2015 and 2017 was decided at the 2014 ESA Ministerial Council meeting, which allocated 800 million euros from ten of its participating member governments (Belgium, Denmark, France, Germany, Italy, the Netherlands, Norway, Spain, Sweden and Switzerland) to cover ISS costs during that time.[97] ESA is expected to continue its contribution to the ISS from 2018 until 2020 by providing an ATV-derived service module for NASA's crewed Orion spacecraft. However the fate of Europe's contribution to the ISS after 2020 was left open, as ESA's new Director General (DG) Johann-Dietrich "Jan" Woerner began to set the Agency's sights on developing an international Moon Village by 2030.[98]

The ESA's focus will centre on the following areas: space applications; science and exploration; transportation; space technology and operations; and administration. While the ESA statement did not clarify how the new organisation would become more efficient and reduce bureaucracy, the reorganization and five focus areas are meant to bring the agency closer to the goal of developing a "United Space in Europe through ESA." Yet in order to reach that goal, ESA—with its growing membership—will need to reassess its geo-return policy, which guarantees that 90 % of a member state's ESA investment will return in the form of industry contracts, in addition to

[95]"ESA Budget 2015 by Domain. 15 Jan. 2015. ESA 4 Mar. 2016 <http://www.esa.int/spaceinimages/Images/2015/01/ESA_Budget_2015_by_domain>.

[96]De Selding, Peter B. "ESA Hikes Budget, Takes Steps To Send Astronaut to Chinese Space Station." 16 Jan. 2015. SpaceNews 10 Dec. 2015 <http://spacenews.com/esa-hikes-budget-takes-steps-to-send-astronaut-to-chinese-space-station/>.

[97]De Selding, Peter B. "Italian Space Increase Falls Short, Clouding Outlook for ESA Ministerial." 1 Dec. 2014. SpaceNews 22 June 2015 <http://spacenews.com/42638italian-space-increase-falls-short-clouding-outlook-for-esa-ministerial/>.

[98]David, Leonard. "Lunar Leap: Europe Is Reaching for a Moon Base by the 2030s." 30 Dec. 2015. Space.com 27 June 2016 <http://www.space.com/31488-european-moon-base-2030s.html>.

ESA's one member-one vote policy, wherein all ESA members have an equal say regardless of the disparity/magnitude of their investment.[99]

Following the substantial schedule delays and cost overruns experienced in developing new lidar technologies, ESA changed its contracting procedures for better coordination of instrument availability and full programme funding. The new procedures will stagger an award for the development of an unproven instrument from the contract award for the overall programme, which should permit advanced technologies to be developed to a sufficient Technology Readiness Level (TRL) to be ready to place on an intended mission platform. ESA's Atmospheric Dynamics Mission (ADM)-Aeolus, and its EarthCare satellite are the latest examples of why such a financing arrangement was needed. Development of ADM-Aeolus began in 2003 with an intended launch in 2007, while development of EarthCare began in 2008 with a launch intended for 2013. Delayed development of the lidar instruments pushed back the launch schedule for both satellites to sometime in 2017 and 2018 respectively. In funding the development of new technologies, ESA would like to see the results of full-model tests in a vacuum before beginning work on the rest of the mission. It plans to also use non-fixed price contracts, and to make additional funding available if technology roadblocks are encountered.[100]

1.2.3.2 European Union

The EU Copernicus flagship programme is entering its operational phase, following the successful launch of the Sentinel-1A satellite on 3 April 2014, and the Sentinel-2A on 22 June 2015. The second Sentinel-1B satellite is expected to launch in the first half of 2016. While funding for the programme had been uncertain during the Commission's development of the 7-year Multiannual Financial Framework (MFF) for 2014–2020, the European Commission managed to keep the Copernicus programme within the MFF budget, albeit with a 2 billion euros reduction, thus amounting to 3.8 billion euros (in 2011 prices).[101] In coordinating the evolution of the Copernicus Space Component, ESA has prepared a long-term plan for the content and associated funding needs, covering the operation of the Sentinels up to 2020, and the procurement of recurrent Sentinel satellites and instruments and access to data available from contributing missions up to 2028.[102]

[99]De Selding, Peter B. "Tough Sledding for Proposed ESA Reorganization." 27 July 2015. SpaceNews 23 Dec. 2015 <http://spacenews.com/tough-sledding-for-proposed-esa-reorganization/>.

[100]"Cost, Schedule Woes on 2 Lidar Missions Push ESA To Change Contract Procedures." 22 May 2015. SpaceNews 11 Jan. 2016 <http://spacenews.com/cost-schedule-woes-on-2-lidar-missions-push-esa-to-change-contract-procedures/>.

[101]"For Europe's Embattled GMES, Good and Bad News." 7 Dec. 2012. SpaceNews 25 Apr. 2014 <http://spacenews.com/article/civil-space/32717for-europe%E2%80%99s-embattled-gmes-good-and-bad-news>.

[102]"Green Light for GMES Copernicus." 4 July 2013. ESA 28 Apr. 2014 <http://www.esa.int/Our_Activities/Observing_the_Earth/Copernicus/Green_light_for_GMES_Copernicus>.

Related to the EU's Copernicus programme, the status of the 500 million euros European Data Relay Satellite (EDRS) programme continued to be in doubt at the beginning of 2015 because of ongoing concerns about its business prospects. The programme was initiated in November 2008 as a Public Private Partnership (PPP) between ESA and Airbus Defence & Space to create a global optical satellite data relay service for civil and military users. The delay in development stemmed both from ESA taking longer than expected to validate EDRS technology and sign a broad Copernicus agreement with the EU, and by the initial resistance from the European Commission to commit to a Service Level Agreement (SLA) that was negotiated between ESA and Airbus D&S and presented to the Commission shortly before the contract's 22 December 2014 deadline.[103] Expected to be included within the EU's Copernicus arrangement with ESA, the SLA assumed that the European Commission would be EDRS's anchor customer, providing 20 million euros per year for 5 years. However, as at 28 January 2015 the Commission seemed willing to provide closer to 12 million euros per year for 5 years for the service.[104] By 20 February 2015, the European Commission, ESA and Airbus D&S reached an agreement on the funding and management of the EDRS programme. While the specific commitments of the agreement were not made public, the European Commission had agreed to spend a little less than 100 million euros for the 2015–2020 period, with funds going toward the development of EDRS ground infrastructure and for data service. Moreover, the EDRS agreement included the formation of a joint ESA-EU-Airbus steering committee to provide all parties with more oversight over the EDRS programme.[105]

The goal of the EDRS programme is to have at least three satellites spread out in GEO orbit, which will receive Earth observation data from low Earth orbiting (LEO) satellites via laser link, and rapidly transmit terabytes of radar and optical data to ground stations. EDRS could also be used to relay UAV data to flight control centres, extending their reach across the globe.[106] The EDRS-A node was launched on 22 June 2015 as a hosted payload aboard the Sentinel 1a satellite. Eutelsat's commercial telecom satellite, Eutelsat 9B, to be launched in early 2016, will carry an EDRS laser terminal as a hosted payload. OHB AG of Bremen is building the third EDRS-C satellite for the Airbus-managed system, which is expected to be launched in late 2016. Moreover, a laser terminal will be launched later as a hosted payload aboard another telecommunications satellite. With the

[103]De Selding, Peter B. "Data-relay Project Exemplifies Uneasy Relationship Between European Commission and ESA." 30 Jan. 2015. SpaceNews 8 Jan. 2016 <http://spacenews.com/data-relay-project-exemplifies-uneasy-relationship-between-european-commission-and-esa/>.

[104]Ibid.

[105]De Selding, Peter B. "Three-way Agreement Extends EDRS Commitment Beyond 2020." 24 Feb. 2015. SpaceNews 8 Jan. 2016 <http://spacenews.com/three-way-agreement-extends-edrs-commitment-beyond-2020/>.

[106]De Selding, Peter B. "Successes in 2014 Prompt Airbus To Reduce Space Layoffs." 20 Jan. 2015. SpaceNews 12 Jan. 2016 <http://spacenews.com/successes-in-2014-prompt-airbus-to-reduce-planned-layoffs-in-space-business/>.

SLA finally secured, Airbus D&S will invest 130 million euros of its own money into the EDRS programme. With the European Commission now vested as EDRS's anchor customer, Airbus would then commercialize the service to civil and military users worldwide. Here, the U.S. DoD seems likely to be a prospective customer, especially following the signing of a U.S. and EU cooperation arrangement on Copernicus Earth observation data on 19 October 2015, and following U.S. and German negotiations on allocating satellite spectrum for UAVs crossing over the Atlantic that were in discussion at the WRC-15 meeting in November 2015.[107]

By the end of 2015, the EU Galileo flagship GNSS programme had managed to recover from the rough start it experienced in 2014, where one of the four in-orbit validation (IOV) satellites launched in 2014 suffered irreparable power degradation from an unknown anomaly, while two satellites were placed in a bad orbit by a Soyuz Fregat upper stage. An additional six Galileo navigation satellites were successfully placed into medium Earth orbit in 2015, enabling the European Commission to debut initial Galileo services (i.e. open service, SAR, and PRS) with the remaining nine fully operational satellites by late 2016. The EU Galileo flagship GNSS programme's 7-year budget of 6.3 billion euros (at 2011 prices) was approved by the European Parliament on 22 November 2013.[108] Switzerland joined the EU's Galileo programme at the beginning of 2014, following an agreement that requires Switzerland to make catch-up payments to the EU totalling 80.05 million euros, for the period 2008–2013, along with annual fees of 27 million euros in return for access to Galileo's restricted Public Regulated Service (PRS) signals. At the end of 2015, the EU was still sorting out access to the PRS signals its 28 member states, in addition to still considering whether to make PRS signals available to Norway and the United States.

On 16 June 2015, during the Paris Air Show, France, Germany, Italy, Spain and Britain agreed to make available their existing optical and radar tracking telescopes for use in the European Union's Space Surveillance and Tracking Framework. The EU initially approved the Framework programme in 2014, and has budgeted 70 million euros between 2015 and 2020. The EU programme extends beyond activities conducted by ESA in space tracking, which mainly focus on near Earth objects and space weather, to also include a military element. Moreover, part of the EU's aim is to reduce dependence on the U.S. Space Surveillance Network in tracking orbital objects to monitor and protect its two space flagship programmes Galileo and Copernicus. The five nation consortium supplying the assets on behalf of the EU will also create a Security Committee to determine who will have access to space surveillance data together with the EU Satellite Centre (EUSC). The

[107]De Selding, Peter B. "U.S., Germany Face Uphill Fight for UAV Spectrum." 13 Nov. 2015. SpaceNews 7 Jan. 2016 <http://spacenews.com/u-s-germany-face-uphill-fight-for-uav-spectrum/>.

[108]"Galileo Funding: A 'Small' Difference of €700 Million." 22 Nov. 2013. The European Sting 25 Apr. 2014 <http://europeansting.com/2013/11/22/galileo-funding-a-small-difference-of-e700-million/>.

division of the 70 million euros among the consortium partners still needs to be determined.[109]

1.2.3.3 EUMETSAT

The European Organisation for the Exploitation of Meteorological Satellites (Eumetsat) is an intergovernmental organisation that supplies weather and climate-related satellite data to the National Meteorological Services of its Member and Cooperating States in Europe, and other users worldwide. Eumetsat's 83rd council meeting took place in Darmstadt, Germany, on 23 and 24 June 2015. At that meeting, the council successfully concluded the approval process for the Eumetsat Polar System Second Generation (EPS-SG) programme securing the commitment of all 30 member states; the meeting also made progress in the approval process for the Jason Continuity of Service (Jason-CS) optional programme, achieving 77.83 % of the financial envelope from participating states.[110] By the 84th council meeting in Darmstadt, Germany, held on 2 December 2015, the financial envelope for the Jason-CS optional programme had increased to 93.07 % with the inclusion of Norway as the 13th participating state[111]; the programme had entered into force on 9 September 2015 following subscriptions from 12 participating states (Austria, France, Germany, Iceland, Italy, Luxembourg, The Netherlands, Portugal, Sweden, Switzerland, Turkey, and the United Kingdom).[112] Also at the 84th meeting, Eumetsat and the U.S. National Oceanic and Atmospheric Administration (NOAA) signed an agreement to establish and operate their Joint Polar System (JPS), which will consist of Eumetsat's Metop-Second Generation (Metop-SG) satellites, NOAA's JPSS satellites and shared ground systems and services in the period from 2020 to 2040.[113]

Eumetsat's current Metop program consists of three identical satellites launched at 6-year intervals between 2006 and 2018, with operations running until 2023. The first polar-orbiting meteorological satellite, Metop-A was launched on 19 October

[109]De Selding, Peter B. "A European Space Surveillance Network Inches Forward." 17 June 2015. SpaceNews 22 Dec. 2015 <http://spacenews.com/a-european-space-surveillance-network-inches-forward/>.

[110]"EUMETSAT Council concludes the approval of the EPS Second Generation programme and achieves progress on Jason-CS." 24 June 2015. Eumetsat 1 June 2016 <http://www.eumetsat.int/website/home/News/DAT_2681070.html?lang=EN&pState=1>.

[111]"Council takes steps to implement the EPS-SG and Jason-CS programmes." 3 Dec. 2015. Eumetsat 1 June 2016 <http://www.eumetsat.int/website/home/News/DAT_2868481.html?lang=EN&pState=1>.

[112]"Jason-CS Programme enters into force – two months ahead of COP 21." 11 Sept. 2015. Eumetsat 1 June 2016 <http://www.eumetsat.int/website/home/News/DAT_2779482.html?lang=EN&pState=1>.

[113]"EUMETSAT and NOAA sign agreement on Joint Polar System." 2 Dec. 2015. Eumetsat 1 June 2016 <http://www.eumetsat.int/website/home/News/DAT_2867890.html?lang=EN&pState=1>.

2006; it was joined by the successful launch of Metop-B on 17 September 2012, which orbits in tandem with Metop-A. The third identical Metop satellite, Metop-C, is expected to launch in October 2018.[114] The entire cost of the Metop program (including development of the three satellites, their launches and the related ground infrastructure) has reached 3.2 billion euros (at 2011 prices), with Eumetsat covering 75 % of the cost, and ESA covering the rest.[115]

The Metop Second Generation (Metop-SG) is the space segment of Eumetsat's Polar System of Second Generation. On 5 October 2015, Eumetsat and ESA signed a cooperation agreement that secures the development of the six-satellite Metop-SG system. The entire program is budgeted at about 4.1 billion euros, with Eumetsat contributing about 80 % of the budget, while ESA will cover the rest; the satellites are intended to operate between 2021 and around 2042.[116] ESA will develop the Metop-SG satellites according to user and system requirements as defined by Eumetsat and is responsible for the procurement of the recurrent satellites on Eumetsat's behalf, while Eumetsat will procure all launch services and develop the ground control systems, in addition to acquiring and processing their data, and delivering products and services to users in response to their evolving needs.[117] The Metop-SG satellites will operate in three pairs, each carrying a different but complementary suite of instruments, and will be manufactured by Airbus Defence and Space under a 1.32 billion euros contract signed on 16 October 2014.[118]

The Meteosat Third Generation (MTG) system is a series of sounding and imaging satellites in geostationary orbit, being developed by France's Thales Alenia Space and Germany's OHB AG, and is aimed at providing services for the 2020–2040 timeframe. On 24 July 2015, the Eumetsat Council approved the contract with Arianespace for the launches of the first three MTG satellites (the MTG-I1, MTG-S1, and one option for MTG-I2).[119] The satellites are scheduled to

[114] "METOP." Eumetsat 1 June 2016 <http://www.eumetsat.int/website/home/Satellites/CurrentSatellites/Metop/index.html>.

[115] "European Weather Satellite Launched after 4-Month Delay." 27 Sept. 2012. SpaceNews 28 Apr. 2014 <http://spacenews.com/article/european-weather-satellite-launched-after-4-month-delay>.

[116] De Selding, Peter B. "ESA, Eumetsat Finalize Contract for Next-gen Weather Satellites." 6 Oct. 2015. SpaceNews 13 Jan. 2016 <http://spacenews.com/esa-eumetsat-finalize-contract-for-next-gen-weather-satellites/>.

[117] "EUMETSAT and ESA sign cooperation agreement on Metop-SG satellites." 5 Oct. 2015. Eumetsat 1 June 2015 <http://www.eumetsat.int/website/home/News/DAT_2804093.html?lang=EN&pState=1>.

[118] De Selding, Peter B. "Airbus Signs $1.7 Billion Contract for Six Metop Weather Satellites." 16 Oct. 2015. SpaceNews 3 June 2016 <http://spacenews.com/42212airbus-signs-17-billion-contract-for-six-metop-weather-satellites/>.

[119] "EUMETSAT and Arianespace announce signature of launch contract for three MTG satellites." 16 July 2015. Eumetsat 3 June 2016 <http://www.eumetsat.int/website/home/News/DAT_2703454.html?lang=EN&pState=1>.

be launched to GEO orbit within the 2019–2023 timeframe, and will operate from 2020 to 2040.[120]

Eumetsat's existing Meteosat programme, consisting of both Meteosat First Generation (MFG) and the Meteosat Second Generation (MSG) satellites operating in geostationary orbit over Europe and Africa, added the MSG-4 to its MSG programme on 15 July 2015. Upon the successful conclusion of the MSG-4's commissioning phase by Eumetsat, the spacecraft will be renamed Meteosat-11 and will be placed into in-orbit storage for 2.5 years, meant for use prior to the deployment of Eumetsat's MTG system.[121] The MSG programme has three other satellites in operation, Meteosat-8 to -10, which are expected to end service in 2019, 2021, and 2022 respectively.

Additionally, Eumetsat's Meteosat-7, the last of the Meteosat First Generation (MFG) satellites, continued to operate throughout 2015, providing imaging and data collection service coverage over the Indian Ocean. Having been in operation since September 1997, and positioned in GEO above the Indian Ocean in July 2006, the spacecraft is expected to reach the end of its life in 2017.[122] With no replacement spacecraft expected, concerns about a gap in weather coverage have been expressed by the U.S. DoD which uses Meteosat-7 data for cloud characterization and weather imagery over the war-wracked region, both essential for maintaining battlespace awareness.[123] The primary role of the Meteosat satellites is to help detect and forecast rapidly developing high impact weather, including thunderstorms, volcanic ash, and fog, up to 6 h ahead through continuous monitoring.[124]

1.2.3.4 National Governments

France

On 5 January 2015, the French space agency (CNES) announced that it had begun a small technology research programme with Germany and other ESA governments

[120]Henry, Caleb. "Arianespace Wins Fresh Eumetsat Contract After MSG 4 Launch." 16 July 2015. Via Satellite 3 June 2016 <http://www.satellitetoday.com/launch/2015/07/16/arianespace-wins-fresh-eumetsat-contract-after-msg-4-launch/>.

[121]"MSG-4 successfully launched." 16 July 2015. Eumetsat 3 June 2016 <http://www.eumetsat.int/website/home/News/DAT_2696903.html?lang=EN&pState=1>.

[122]"Meteosat-7 becomes EUMETSAT's longest-serving operational satellite." 23 Jan. 2015. Eumetsat 3 June 2016 <http://www.eumetsat.int/website/home/News/DAT_2526201.html?lang=EN&pState=1>.

[123]Gruss, Mike. "U.S. Military Faces Weather Coverage Gap over Hot Zones." 13 Mar. 2015. SpaceNews 22 Dec. 2015 <http://spacenews.com/u-s-central-command-faces-weather-coverage-gap-over-hot-zones/>.

[124]"There are two generations of active Meteosat satellites, Meteosat First Generation (MFG) and Meteosat Second Generation (MSG), providing images of the full Earth disc, and data for weather...." Eumetsat 30 May 2014 <http://www.eumetsat.int/website/home/Satellites/CurrentSatellites/Meteosat/index.html>.

to develop the technology needed for a reusable rocket stage. CNES had paused on the idea of reusability a decade ago, after a joint-study with Russia (called Baikal) revealed that a first stage launcher would need a rhythm of 40 launches per year to bring a return on the investment.[125] That rhythm estimate has since been reduced to 30 per year to complete the business case, as it would also include the cost of return and refurbishment, smaller economies of scale, and therefore higher per unit costs.[126] With interest reignited in light of the progress made by SpaceX's Falcon 9 launcher, CNES, DLR and ESA could use their collective experience in reusability technology to produce a consensus on a future technology roadmap, and hopefully conduct a technology demonstration after 2026. By October 2015, CNES and France's ONERA aerospace research institute announced that they had begun studies on digital aerothermodynamic simulations, but would decide on further deepening collaboration in 2016.[127]

In contrast to CNES' aim to return an entire stage to Earth for reuse, on 5 June 2015, Airbus Defence & Space revealed its own design for a reusable first-stage engine and avionics package intended for a next generation Ariane rocket. Named Adeline (or Advanced Expendable Launcher with Innovative engine Economy), the design was said to have resolved issues related to an engine's exposure to high speed stresses experienced when descending through the atmosphere before landing back on Earth. The Airbus design simplifies that used by SpaceX by separating the propulsion bay (accounting for up to 80 % of a rocket's cost) from the rest of the stage, protecting only the motor on re-entry, and returning horizontally on a conventional airstrip using winglets and turbofans. Over the course of 5 years, Airbus has invested 5 million euros in the design, but it would require substantial government support if it is needed by 2025.[128] While the Adeline is a proposal for the planned two-stage Ariane 6 launcher, its liquid fuelled engine could also be used with other rockets.

Prior to Airbus Safran Launcher's (ASL) formal bid on 7 May 2015 to ESA for the production of the Ariane 6 launcher, questions remained over the division of responsibilities between industry and government. The 2014 ESA Ministerial Council meeting left open issues such as whether ASL would be required to invest upwards of 400 million euros of its own capital in the Ariane 6 launcher

[125]De Selding, Peter B. "With Eye on SpaceX, CNES Begins Work on Reusable Rocket Stage." 5 Jan. 2015. SpaceNews 30 Dec. 2015 <http://spacenews.com/with-eye-on-spacex-cnes-begins-work-on-reusable-rocket-stage/>.

[126]De Selding, Peter B. "With Revenue Looking Up, Arianespace Seeks To Bring Ariane 5 Costs Down." 21 Oct. 2015. SpaceNews 13 Jan. 2016 <http://spacenews.com/with-revenue-looking-up-arianespace-seeks-to-bring-ariane-5-costs-down/>.

[127]De Selding, Peter B. "French Agencies To Collaborate on Reusable Rocket." 2 Oct. 2015. SpaceNews 5 Jan. 2016 <http://spacenews.com/french-agencies-to-collaborate-on-reusable-rocket/>.

[128]De Selding, Peter B. "Meet Adeline, Airbus' Answer To SpaceX Reusability." 5 June 2015. SpaceNews 5 Jan. 2016 <http://spacenews.com/meet-adeline-airbus-response-to-reusable-spacex-rocket/>.

programme, and whether an additional 200 million euros would be temporarily removed by ESA from the project that was initially valued at 3.215 billion euros.[129]

An ESA concern was that if ASL were forced to pay the 400 million euros, it might spread that cost over future commercial launches. The effect on Ariane 6's competitiveness might come into question, especially following SpaceX's streak in the commercial market, and Japan considering lowering its launcher cost by 50 % to win additional commercial business. Moreover, the move could dampen the effect of the favourable exchange rate that is currently developing, allowing Europe's launcher to compete with U.S. launchers at near par-value currencies.[130]

Separately, ASL raised its desire for a complete paradigm shift in governance before it would consider investing its own resources into the new launcher. That is, in order for the Ariane 6 to become more competitive, ASL wanted oversight over the design, production, commercialization and operations of the launcher. A second issue that threatened to stall negotiations was the ownership of Arianespace, in which ASL sought to increase its 39 % stake in the launch company to include the near 34.7 % stake held by the French government through CNES, resulting in the removal of CNES's minority blocking power in launcher development.[131]

The French government agreed to sell its stake in Arianespace to ASL by the end of 2015, but reserved informal oversight over the company, which will remain intact at its Evry headquarters, and will utilize the Ariane 6 launch platform CNES is building in French Guiana.[132] Following negotiations between ESA and ASL, it was agreed that ASL would contribute 400 million euros to the Ariane 6 development contract, while 200 million euros will be cut out partly by shaving the cost of certain buildings and facilities that will integrate the launcher horizontally, and an additional 200 million euros will be removed as unnecessary expenditures. ESA signed the Ariane 6 development contract with ASL on 12 August 2015. The contract, now valued at around 2.4 billion euros, will cover development of the launcher from 2015 to its inaugural flight in 2020.

A related industry concern that remains to be resolved is whether the ASL launch contractor configuration will be neutral among European satellite manufacturers. While Airbus D&S is a 50 % shareholder in ASL, it also directly competes with Thales Alenia Space and smaller prime contractor OHB, in manufacturing

---

[129]De Selding, Peter B. "ESA, Industry at Odds over Ariane 6 Funding Responsibilities." 3 Apr. 2015. SpaceNews 13 Jan. 2016 <http://spacenews.com/esa-industry-at-odds-over-ariane-6-funding-responsibilities/>.

[130]De Selding, Peter B. "Arianespace Assures French Parliament it Can Outcompete SpaceX." 13 May 2015. SpaceNews 5 Jan. 2016 <http://spacenews.com/arianespace-assures-french-parliament-it-can-outcompete-spacex/>.

[131]De Selding, Peter B. "New Airbus-Safran Venture Eyes Full Control of Arianespace." 8 Jan. 2015. SpaceNews 12 Jan. 2016 <http://spacenews.com/new-airbus-safran-venture-eyes-full-control-of-arianespace/>.

[132]De Selding, Peter B. "France Giving up Arianespace Ownership, but not Oversight." 19 June 2015. SpaceNews 22 Dec. 2015 <http://spacenews.com/france-giving-up-arianespace-ownership-but-not-oversight/>.

commercial satellites. Hence, a valid concern among European competitors is the possibility that an Airbus-controlled Ariane 6 might form a preference for its satellites through commercial or design biases.

Germany

On 18 June 2015, the German Aerospace Centre (DLR) elected Pascale Ehrenfreund as the new Chair of the DLR Executive Board, replacing Jan Woerner who began his new role as ESA's Director General on 1 July 2015.[133] Pascale Ehrenfreund, who is the DLR's first female Chair, began her role on 17 August 2015. Pascale was born in Vienna, Austria, where she obtained degrees in Astronomy and Biology at the University of Vienna, in addition to her doctorate and various other degrees. Prior to her appointment at the DLR, Pascale was a Professor of Space Policy and International Affairs at the Elliott School of International Affairs, George Washington University, and Lead Investigator at the NASA Astrobiology Institute, in addition to accepting the honorary position of President of the Austrian Science Fund (FWF) in 2013, and being a visiting researcher at the European Space Policy Institute.[134]

On 8 December 2015, German and French officials at the COP21 meeting in Paris reaffirmed a joint commitment to develop the "Merlin" greenhouse gas monitoring satellite. In the wake of the financial crisis of 2007, the project that was originally endorsed in 2010 had been shelved by the two countries for financial reasons, and because the satellite's lidar technology still needed to be developed. The methane monitoring satellite is aimed to be a globally recognised system to verify government commitments to reducing greenhouse gas emissions. The system is expected to cost 250 million euros ($266 million) covering the satellite's construction and launch, in addition to the associated infrastructure and 3 years of operations. Germany's 150 million euros contribution to the Merlin programme will go mainly toward the development of the lidar instrument, while France's 100 million euros will be put toward the development of a next generation Myriade satellite platform and its ground segment. Completion of the satellite is expected sometime in 2019.[135]

Italy

Italy's Avio SpA, the industrial prime contractor for the Vega launcher, was in search of a new strategic owner in 2015, as its current 81 % majority stakeholder,

---

[133] "First woman to head a major German research facility." 18 June 2015. DLR 29 June 2016 <http://www.dlr.de/dlr/en/desktopdefault.aspx/tabid-10081/151_read-13955/#/gallery/19767>.

[134] "Pascale Ehrenfreund – Chair of the DLR Executive Board." 7 Oct. 2015. DLR 29 June 2016 <http://www.dlr.de/dlr/en/desktopdefault.aspx/tabid-10329/510_read-14467>.

[135] De Selding, Peter B. "France, Germany Pressing Ahead with Greenhouse Gas Monitoring Satellite." 8 Dec. 2015. SpaceNews 11 Jan. 2016 <http://spacenews.com/france-germany-pressing-ahead-with-greenhouse-gas-monitoring-satellite/>.

Cinven sought to sell its shares in the launcher. Italy's Finmeccanica owns a 14 % share in Avio, while the remaining 5 % share belonged to Avio's company management. The newly established Airbus Safran Launchers (ASL) had its eye on acquiring the company to have Vega join its Ariane 5 and 6 launchers being developed under the Arianespace launch consortium but officials in the Italian government and its industry preferred to keep Avio under Italian control. Those stakeholders would likely welcome a takeover bid by Finmeccanica, provided it meets Cinven's transaction requirements. The Vega launcher is delivered by Avio to Arianespace under a joint venture with the Italian Space Agency (ASI) called ELV SpA (Avio has a 70 % stake in ELV, while ASI holds the remaining 30 % share). As in the case of the French government selling its shares in Arianespace to ASL to provide industry with more control, one scenario being discussed is whether ASI should sell its share in the ELV joint venture to ASL as well, thereby keeping Avio Italian owned, while attaining market balance among companies in Europe's launch sector.[136]

Relatedly, on 12 August 2015, ESA signed a 395 million euros contract with ELV and ASI to develop the upgraded Vega-C rocket, which includes the cost of the launcher's inaugural flight in 2018.[137] Moreover, 225 million euros will be dedicated to developing Vega's P120 first stage engine; and will also serve as a strap-on booster for the Ariane 6 launcher, in its 62 and 64 configurations. ELV aims to build a production facility capable of manufacturing 4 Vega launchers per year, along with an additional 31 P120 stages per year for the Ariane 6 launcher.

Despite committing to funding nearly 50 % of the next generation Vega launcher at the 2014 Ministerial Council meeting, the Italian government was also able to allocate 30 million euros per year from 2015 to 2018, to put forward a second generation Cosmo-SkyMed military and civilian radar Earth observation programme. Cosmo-SkyMed is part of a bilateral agreement with the French Defence Ministry, wherein Italy contributes radar data to France, while France provides Italy with products from its optical reconnaissance satellites. Valued at 500 million euros, the government's contract with ASI and Thales Alenia Space will be for two next generation satellites, and is expected to employ nearly 400 engineers and technicians at Thales Alenia Space's facilities in Italy (i.e. Rome, L'Aquila, and Turin). The new generation will be launched into orbit between 2017 and 2018.

The United Kingdom

Effective as of 1 October 2015, the UK Outer Space Act limited the liability of UK-licenced satellite operators to 60 million euros depending on the results of a

[136]De Selding, Peter B. "Vega Prime Contractor Avio Expects To Find Strategic Owner This Year." 19 June 2015. SpaceNews 5 Jan. 2016 <http://spacenews.com/vega-prime-contractor-avio-expects-to-find-strategic-owner-this-year/>.

[137]De Selding, Peter B. "ESA Inks $3.8 Billion in Contracts for Ariane 6, Vega-C and Spaceport Upgrades." 12 Aug. 2015. SpaceNews 13 Jan. 2016 <http://spacenews.com/esa-inks-3-8-billion-in-contracts-for-ariane-6-vega-c-and-spaceport-upgrades/>.

risk analysis performed for each new license application; higher-risk missions may expect to be liable for a higher amount. The move was made to encourage growth in Britain's commercial space sector, and to put the United Kingdom's requirements in line with other spacefaring nations. Operators with existing licenses may also apply for the liability cap for their licenses by submitting a formal request to the UK Space Agency, but that option is for now available only to domestic licenses, not including British dependencies and overseas territories. The new policy does not affect the UK's additional requirement for operators to also purchase 60 million euros in third party liability insurance as a licence requirement regardless of the use of the satellite.[138]

Yet while the UK stands out in Europe for its commercial focus and development in the space sector, it might need to loosen its third party liability insurance requirement on licence applicants even more if it aims to be more competitive with the United States. Even with recent policy changes geared to provide more incentive for small satellite developers by not charging fees for licence applications, and capping the amount of third party insurance needed to a slightly lower amount, those actions may not be enough to kick start its small satellite space sector. For instance, regardless of the size of a satellite being registered in the UK, its owner would likely have to pay an annual premium of about 1 %, or 600,000 euros each year on third party liability insurance. In the case of small satellite makers, that requirement may be still too prohibitive for start-up companies whose total budgets might be about the same size as the amount needed to be paid in additional insurance.[139] By contrast, the cost of licensing a satellite in the U.S. depending on the category might be anywhere from $0 for amateur satellites, to $60 for experimental, to $430,000 for an operational satellite. And aside from requiring third party liability insurance for the 30 days following launch of a payload, or 30 days following the initiation of re-entry, the U.S. has no third-party liability insurance requirement for a satellite's on-orbit activities.[140] Here, as can be seen in recent years, small satellite makers seem to be drawn to this arrangement, as the most popular categories of small satellites have been either amateur or experimental.[141]

---

[138]De Selding, Peter B. "U.K. Agrees to Liability Cap for Satellite Operators." 27 July 2015. SpaceNews 23 Dec. 2015 <http://spacenews.com/u-k-agrees-to-liability-cap-for-satellite-operators/>.

[139]De Selding, Peter B. "Smallsat Designer Flees UK Ofcom for US FCC." 17 Mar. 2015. SpaceNews 13 Jan. 2016 <http://spacenews.com/smallsat-designer-flees-uk-ofcom-for-us-fcc/>.

[140]*See* 14 C.F.R. §§ 440.9, 440.11, 440.12 (2014); *See also*, Schaefer, Matthew. "The Need for Federal Preemption and International Negotiations Regarding Liability Caps and Waivers of Liability in the U.S. Commercial Space Industry." Berkeley Journal of International Law 33.1.8 (2015): 223–273 <http://scholarship.law.berkeley.edu/bjil/vol33/iss1/8>.

[141]De Selding, Peter B. "News from the ITU Symposium on Small Satellite Regulation." 16 Mar. 2015. SpaceNews 12 Jan. 2016 <http://spacenews.com/news-from-the-itu-symposium-on-small-satellite-regulation/>.

The United Kingdom is also getting closer to determining where to build a spaceport to mainly conduct suborbital launch activities. On 3 March 2015, the British government, its UK Space Agency, and other agencies issued a joint public statement on the outcome of a consultation prepared by the UK's Civil Aviation Authority (CAA) on criteria for a commercial spaceport. Having reviewed the eight potential sites that were recommended by the report, the government reduced the list to six after receiving feedback from industry, government, and public stakeholders. The potential sites vary from military airfields to smaller commercial airports, with a majority of options located in Scotland. Four of the six sites are located in Scotland, while one is in Wales, and the other one is in England. The next step in the spaceport's development is for the British government's Department for Transport to publish detailed technical specifications of spaceport requirements, with a particular focus on suborbital vehicles and/or air launch systems that take off on runways.[142]

1.2.4 The United States

On 2 February 2015, the White House submitted its 2016 NASA budget proposal to the U.S. Congress, seeking $18.53 billion for 2016, an increase of 2.9 % from the $18.01 billion it received in 2015. The proposal contained few significant changes to the agency's programmes aside from a new initiative for the development of new Earth science missions, including the development of the Landsat 9 satellite and a Thermal-Infrared Free Flyer satellite, under the agency's Sustainable Land Imaging (SLI) multi-decade initiative. The Landsat 9 satellite, planned for launch in 2023, will be based on the Landsat 8 launched in 2013, but will correct design flaws in one of Landsat 8's instruments. The Thermal-Infrared Free Flyer, planned for launch in 2019, is meant to be a data gap-filler for the Landsat 7 mission which is expected to run out of propellant in that same year. Funding for other initiatives such as the Asteroid Redirect Mission (ARM)—initially proposed in 2014 but in stasis at the beginning of 2015 as planners were undecided on whether to redirect an entire 10 m asteroid or capture a small boulder off the surface of a larger asteroid—amounted to $220 million spread among the science, human exploration and operations, and space technology mission directorates. Funding for NASA's Orion programme decreased by 8.4 % to $1.1 billion from $1.19 billion in 2015, while funding for its SLS programme dropped by 20 % to 1.36 billion from 1.7 billion in 2015. However, the funding requested for NASA's Commercial Crew Programme received a significant boost reaching $1.24 billion for 2016, an increase of 54.5 % from $805 million in 2015; that increase is tied to milestones NASA

[142] Foust, Jeff. "British Government Narrows List of Potential Spaceport Sites." 3 Mar. 2015. SpaceNews 14 Dec. 2015 <http://spacenews.com/british-government-narrows-list-of-potential-spaceport-sites/>.

awarded to Boeing and SpaceX in September 2014. The budget request also cut funding for the Lunar Reconnaissance Orbiter (LRO) and Mars rover Opportunity in 2016.[143] The U.S. House Science Committee Chair expressed disapproval, finding too much funding going toward Earth science while not enough went to the Orion and SLS programmes[144]; the U.S. Senate Commerce space subcommittee voiced similar concerns in March 2015, with the essential issue boiling down to whether NASA's core mission of space exploration included Earth science rather than this being a competing priority.[145]

On 18 December 2015, U.S. President Obama signed the 2016 omnibus spending bill into law, including an adjusted NASA budget for 2016 amounting to $19.285 billion; an increase of $785 million above the White House budget request, and $1.285 billion more than NASA's budget in 2015.[146] The approved bill was a windfall for planetary science, and the Orion and SLS programmes, which drew substantially more funding than the initial budget request, while funding for the Earth Science SLI programme's Landsat 9 satellite and the Thermal-Infrared Free Flyer satellite appeared to survive the May 2015 budget cuts by the House Appropriations Committee.[147]

On 25 November 2015, U.S. President Obama signed into law the U.S. Commercial Space Launch Competitiveness Act (CSLCA) of 2015.[148] The CSLCA is the final version of a bill that reconciled a Senate promoted bill by the same name approved in August 2015 with the House bill known as the Spurring Private Aerospace Competitiveness and Entrepreneurship (SPACE) Act that was approved in May 2015.[149] The CSLCA is bound to have reverberating effects on the domestic U.S. space industry, in addition to the global space community, as it brings into question whether the act contravenes international

[143]Foust, Jeff. "For NASA, a New Budget But Old Battles Continue." 6 Feb. 2015. SpaceNews 10 Dec. 2015 <http://spacenews.com/for-nasa-a-new-budget-but-old-battles-continue/>.

[144]Ferster, Warren. "Editorial | NASA's Groundhog Day Budget Foreshadows More Stalemate." 16 Feb. 2015. SpaceNews 10 Dec. 2015 <http://spacenews.com/editorial-nasas-groundhog-day-budget-foreshadows-more-stalemate/>.

[145]Foust, Jeff. "Senators, Bolden Clash over the "Core Mission" of NASA." 12 Mar. 2015. SpaceNews 22 Dec. 2015 <http://spacenews.com/senators-bolden-clash-over-the-core-mission-of-nasa/>.

[146]Smith, Marcia S. "NASA Gets Big Boost in Final FY2016 Appropriations Bill." 16 Dec. 2015. Space Policy Online 30 June 2016 <http://www.spacepolicyonline.com/news/nasa-gets-big-boost-in-final-fy2016-appropriations-bill>.

[147]C.f. Foust, Jeff. "House Budget Cuts NASA Earth Science By More Than $250 Million." 19 May 2015. SpaceNews 22 Dec. 2015 <http://spacenews.com/house-budget-cuts-nasa-earth-science-by-more-than-250-million/>.

[148]"President Obama Signs Bill Recognizing Asteroid Resource Property Rights into Law." 25 Nov. 2015. Planetary Resources 26 Nov. 2015 <http://www.planetaryresources.com/2015/11/president-obama-signs-bill-recognizing-asteroid-resource-property-rights-into-law/>.

[149]Foust, Jeff. "U.S. Senate Passes Compromise Commercial Space Bill." 11 Nov. 2015. SpaceNews 26 Nov. 2015 <http://spacenews.com/u-s-senate-passes-compromise-commercial-space-bill/>.

space law treaties in recognising asteroid mining property rights for private commercial actors. The CSLCA also extends federal indemnification of commercial launches for third-party damages, and extends a restriction on the Federal Aviation Administration's (FAA) ability to enact safety regulations for people flying on commercial spacecraft. Moreover, the CSLCA also extends the operation and utilisation of the International Space Station (ISS), introduces a new type of astronaut, and seeks to identify appropriate oversight for the commercial development of space, and to streamline the FAA launch license process. Relatedly, from a previous estimate of the Senate's precursor to the bill, its implementation is likely to cost less than $500,000 over the 2016–2020 period, but it will jump to upwards of $14.3 billion between 2020 and 2025, assuming appropriation of the necessary funds.[150]

The most surprising development of the CSLCA concerns the inclusion of a 'ground-breaking' bill that permits prospecting and development of space resources. Intended to create more certainty and clarity for current and future U.S. investors in this burgeoning industry, Title IV of the CSLCA, on Space Resource Exploration and Utilization, promotes the right of U.S. citizens and companies to engage in commercial exploration and commercial recovery of space resources free from harmful interference. With regard to resource rights, the CSLCA grants commercial U.S. prospectors an entitlement to any asteroid resource or space resource obtained, including "to possess, own, transport, use, and sell the resource obtained". Aside from the extracted resources, the CSLCA makes sure to disclaim that the United States does not thereby assert sovereignty or sovereign or exclusive rights or jurisdiction over, or the ownership of, any celestial body.[151] However, the U.S. interpretation of Articles I, II, and VI of the international Outer Space Treaty[152] is likely to draw a substantial rebuff from some members within the international community that might view the U.S. asteroid mining property rights for private commercial actors as a violation of international law.

A more domestically-contentious issue in the lead-up to the CSLCA focussed on the duration of third-party launch indemnification. Originally introduced in 1988, with the latest indemnification authority set to expire on 31 December 2016, the indemnification provision means that the U.S. federal government will cover any third-party damages from a commercial launch accident in excess of the "maximum probable loss" insurance level that the U.S. company holding the launch license is

---

[150] "S. 1297, U.S. Commercial Space Launch Competitiveness Act." 17 June 2015. Congressional Budget Office 26 Nov. 2015 <https://www.cbo.gov/publication/50323>.

[151] "President Obama Signs Bill Recognizing Asteroid Resource Property Rights into Law." 25 Nov. 2015. Planetary Resources 26 Nov. 2015 <http://www.planetaryresources.com/2015/11/president-obama-signs-bill-recognizing-asteroid-resource-property-rights-into-law/>.

[152] Treaty on Principles governing the Activities of States in the Exploration and Use of Outer Space, including the Moon and Other Celestial Bodies (Outer Space Treaty), London/Moscow/Washington, done 27 January 1967, entered into force 10 October 1967, 610 UNTS 205, 6 ILM 386 (1967).

responsible for, up to a level of approximately $3 billion.[153] While the initial House bill offered an 8-year extension through 2023, some members found it to be too generous to the priorities of the commercial space industry, wanting to shorten it to a 5-year period, in addition to other cutbacks; similarly, the initial Senate bill had launch indemnification run through 2020.[154] Yet following several months of deliberation within the House and Senate, the extension period was increased through September 2025.[155]

Another evolving issue regarded the FAA's ability to enact safety regulations for people flying on commercial spacecraft. With the passing of the Commercial Space Launch Amendments Act of 2004, the commercial spaceflight industry was granted an 8-year 'learning period'—later extended by an additional 3 years—that restricted the FAA from enacting regulations regarding the safety of people who might actually fly on commercial spacecraft until 1 October 2015. While the original version of the House bill allowed for an 8-year extension of the learning period through 2023, echoing the indemnification issue, some House representatives wished to scale it back to 5 years, running through 2020; similarly, the initial Senate bill had launch indemnification running through 2020.[156] Yet to ensure investment in the industry, the learning period provision in the final CSLCA was extended to 1 October 2023; limiting the FAA's safety regulating ability to cases where there is an accident that causes injury, death, or extensive property damage.[157]

On less contentious issues, the CSLCA extends the operation and utilization of the ISS through 2024, helping to ensure mission certainty to the station even in case of a more financially restrictive future presidential administration.[158] The CSLCA also established a new 'Government Astronaut' class; a separate class from crew and space flight participants. This new class is in regard to government employees that are transported to space on commercial vehicles, such as with the upcoming SpaceX crewed Dragon capsule.[159] Other provisions within the act include a request to the Secretary of Transportation, in consultation with NASA, the

---

[153]Foust, Jeff. "Congress launches commercial space legislation." 26 May 2015. The Space Review 26 Nov. 2015 <http://www.thespacereview.com/article/2759/1>.

[154]Foust, Jeff. "Congress launches commercial space legislation." 26 May 2015. The Space Review 26 Nov. 2015 <http://www.thespacereview.com/article/2759/1>.

[155]Foust, Jeff. "U.S. Senate Passes Compromise Commercial Space Bill." 11 Nov. 2015. SpaceNews 26 Nov. 2015 <http://spacenews.com/u-s-senate-passes-compromise-commercial-space-bill/>.

[156]Foust, Jeff. "Congress launches commercial space legislation." 26 May 2015. The Space Review 26 Nov. 2015 <http://www.thespacereview.com/article/2759/1>.

[157]Dillow, Clay. "Obama is About to Give Private Space Companies a Big Break." 24 Nov. 2015. Fortune.com 26 Nov. 2015 <http://fortune.com/2015/11/24/obama-commercial-space-break/>.

[158]Dillow, Clay. "Obama is About to Give Private Space Companies a Big Break." 24 Nov. 2015. Fortune.com 26 Nov. 2015 <http://fortune.com/2015/11/24/obama-commercial-space-break/>.

[159]Press Releases. "Senate Approves U.S. Commercial Space Launch Competitiveness Act." 4 Aug. 2015. U.S. Senate Committee on Commerce, Science, & Transportation 26 Nov. 2015 <http://www.commerce.senate.gov/public/index.cfm/2015/8/senate-approves-u-s-commercial-space-launch-competitiveness-act>.

U.S. Secretary of State and other federal agencies, and the commercial space sector to identify an appropriate oversight mechanism that would protect public health and safety, safety of property, national security interests, and the foreign policy interests of the United States, in addition to eliminating duplicative requirements relevant to any commercial space launch or re-entry of a vehicle.[160] Here the FAA launch licence process allows a company to hold both an experimental permit and launch licence concurrently for the same vehicle, i.e. a licence for commercial missions and a less-restrictive permit for test flights.[161]

On April 2015, the U.S. Congress began to consider whether to renew the authorization of the U.S. Export-Import Bank (Ex-Im Bank). The Ex-Im Bank, and its regular partner Coface of France, entered the spotlight in 2015 while struggling with their near $400 million backing of the Australian start-up satellite operator, NewSat.[162] A decision of the U.S. Bankruptcy Court of Delaware on 22 May 2015 resulted in the cancellation of NewSat's contract with Lockheed Martin, allowing Lockheed to take possession of the nearly completed Jabiru-1 satellite. That decision put Ex-Im Bank's contract with NewSat in limbo, as Ex-Im Bank appeared to be in danger of having an unrecoverable loss of over $100 million.[163] To compound the situation, the decision arrived about a month prior to the set expiry of the Ex-Im Bank's charter (i.e. 30 June 2015), with industry opinions varying as to whether it would tip Congress' opinion in opposition of reauthorizing the bank, or maybe increase the amount of scrutiny over its investments.[164] The Ex-Im Bank, like its counterpart agencies in France, China, Russia, and elsewhere, finance overseas projects that provide export business for its domestic companies; in the space sector, the Ex-Im Bank often pairs up with France's Coface to launch U.S. built telecommunications satellites into orbit on Europe's Ariane 5 launcher. Similarly, China's government-backed loans, and soon Russia's as well, stimulate the sale of their commercial satellites to other countries by offering bundled packages to place them into orbit.[165] Yet, these practices have

[160] "H.R.2262 – U.S. Commercial Space Launch Competitiveness Act | Text: H.R.2262 — 114th Congress (2015–2016)." 25 Nov. 2015. Congress.gov 2 Dec. 2015 <https://www.congress.gov/bill/114th-congress/house-bill/2262>.

[161] Foust, Jeff. "Senate Passes Commercial Space Bill." 5 Aug. 2015. SpaceNews 26 Nov. 2015 <http://spacenews.com/senate-passes-commercial-space-bill/>.

[162] De Selding, Peter B. "As Ex-Im Battle Brews, Novel Satellite Export Deals Stir Pot." 7 Apr. 2015. SpaceNews 13 Jan. 2016 <http://spacenews.com/as-ex-im-battle-brews-novel-satellite-export-deals-stir-pot/>.

[163] De Selding, Peter B. "NewSat Bankruptcy is Costly for U.S. Ex-Im Bank." 28 May 2015. SpaceNews 13 Jan. 2016 <http://spacenews.com/newsat-bankruptcy-is-costly-for-u-s-ex-im-bank/>.

[164] De Selding, Peter B. "Will NewSat's Failure Dim Ex-Im's Appetite for Satellite Deals?" 1 June 2015. SpaceNews 13 Jan. 2016 <http://spacenews.com/will-newsats-failure-dim-ex-ims-appetite-for-satellite-deals/>.

[165] De Selding, Peter B. "As Ex-Im Battle Brews, Novel Satellite Export Deals Stir Pot." 7 Apr. 2015. SpaceNews 13 Jan. 2016 <http://spacenews.com/as-ex-im-battle-brews-novel-satellite-export-deals-stir-pot/>.

drawn criticism by some representatives in the U.S. Congress who deride it as a form of market-distorting corporate welfare.[166]

On 1 July 2015, the Ex-Im Bank's authority lapsed, closing its doors to new satellite and other export business, but continuing all pre-existing loans, guarantees, and insurance policies into their maturity.[167] Nevertheless, even with other export credit agencies leaping in to fill the gap in financing, the effect of the Ex-Im Bank closure meant that there was not enough money available in the market to finance every new order of telecommunications satellites.[168] One casualty of the Ex-Im Bank closure was ABS of Hong Kong, which was forced to suspend its order for the ABS-8 satellite from Boeing, as the contract was in the midst of arranging financing when the lapse occurred[169]; and another deal was said to be lost by Orbital ATK who had been the frontrunner bidder to construct a second Azerspace-2 satellite for the government of Azerbaijan.[170]

On 27 October 2015, lawmakers in the U.S. House of Representatives gave bipartisan approval to reauthorize the Ex-Im Bank after a clutch of representatives succeeded in forcing its vote using the rarely invoked discharge petition mechanism earlier in the month.[171] With the bill having gained Senate approval shortly thereafter, U.S. President Obama signed the legislation into law on 4 December 2015, extending the bank's life through September 2019. However, a lack of quorum in the Ex-Im Bank's five-seat board meant that satellite financing continued to be stalled by the end of 2015; such loans must await confirmation of President Obama's nominees by the Senate.[172] Another open question is the legal mechanism the Ex-Im Bank could use to recoup some of its NewSat investment.[173]

[166]SpaceNews Editor. "Editorial | NewSat's Failure and the Ex-Im Bank." 15 June 2015. SpaceNews 13 Jan. 2016 <http://spacenews.com/editorial-newsats-failure-and-the-ex-im-bank/>.

[167]De Selding, Peter B. "Ex-Im is Closed to New Business; Will the U.S. Space Industry Suffer?" 2 July 2015. SpaceNews 13 Jan. 2016 <http://spacenews.com/ex-im-is-closed-to-new-business-will-the-u-s-space-industry-suffer/>.

[168]De Selding, Peter B. "MDA Corp. Explains Drop in Satellite Orders, Says Move Out of U.S. Possible." 3 Aug. 2015. SpaceNews 13 Jan. 2016 <http://spacenews.com/mda-corp-explains-drop-in-satellite-orders-says-move-out-of-u-s-possible/>.

[169]De Selding, Peter B. "Boeing, Lockheed Gaze Abroad as Ex-Im Closure Costs them Satellite Sales." 18 Sept. 2015. SpaceNews 13 Jan. 2016 <http://spacenews.com/boeing-lockheed-gaze-abroad-as-ex-im-closure-costs-them-satellite-sales/>.

[170]Foust, Jeff. "The Import of the Satellite Exporter's Bank." 29 Sept. 2015. SpaceNews 13 Jan. 2016 <http://spacenews.com/the-import-of-the-satellite-exporters-bank/>.

[171]Timiraos, Nick. "House Votes to Reauthorize U.S. Export-Import Bank." 27 Oct. 2015. The Wall Street Journal 11 Mar. 2016 <http://www.wsj.com/articles/house-votes-to-reauthorize-u-s-export-import-bank-1445986019>.

[172]Calmes, Jackie. "Ex-Im Bank Is Reopened, but Big Loans Are Stalled." 7 Dec. 2015. The New York Times 11 Mar. 2016 <http://www.nytimes.com/2015/12/08/business/ex-im-bank-is-reopened-but-big-loans-are-stalled.html>.

[173]De Selding, Peter B. "Ex-Im Hasn't Given Up on Recovering $100 Million NewSat Loss." 9 June 2015. SpaceNews 13 Jan. 2016 <http://spacenews.com/ex-im-hasnt-given-up-on-recovering-100-million-newsat-launch/>.

1.2.5 Canada

On 27 February 2015, Canada's Prime Minister Stephen Harper announced the selection of Sylvain Laporte as president of the Canadian Space Agency (CSA). Effective 9 March 2015, Laporte replaced Luc Brûlé who had been sitting as the acting president of the CSA following the departure of the previous CSA president, Walter Natynczyk in November 2014. Laporte's experience stems from a broad background within Industry Canada, Canada Post Corp., and the Royal Canadian Air Force.[174]

The CSA fiscal budget for 2015, extending from 1 April 2015 to 31 March 2016, increased by 27.65 % to C$483.43 from C$378.72 million the previous 2014 fiscal year.[175] Moreover, on 21 April 2015, Canada announced its commitment to increasing its member contribution to the European Space Agency (ESA). Under its Economic Action Plan, Canada plans to increase its spending by an additional C$30 million for ESA's ARTES programme, distributed over the period of 2016–2019. Canada also committed to extending its participation in the ISS to 2024, and has historically provided 2.3 % of the ISS's common operating costs.[176] Canada's contribution to ESA was 15.5 million euros (C $16.7 million) in 2015.[177]

With the U.S. Export-Import Bank (Ex-Im Bank) shut down for financing satellite contracts for part of 2015, business across its border and across the Atlantic Ocean looked to gain a noticeable uptick; here, Canada's Export Development Canada (EDC), and France's Coface export credit agency sought to fill the void left by the Ex-Im Bank. One company that is in a favourable position is Canada's MacDonald, Dettwiler and Associates Ltd. (MDA), the parent company of top prime contractor Space Systems/Loral (SS/L). In this case, because SS/L's parent is a Canadian company, under the EDC credit agency the U.S. based SS/L would be considered a Canadian company as well, and would be eligible for EDC backing without being required to provide work in Canada in return. Unlike with other top prime contractors in the U.S. and in Europe, with a Canadian identity SS/L is able to keep its entire operation in the U.S. with Canadian financing. In contrast, if Boeing had turned to the EDC or France's Coface export credit agency when its commercial satellite contract with ABS of Hong Kong stalled in mid-2015 following the

---

[174]Foust, Jeff. "Laporte Named President of Canadian Space Agency." 2 Mar. 2015. SpaceNews 14 Dec. 2015 <http://spacenews.com/laporte-named-president-of-canadian-space-agency/>.

[175]"2015–16 Report on Plans and Priorities." 31 Mar. 2015. Canadian Space Agency 3 July 2015: 12 <http://www.asc-csa.gc.ca/pdf/eng/publications/rpp-2015.pdf>.

[176]SpaceNews Editor. "Editorial | Canadian Commitment Builds Momentum for Space Station Extension." 5 May 2015. SpaceNews 11 Jan. 2016 <http://spacenews.com/editorial-canadian-commitment-builds-momentum-for-space-station-extension/>.

[177]De Selding, Peter B. "Canada's New Space Budget Extends ISS Commitment to 2024." 22 Apr. 2015. SpaceNews 10 Jan. 2016 <http://spacenews.com/canadas-new-space-budget-extends-iss-commitment-to-2024/>.

shutdown of the Ex-Im Bank, it is likely that part of the work would have had to be conducted within those countries as an additional requirement for funding.[178]

1.2.6 Russia

Russia's merger of its Federal Space Agency (Roscosmos) and the newly-founded state-owned holding company United Rocket and Space Corporation (URSC) was approved by President Vladimir Putin on 22 January 2015. While the two organisations were initially meant to co-exist in an arms-length customer and contractor relationship, merging the two into one entity was seen as a step in raising Russia's competitiveness, both in gaining market share, and in securing parity and advantage over geopolitical opponents.[179] The merger might help to strengthen Russia's struggling space industry, which has a seen a number of high-profile failures in recent years.

While in the midst of its restructuring, Russia's space industry experienced a 13 % decrease in export revenue, earning 4.374 billion roubles ($76.99 million) for the year in 2014; most of that decline occurred prior to the rouble's fall late in the year.[180] The rouble fell even further in 2015, bottoming out at US¢14.09 to 1 rouble near the end of the year—less than half its value in 2013 when Russia planned to spend 2.1 trillion roubles (~$63 billion in 2013 prices) including extra-budgetary sources, to boost the development of its national space activities from 2013 to 2020. The rouble's free-fall led to Russia's reassessment of its draft Federal Space Program 2016–2025 proposal, which initially requested 3.4 trillion roubles ($52.5 billion) in March 2015, but was granted just 1.5 trillion roubles ($22.5 billion) as at 1 December 2015 for the 10 year period.[181]

---

[178] De Selding, Peter B. "Boeing, Lockheed Gaze Abroad as Ex-Im Closure Costs them Satellite Sales." 18 Sept. 2015. SpaceNews 24 Dec. 2015 <http://spacenews.com/boeing-lockheed-gaze-abroad-as-ex-im-closure-costs-them-satellite-sales/>.

[179] Bodner, Matthew. "Putin Approves Roscosmos Merger with Conglomerate." 23 Jan. 2015. SpaceNews 12 Jan. 2016 <http://spacenews.com/putin-approves-roscosmos-merger-with-conglomerate/>.

[180] De Selding, Peter B. "Roscosmos Details Russia's Struggling Space Sector." 20 Apr. 2015. SpaceNews 13 Jan. 2016 <http://spacenews.com/roscosmos-details-russias-struggling-space-sector/>.

[181] "Roscosmos 10-Year Budget Cut For Third Time." 1 Dec. 2015. The Moscow Times 1 July 2016 <http://www.themoscowtimes.com/business/article/roscosmos-10-year-budget-cut-for-third-time/551337.html>.

1.2.7 Japan

Japan's Space Policy Commission published a third iteration of its Basic Plan for Space Policy on 9 January 2015. Marking a shift in priorities from the two previous iterations published in 2009, and updated early in 2013, the new 10-year roadmap focuses on security and commerce from its earlier emphasis on the peaceful use of outer space. Another departure in this policy is in its naming of China as a destabilizing factor in global security, particularly in its growing counter-space capability and development of anti-satellite weapons.[182] The new policy puts greater focus on developing Japan's Information Gathering Satellites (IGS) to further enable the country's surveillance and reconnaissance competencies. Moreover, it looks to increasing its cooperation with the U.S. on an equal basis, while also maintaining and strengthening its own industrial and science and technology sector.[183] A revised version of the plan was released on 11 November 2015, this time recommending that the IGS system be expanded from the original two optical and two radar satellites (plus one on-orbit spare), to eight satellites (plus two relay satellites to support the constellation).[184] The current constellation has four optical satellites, and three radar satellites (including one spare) in orbit; Japan launched its IGS-Radar Spare on 1 February 2015, and its IGS-Optical 5 on 26 March 2015.

In line with its new policy, Japan's combined space budget was increased by 18.5 %, amounting to ¥324.5 billion for the fiscal year 2015 (beginning on 1 April 2015 and ending 30 March 2016).[185] Its 2014 budget in the same fiscal period is estimated to be around ¥274 billion.[186] The new budget, encompassing the space activity of 11 government ministries, saw a 68 % increase in funding (i.e. ¥22.3 billion) moving toward the government's Cabinet Office (with most of the funding going toward its Quasi-Zenith regional navigation system); its Cabinet Secretariat, tasked with controlling Japan's IGS system, received a 14 % funding increase (i.e. ¥69.7 billion). The Education Ministry, governing the Japan Aerospace Exploration Agency (JAXA), also had a 19 % funding increase (i.e. ¥182 billion); within

---

[182]Kallender-Umezu, Paul. "Japan Boosts Space Spending In Support of Security Focus." 2 Feb. 2015. SpaceNews 10 Dec. 2015 <http://spacenews.com/japan-boosts-space-spending-in-support-of-security-focus/>.

[183]Rajagopalan, Rajeswari Pillai. "Op-ed | Japan's Space Policy Shift Reflects New Asian Realities." 23 Feb. 2015. SpaceNews 10 Dec. 2015 <http://spacenews.com/op-ed-japans-space-policy-shift-reflects-new-asian-realities/>.

[184]Kallender-Umezu, Paul. "What's Behind Japan's Sudden Thirst for More Spy Satellites." 13 Nov. 2015. SpaceNews 7 Jan. 2016 <http://spacenews.com/whats-behind-japan-sudden-thirst-for-more-spy-satellites/>.

[185]Kallender-Umezu, Paul. "Japan Boosts Space Spending In Support of Security Focus." 2 Feb. 2015. SpaceNews 10 Dec. 2015 <http://spacenews.com/japan-boosts-space-spending-in-support-of-security-focus/>.

[186]Euroconsult 2015. Profiles of Government Space Programs.

JAXA's allocation, its next-generation H-3 launcher (to be completed in 2020) received a 78 % increase to ¥12.5 billion.[187]

And in line with its enhanced cooperation with the U.S., Japan indicated that it would agree to extend its participation in the International Space Station (ISS) through 2024. Yet by the end of 2015, no formal agreement on extending the station was in place, as the Japanese and U.S. governments were still coordinating a new cooperation framework. Japan spends about ¥40 billion on the ISS annually; however sustaining that contribution might call for greater participation in the ISS, in addition to its symbolic value showing bilateral cooperation with the U.S.[188]

1.2.8 China

The year 2015 ended the cycle of China's 12th Five Year Plan (2011–2015); its 13th Five Year Plan (2016–2020) was preliminarily approved by China's Communist Party on 29 October 2015 with its details to be finalised in March 2016.[189] Nevertheless, the 13th Plan is expected to address the following four economic thematic areas: (1) secure a high growth target for 2020; (2) place more emphasis on domestic consumption with less focus on investment and exports; (3) further embrace competitive market mechanisms as part of its structural reform; and (4) enhance individual well-being through social welfare and healthcare reforms.[190] In that last thematic area, China is expected to end its 35-year-running one-child family planning policy, which has resulted in a top-heavy aging population.[191] However, some of the rhetoric behind this 'new normal' benchmark for China has raised some mixed concerns; while it promises a real-time environmental monitoring system, increased low-carbon public transportation, and more alternative energy vehicles—it also conveys an aim to tighten controls over the internet, art

[187]Kallender-Umezu, Paul. "Japan Boosts Space Spending In Support of Security Focus." 2 Feb. 2015. SpaceNews 10 Dec. 2015 <http://spacenews.com/japan-boosts-space-spending-in-support-of-security-focus/>.

[188]Foust, Jeff. "Japan Seeks To Become Full Partner with U.S. in Space." 11 Dec. 2015. SpaceNews 7 Jan. 2016 <http://spacenews.com/japan-seeks-to-become-full-partner-with-u-s-in-space/>.

[189]Magnier, Mark. "China's Communist Party Approves Five-Year Plan." 29 Oct. 2015. The Wall Street Journal 1 Feb. 2016 <http://www.wsj.com/articles/chinas-communist-party-approves-five-year-plan-1446124597>.

[190]Reeves, Martin, and David He. "What China's 13th Five-Year Plan Means for Business." 7 Dec. 2015. Harvard Business Review 1 Feb. 2016 <https://hbr.org/2015/12/what-chinas-13th-five-year-plan-means-for-business>.

[191]Moody, Andrew. "Dissecting China's Five-Year Plan." 23 Nov. 2015. The Telegraph 1 Feb. 2016 <http://www.telegraph.co.uk/sponsored/china-watch/politics/12006280/china-five-year-plan.html>.

and culture, and increase its say in areas including the exploration of space, the deep sea and polar regions.[192]

The year also brought new prospects for space cooperation between China and other major space actors. Europe signed a framework agreement with the China Manned Space Agency (CMSA) in December 2014, thereafter establishing three working groups: space experiments and utilisation; astronaut selection, training and flight; and space infrastructure to analyse and propose concrete cooperation areas of mutual interest.[193] Russia has also signed an initial space station cooperation agreement with China.[194] And there are also indications that China and the U.S. will increase cooperation in the near future. While the export controls entrenched in the U.S. Strom Thurmond National Defense Authorization Act of 1999 restrict U.S. manufacturers from selling launchers, satellites and their components to China over concerns of technology transfer, it was the U.S. Consolidated and Further Continuing Appropriations Act of 2012 (signed into law on 18 November 2011) that banned any funding meant for bilateral cooperation or communication between NASA and the White House Office of Science and Technology Policy (OSTP) with China.[195] Since then, it has become increasingly apparent to both NASA and the OSTP that the U.S. is at risk of being left behind if it continues to exclude China in space exploration projects. However in mid-2015 meetings between China and the U.S. State Department (not barred by those restrictions), the two bodies agreed to consult with one another as part of a new U.S.-China Civil Space Cooperation Dialogue, for the purpose of satellite collision avoidance and the long-term sustainability of space.[196] The first meeting took place on 28 September 2015 in Beijing, China; a second dialogue meeting was expected to take place in Washington, D.C., in 2016.[197] Additionally, a Chinese experiment may be sent to Japan's Kibo module on the ISS in 2016, as

---

[192]Shi, Ting. "China's Big Development Plan Has Xi's Fingerprints All Over It." 3 Nov. 2015. Bloomberg.com 1 Feb. 2016 <http://www.bloomberg.com/news/articles/2015-11-03/china-s-big-development-plan-has-xi-s-fingerprints-all-over-it>.

[193]David, Leonard. "US-China Cooperation in Space: Is It Possible, and What's in Store?" 16 June 2015. Space.com 22 Jan. 2016 <http://www.space.com/29671-china-nasa-space-station-cooperation.html>.

[194]De Selding, Peter B. "China's Space Station Planners Put out Welcome Mat." 13 Oct. 2015. SpaceNews 11 Jan. 2016 <http://spacenews.com/chinas-space-station-planners-put-out-welcome-mat/>.

[195]Section 539 Consolidated And Further Continuing Appropriations Act, 2012, Pub. L. no 112-55, 125 Stat. 639 (18 Nov. 2011) <https://www.gpo.gov/fdsys/pkg/PLAW-112publ55/pdf/PLAW-112publ55.pdf>.

[196]Smith, Marcia S. "U.S., China Agree to Bilateral Civil Space Cooperation Dialogue." 26 June 2015. SpacePolicyOnline.com 22 Jan. 2016 <http://www.spacepolicyonline.com/news/u-s-china-agree-to-bilateral-civil-space-cooperation-dialogue>.

[197]"The First Meeting of the U.S.-China Space Dialogue." 28 Sept. 2015. U.S. Department of State 22 Jan. 2015 <http://www.state.gov/r/pa/prs/ps/2015/09/247394.htm>.

part of a commercial arrangement brokered by the private U.S. company, NanoRacks.[198]

China also released its Middle and Long Term Development Plan for State Civil Space Infrastructure (2015–2025) on 26 October 2015.[199] Under this new roadmap, satellite remote sensing, communications, broadcasting, and navigation systems will be built during the 13th Five Year Plan to establish the state civil space infrastructure system. Using those systems, the 2015–2025 space development plan aims to produce comprehensive application demonstrations in 12 fields (including territory, mapping, energy, communications, and environmental protection) to provide core business with timely, accurate, and stable space information service. Additionally, the 2015–2025 plan stresses the importance of investment in its domestic industry, and calls for more investment of private capital.[200] For instance, it foresees 100 launches of its Long March launcher family during the 2015–2025 period to meet domestic demand; it also aims to court commercial launch contracts by providing commercial launch services outside of its territory.[201]

In the month preceding the 2015 United Nations Climate Change Conference (COP 21) held in Paris, France on 30 November–12 December 2015, Chinese news sources reported that China planned on launching a global carbon dioxide observatory satellite into orbit in 2016. The so-called "CO_2 Observatory" satellite will carry carbon dioxide sensors that will have an accuracy better than four parts per millions.[202] The system would allow China to monitor its own carbon dioxide emissions, along with emissions in other major regions across the globe. Across the Pacific, an open question remains as to whether the U.S. NOAA or any other U.S. institution will rely on Chinese environmental data, despite anticipating a gap in U.S. weather coverage as early as 2017 for parts of Asia, including the Middle East and Afghanistan.[203]

And in anticipation of the inevitability of reusable launch services, China looks eager to compete on the global commercial launch market. At the 66th International

[198]David, Leonard. "US-China Space Freeze May Thaw with Historic New Experiment." 21 Aug. 2015. Space.com 22 Jan. 2016 <http://www.space.com/30337-chinese-experiment-international-space-station.html>.

[199]Google Translated. "关于印发国家民用空间基础设施中长期发展规划(2015–2025年)的通知." 26 Oct. 2015. National Development and Reform Commission 2 Feb. 2016 <http://www.sdpc.gov.cn/zcfb/zcfbghwb/201510/t20151029_756376.html>.

[200]"Middle and Long Term Development Plan for Civil Space Infrastructure Issued." 13 Jan. 2016. LexisNexis 2 Feb. 2016 <https://hk.lexiscn.com/latest_message.php?access=show_detail&id=184408>.

[201]De Selding, Peter B. "With Naga-L Rocket, China Would Turn Tables on U.S. Export Ban." 15 Oct. 2015. SpaceNews 24 Dec. 2015 <http://spacenews.com/with-naga-l-rocket-china-would-turn-tables-on-u-s-export-ban/>.

[202]"China plans to launch CO2 monitoring satellite in 2016." 23 Oct. 2015. English.News.Cn 22 Jan. 2016 <http://news.xinhuanet.com/english/sci/2015-10/23/c_134743536.htm>.

[203]Gruss, Mike. "U.S. Military Faces Weather Coverage Gap over Hot Zones." 13 Mar. 2015. SpaceNews 22 Dec. 2015 <http://spacenews.com/u-s-central-command-faces-weather-coverage-gap-over-hot-zones/>.

Astronautical Congress (IAC) held on 12–16 October 2015 in Jerusalem, Israel, the China Academy of Launch Vehicle Technology (CALT) announced that it was designing a new rocket that would be launched at sites outside of China which would allow it to bypass some U.S. ITAR restrictions and launch international commercial satellites.[204] The two-stage Naga-L launcher will be able to lift small satellites into LEO orbit (i.e. 1550 kg to 400 km LEO, 820 kg to 500 km SSO, and 620 kg to 700 km) at a cost of $10 million per flight. Its first flight is expected to take place at the end of 2017.[205] CALT has already begun negotiations with authorities in Indonesia, Sweden, and Tanzania to host its launcher. Previously, in July 2015, the Chinese government had been in talks with Russia's Energia over the purchase of the struggling commercial launch service provider, Sea Launch AG. As with the Naga-L launcher, China was considering stationing the launcher in international waters outside its territory to skirt ITAR restrictions and court the global commercial launch market. Yet the sale of the multi-national Sea Launch to China would likely come with an array of regulatory and political hurdles.[206] China may have lost interest in the purchase in the following months, as Energia was still in search of a buyer by the end of 2015.[207]

1.2.9 India

The Indian government is in the middle of its 12th 5-year plan (2012–2017), which puts emphasis on speeding up the implementation of infrastructure projects, and aims to achieve an annual average economic growth rate of 8.2 %.[208] India's space programme aims to strengthen its operational services in communications and navigation; enhance its remote sensing imaging capability; continue space science missions for better understanding of the solar system and the universe; and to build on its experience in planetary exploratory missions.

On 9 March 2015, India's parliament allocated 73.9 billion rupees ($1.18 billion) to the Indian Space Research Organisation (ISRO) for the fiscal year 2015–2016,

---

[204] De Selding, Peter B. "With Naga-L Rocket, China Would Turn Tables on U.S. Export Ban." 15 Oct. 2015. SpaceNews 24 Dec. 2015 <http://spacenews.com/with-naga-l-rocket-china-would-turn-tables-on-u-s-export-ban/>.

[205] Messier, Doug. "China Developing New Smallsat Launch Vehicle." 15 Oct. 2015. Parabolic Arc 29 Jan. 2016 <http://www.parabolicarc.com/2015/10/15/china-developing-smallsat-launch-vehicle/>.

[206] De Selding, Peter B. "China Eyes Purchase of Sea Launch Assets." 17 July 2015. SpaceNews 13 Jan. 2016 <http://spacenews.com/china-eyes-purchase-of-sea-launch-assets/>.

[207] "Sea Launch On the Sales Block?" 3 Dec. 2015. Satnews 29 Jan. 2016 <http://www.satnews.com/story.php?number=25117559>.

[208] "Government Approves 12th Five Year Plan." 4 Oct. 2012. The Times of India 3 May 2014 <http://timesofindia.indiatimes.com/business/india-business/Government-approves-12th-five-year-plan/articleshow/16672927.cms>.

beginning on 1 April 2015. The budget is a modest increase from the 72.4 billion rupees ($1.21 billion) initially allocated in the previous fiscal year. However, as with the previous fiscal year budget which was later revised to 52 billion rupees, it is possible that ISRO's spending budget for 2015–2016 will be less than initially allocated.[209] About 35.5 % of the new fiscal budget was allocated toward launch vehicle technology, covering the continued development and operation of India's Geosynchronous Satellite Launch Vehicle (GSLV)-Mark 3 launcher, its workhorse Polar Satellite Launch Vehicle (PSLV), and other related technologies. Another 17.9 % was allocated toward ISRO's INSAT constellation of communications and meteorological satellites, while 15.0 % went toward ISRO's satellite technology budget. Moreover, 13.0 % of the budget will be put toward India's space applications, while another 11.7 % will be put toward launch support, tracking network and range facilities such as increasing India's satellite launching capacity by increasing ISRO's Satish Dhawan Space Centre launch range infrastructure. Just 4.1 % of the budget will go toward ISRO's space sciences, a notable decrease from the 5.8 % initially allocated in the previous fiscal year; and lastly, direction and administration including other programmes amount to the remaining 1.6 % of the fiscal year budget.[210]

On 21 May 2015, India's office of the Prime Minister announced that it had approved an order to build 15 additional PSLV launchers to be used between 2017 and 2020, at a rate of 4–5 launches per year. Upon the completion of that order, the ISRO's docket of PSLV launchers will have reached 50 in total. While the launch system has mainly been used for India's institutional missions, the PSLV has also provided low-cost commercial launch services to governments and commercial operators from Europe, Canada, Algeria, and the U.S. (when granted by a waiver).[211] As the U.S. government reconsiders its policy on discouraging the use of the PSLV toward U.S. commercial and government operators based on its earlier view of unfair market-based pricing, ISRO's commercial arm Antrix may be in a position to win additional commercial launch service contracts for both its PSLV and GSLV launchers.[212]

At the start of 2015, Shailesh Nayak, India's Ministry of Earth Sciences secretary, was asked to also be the interim secretary of the Department of Space following the retirement of ISRO's Chairman Koppillil Radhakrishnan on 31 December 2014. The Department of Space secretary typically has the dual function of also serving as the Chairman of ISRO.[213] Nearly 2 weeks into his

[209] Jayaraman, K.S. "India Allocates $1.2 Billion for Space Activities." 9 Mar. 2015. SpaceNews 14 Dec. 2015 <http://spacenews.com/india-allocates-1-2-billion-for-space-activities/>.

[210] "Department of Space | No.93/Department of Space." 1 Mar. 2015. Indiabudget.nic.in 11 Feb. 2016 <http://indiabudget.nic.in/ub2015-16/eb/sbe93.pdf>.

[211] De Selding, Peter B. "India OKs Budget for Building, Launching 15 PSLV Rockets by 2020." 22 May 2015. SpaceNews 5 Jan. 2016 <http://spacenews.com/india-to-build-and-launch-15-pslv-rockets-by-2020/>.

[212] SpaceNews Editor. "Getting the Cubesat Revolution Out of Low Gear." 24 Nov. 2015. SpaceNews 6 Jan. 2016 <http://spacenews.com/getting-the-cubesat-revolution-out-of-low-gear/>.

[213] Jayaraman, K.S. "Nayak Named Interim Chief at ISRO." 6 Jan. 2015. SpaceNews 10 Dec. 2015 <http://spacenews.com/nayak-named-interim-chief-at-isro/>.

role, on 13 January 2015, the Indian government announced that Alur Seelin Kiran Kumar would relieve Nayak, and run ISRO for 3 years beginning in 2015.[214]

Following the 14 September 2015 International Criminal Court (ICC) ruling directing ISRO's commercial arm Antrix Corp. to pay $672 million in damages to Devas Multimedia (an Indian company backed by German and U.S. investors), the Indian government announced that it will challenge the ruling and apply for remedy. In January 2005, Antrix and Devas agreed to have ISRO build and launch the GSAT-6 and GSAT-6A S-band communication satellites, and lease 90 % of its transponder capacity to Devas for 12 years, in exchange for $300 million. The deal was nullified by the Indian government on 25 February 2011, when the S-band spectrum used by the satellites was reallocated to India's defence forces. Devas subsequently turned to the ICC seeking $1.6 billion in damages, arguing that the cancellation had been arbitrary and illegal.[215]According to the arbitration statement, interest on the award will accrue at 18 % per year until damages are settled.[216]

1.2.10 Brazil

Despite long-standing obstacles for Brazil's space and R&D development, and the lack of expertise in its workforce, Brazil is still an important space player of the future due to its geographic position on the equator, and its growing economy.

In terms of partnerships with space powers around the globe, Brazil has formed a number of cooperation arrangements including cooperation with China on Earth observation through the China-Brazil Earth Resources Satellite (CBERS) project since 1988, hosting three monitoring stations for Russia's Glonass GNSS system, small satellite development with Argentina and Japan, and space technology agreements involving manufacturers in Canada and France, among others. However, the broad application of the United States' International Traffic in Arms Regulations (ITAR) export controls, even after modifications entered into force in November 2014, has been a main factor in Brazil's resistance in forming closer ties with the U.S. space sector. While the U.S. is the main source of components in the global

---

[214]Jayaraman, K.S. "India Taps ISRO Veteran To Run Space Program." 13 Jan. 2015. SpaceNews 10 Dec. 2015 <http://spacenews.com/india-taps-isro-veteran-to-run-space-program/>.

[215]Jayaraman, K. S. "Antrix To Appeal $672M Judgment for Nixed Satellite Contract." 30 Sept. 2015. SpaceNews 13 Jan. 2016 <http://spacenews.com/antrix-to-appeal-672m-judgment-for-nixed-satellite-contract/>.

[216]Jai Krishna, R. "ISRO's Antrix Corp asked to pay $562 million in damages in Devas Multimedia dispute." 30 Sept. 2015. The Financial Express 12 Feb. 2016 <http://www.financialexpress.com/article/industry/companies/antrix-corp-asked-to-pay-562-million-in-damages-in-devas-multimedia-dispute/143518/>.

space sector, Brazilian authorities tend to favour competitive bids that have the fewest U.S. parts to avoid potential ITAR restrictions.[217]

Yet in the midst of a slumping economy, Brazil's government has decided to end its decade-long space launch cooperation agreement with Ukraine to operate Ukraine's Cyclone-4 rocket from Brazil's Alcântara Launch Centre (CLA) due to budget constraints, cost overruns, and uncertainty about the future market success of the arrangement.[218] Brazil and Ukraine had created the binational company "Alcântara Cyclone Space" (ACS), with financing divided equally, for commercial launches using the Ukrainian vehicle Cyclone-4 from the CLA spaceport whose proximity to the equator gives the launch vehicle more capacity to orbit than can be offered from Russian spaceports.[219] The launcher development program cost an estimated 1.58 billion reals over the 10-year period, while the Cyclone-4 program was allocated an additional 459.8 million reals during that time. Another 1.9 billion reals had been allocated for space infrastructure improvements, mostly going toward development of the CLA.[220] It is unclear whether Brazil will have to pay Ukraine any financial penalties for the unilateral cancellation of their agreement.[221]

1.3 Worldwide Space Budgets and Revenues

Total government space expenditure was $76.52 billion in 2015, down from $80.42 billion in 2014—the reduction is explained in part by currency fluctuations resulting from the strengthened U.S. dollar to the currencies of many spacefaring countries.[222] Total government expenditure for civil space programs decreased by 3.4 % to $42.37 billion from $43.84 billion in 2014. The compound annual growth rate (CAGR) of the entire space industry, including commercial revenues and government expenditure appeared to decrease by 2.1 %, masking the growth experienced by most of the global space sector whose financial activity is conducted

[217]De Selding, Peter B. "Brazil Bypassing the U.S. as It Builds out a Space Sector." 16 Apr. 2015. SpaceNews 22 Dec. 2015 <http://spacenews.com/brazil-bypassing-the-us-as-it-builds-out-a-space-sector/>.

[218]De Selding, Peter B. "Brazil Pulling Out of Ukrainian Launcher Project." 16 Apr. 2015. SpaceNews 22 Dec. 2015 <http://spacenews.com/brazil-pulling-out-of-ukrainian-launcher-project/>.

[219]"News from the 64th International Astronautical Congress | Backers Insist Brazilian Spaceport Is Nearing Launch Readiness." 4 Oct. 2013. SpaceNews 3 May 2014 <http://www.spacenews.com/article/launch-report/37550news-from-the-64th-international-astronautical-congress-backers-insist>.

[220]"Brazil Scales Back Launch Vehicle Plans." 10 Feb. 2013. Parabolic Arc 3 May 2014 <http://www.parabolicarc.com/2013/02/10/brazil-scales-back-launch-vehicle-plans/>.

[221]De Selding, Peter B. "Brazil Pulling Out of Ukrainian Launcher Project." 16 Apr. 2015. SpaceNews 22 Dec. 2015 <http://spacenews.com/brazil-pulling-out-of-ukrainian-launcher-project/>.

[222]The Space Report 2016. Colorado Springs: The Space Foundation, 2016: 16.

in other currencies.[223] The following section provides a more detailed analysis of institutional budgets.

The Space Report 2016 noted that the total revenue of commercial satellite services, including telecommunications, Earth observation and positioning services, increased by 3.65 %, to $126.33 billion in 2015 from $121.88 billion in 2014. However, revenue from space-related commercial infrastructure, including manufacturing of spacecraft and in-space platforms, launch services as well as ground equipment decreased by 5.22 % to $120.09 billion in 2015 from $126.71 billion in 2014; the decrease mainly came from the revenue of GNSS receivers which is benchmarked in euros hence subject to the same currency fluctuation. Overall, total commercial space revenue decreased by 0.9 % to $246.42 billion in 2015 from $248.59 billion in 2014.[224]

1.3.1 Overview of Institutional Space Budgets

From the Space Report 2016, total institutional spending on space programs in 2015, including that of intergovernmental organisations, decreased by 4.9 % to $76.52 billion from $80.42 billion in 2014; the decrease is attributable to dollar exchange rates, which saw investment by non-U.S. government space actors decrease by 14.2 % to $31.95 billion from $37.24 billion in 2014.[225] Around 55.4 % of the total institutional space expenditure went toward civil expenditure ($42.37 billion), while 44.6 % of the spending went to defence expenditure ($34.15 billion), showing no change in the ratio of civil and defence spending from 2014.[226]

The Space Report 2016 also estimates that worldwide defence related expenditure lowered to $34.15 billion in 2015, with the United States accounting for 69.0 % (or $23.75 billion) for space security programmes under its Department of Defence (DoD). The U.S. DoD's space budget funds its military space programmes, in addition to organisations such as the National Reconnaissance Office (NRO) and the National Geospatial-Intelligence Agency (NGA). Defence spending by non-U. S. government space actors accounted for the remaining 31.0 % (or $10.58 billion), a notable drop from its 36.4 % share in 2014, attributable to the fluctuations in currency values.[227] Moreover, it should be noted that a degree of uncertainty exists regarding expenditures on defence space activities as not all relevant funding is made public.

While spending among space actors saw significant changes in 2015, they should not be ranked against each other given the uncertainties due to exchange rates

[223]Ibid.

[224]Ibid.

[225]Ibid. at 38.

[226]Ibid.

[227]Ibid.

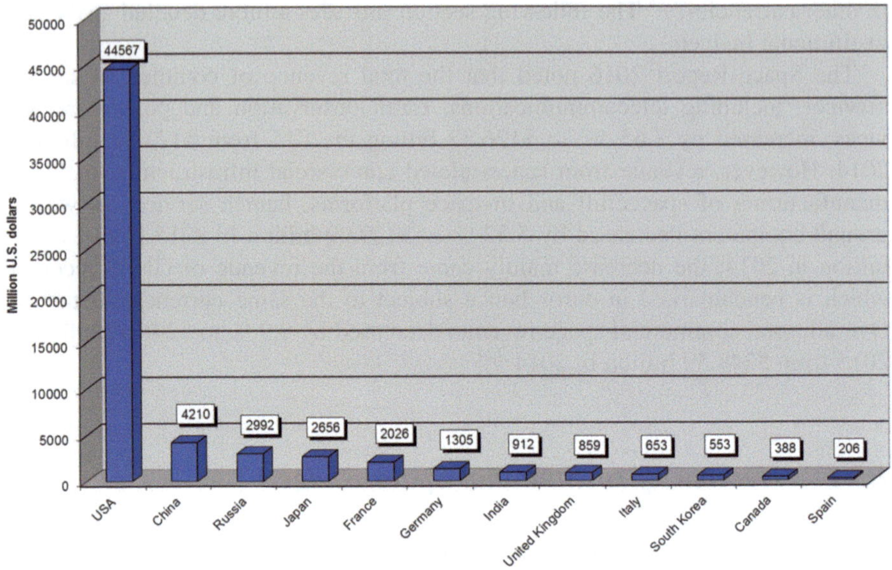

Fig. 1.1 Public space budgets of major space powers in 2015 (Based on Space Report 2016 and ASD-Eurospace data)

(Fig. 1.1).[228] The United States continued to have the largest space budget, increasing its civilian space spending to \$20.995 billion, and defence spending to \$23.572 billion, while China's space budget is based on a modest expenditure estimate of at least 26.1 billion Yuan (\$4.21 billion) according to the Space Report 2016.[229] Next, with the weaker value of the rouble in 2015, Russia's budget of 165.814 billion roubles (\$2.992 billion) in 2015 stayed level with the 165.814 billion roubles (\$4.88 billion) spent in 2014.[230] And Japan's space budget was ¥324.5 billion (\$2.656 billion), followed by France, Germany and India.[231] Estimates for European countries are slightly more conservative than in previous years, which is more reflective of a change in data authorities than in yearly spending by the individual countries.

The 2015 budget of the European Space Agency increased by 8.1 % to 4.433 billion euros (\$4.944 billion) from 4.102 billion euros (\$5.65 billion) in 2014,

[228]N.B.: Figures in this section are based on ASD-Eurospace estimates for European countries (at an €:\$ exchange rate of 1:1.1153) and Space Report 2016 data for all non-European spacefaring countries (i.e. USA, Russia, Japan, China, India, Canada, and South Korea). Moreover, the different currencies in other sections of this chapter were not converted to a baseline currency (except for comparison purposes) as recent currency fluctuations skewed the changes in the spending by other countries.

[229]The Space Report 2016. Colorado Springs: The Space Foundation, 2016: 38.

[230]Ibid.

[231]Ibid.

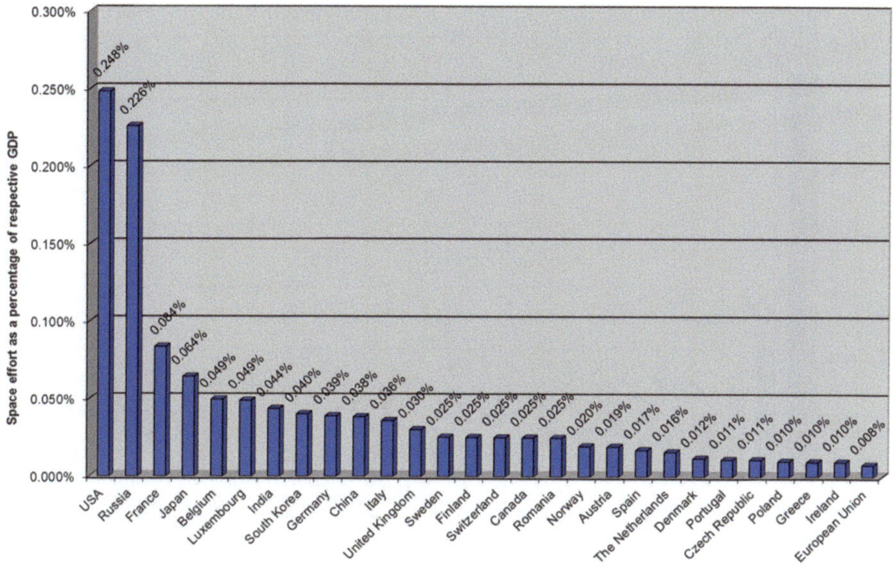

Fig. 1.2 Public space budgets (selection) as a share of nom. GDP in 2015 (Source: The Space Report/Eurospace/IMF)

following a substantial 65.2 % increase in spending by the EU. ESA member state spending decreased by 2.9 % in 2015, reaching 3.241 billion euros ($3.61 billion from $3.339 billion ($4.60 billion) in 2014.[232] Among ESA member states, the five biggest contributors to the total ESA budget were Germany 18.0 %, France 16.2 %, Italy 7.4 %, the UK 7.3 %, and Belgium 4.3 %. Switzerland was the next highest contributor at 3.0 %, followed closely by Spain for seventh position in the 2015 budget.[233]

Additional perspective can be gained by measuring the investment of countries in the space sector with regard to GDP generated in the 2015 (Fig. 1.2).

The U.S. maintained its spot as pack leader in space spending as a share of GDP in 2015, increasing to 0.2483 % from 0.2466 % in 2014. Russia's spending as a share of its GDP kept it in second position, but had decreased to 0.2259 % from 0.2372 % in 2014. France remained in the third position in 2015, with spending at 0.0837 %, while Japan maintained its fourth spot with 0.0644 % in space spending as a share of GDP in 2015. Other leading space countries in Europe and the rest of

[232]"ESA Budget for 2015." 24 Feb. 2015. ESA 28 June 2016 <http://www.esa.int/For_Media/Highlights/ESA_budget_2015>.

[233]Ibid.

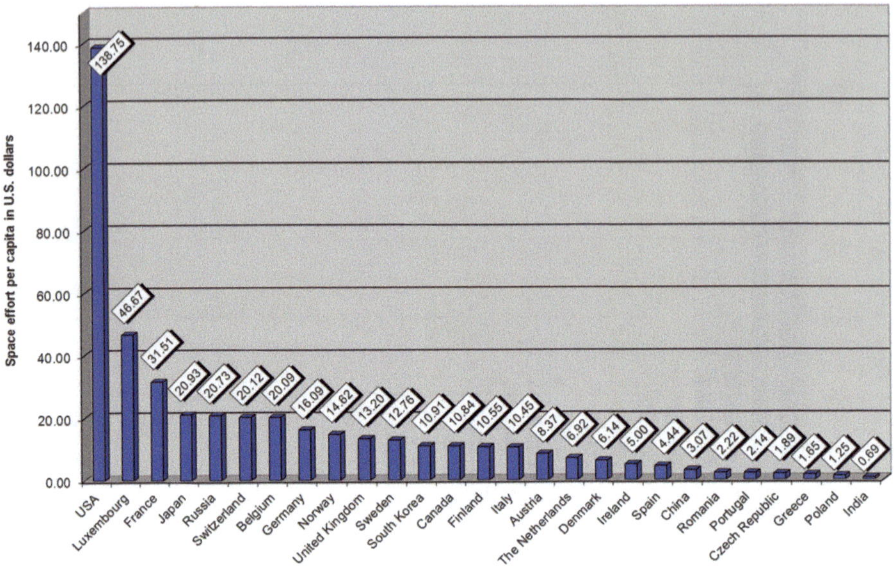

Fig. 1.3 Public space budgets per capita (selection) in 2015 (Source: The Space Report/Eurospace/PRB)

the globe invested less than 0.05 % of their GDP in space activity, while the European Union overall spent an estimated 0.00783 % of its GDP on space.

When looking at space spending in terms of per capita investment, a different ordering becomes visible (Fig. 1.3). U.S. per capita spending increased by 2.6 % in 2015, reaching $138.75 from $135.21 in 2014. The United Kingdom also had an increase of 7.5 % in its spending, while South Korea's spending grew by 19.8 %, and Portugal increased by 5.8 %. France's per capita expenditure reduced to $31.51 in 2015, while Germany's spending lowered further to $16.09 partially due to the change in currency values. Japan's expenditure overtook Russia, with $20.93 per capita compared to $20.73 respectively. Moreover, per capita spending by Switzerland and Belgium were nearly on par, with Switzerland edging ahead by 0.03 at $20.12 in 2015.

Contrasting the GDP share of public space funds and per capita public space funds provides another picture of institutional investment in space, (see Figs. 1.4 and 1.5). Here, the U.S. continues to excel by a significant margin, matched by Russia in the percentage of its GDP spent on the space sector, but outdoing all countries in terms of space spending per capita and in its overall space budget. France holds the third position both in space budget per GDP and space budget per capita, but trails the U.S., China, Russia, and Japan in its total space budget. Yet, some caution is needed when considering these figures, due to fluctuating exchange rates, and the uncertainty of reported values.

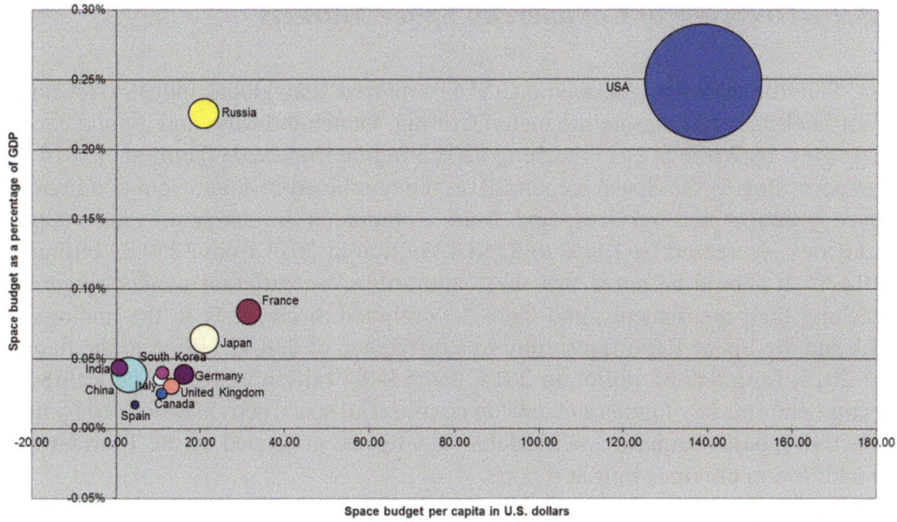

Fig. 1.4 Public space budgets as share of GDP mapped against space budgets per capita in 2015. The bubble size indicates the absolute space budget (Based on the Space Report 2016, Eurospace, and publicly available data)

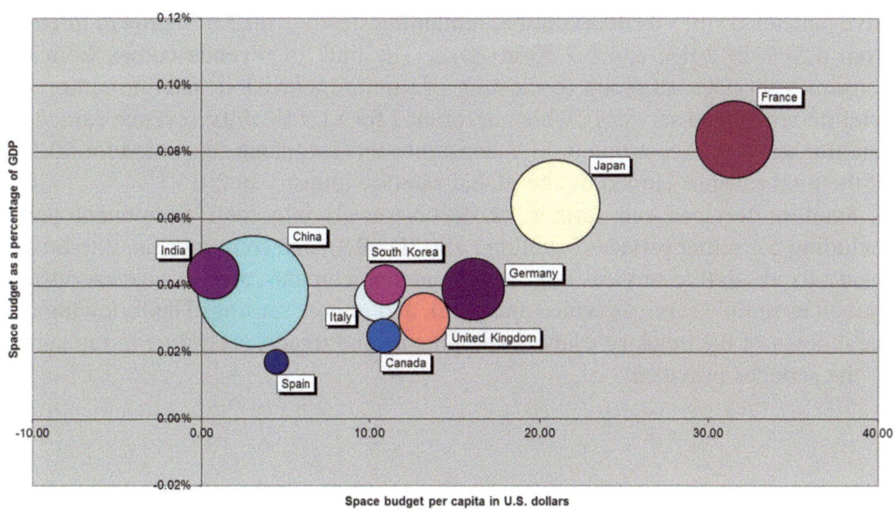

Fig. 1.5 Magnification of public space budgets as a share of GDP mapped against space budgets per capita in 2015, not including the U.S. and Russia. The bubble size indicates the absolute space budget (Based on the Space Report 2016, Eurospace, and publicly available data)

1.3.2 Overview of Commercial Space Markets

The Satellite Industry Association (SIA) reported that global industry revenues from satellite services, satellite manufacturing, launch industry, and ground equipment grew by 2.6 % in 2015 reaching $208.3 billion from $203.0 billion in 2014.[234] Yet according to the Space Report 2016, the combined revenue from commercial space products and services, and from commercial infrastructure and support industries, decreased by 1.8 % to $246.42 billion in 2015 from $250.83 billion in 2014.[235] It should be noted that these authorities use different methodologies in reaching their assessments, and there is continued discrepancy in the findings of SIA and the Space Report, resulting in a difference of $38.12 billion in the figures for 2015, from $47.83 billion in 2014, and $44.87 billion in 2013. The following section presents key figures and data on commercial space activities divided by field of activity, based primarily on available SIA figures generated by the Tauri Group, in addition to previous Futron reports.

Satellite Services
A sustained expansion of satellite capacity and corporate revenue was fuelled by continued demand from emerging economies and developing regions. The revenue earned from satellite services grew by 3.7 % to $127.4 billion, from the $122.9 billion earned in 2014.[236] The rate of revenue growth in satellite services appears to have plateaued since its deceleration, remaining steady with 3.6 % growth in 2014, from 4.5 % in 2013, and 5.2 % in 2012. The bulk of revenue comes from the consumer services subgroup (consisting of satellite television, satellite radio, and satellite broadband services), which accounted for 81.9 % of the revenue earned by satellite services. According to SIA, consumer services alone accounted for 50.1 % of the total revenue earned by the global satellite industry in 2015.[237]

Satellite services can further be deconstructed into their component parts, including consumer services (satellite radio (DARS), and consumer satellite broadband), fixed satellite services (e.g. transponder agreements, and managed services), as well as mobile services (voice and data), and remote sensing. The following is a breakdown of the industry's key developments and trends, according to the nature of the services provided.

---

[234] "2016 State of the Satellite Industry Report." 2 June 2016. Satellite Industry Association and The Tauri Group 5 Sept. 2016 <http://www.sia.org/wp-content/uploads/2016/06/SSIR16-Pdf-Copy-for-Website-Compressed.pdf>.

[235] C.f. The Space Report 2016. Colorado Springs: The Space Foundation, 2016: 16, and The Space Report 2015. Colorado Springs: The Space Foundation, 2015, 14.

[236] "2016 State of the Satellite Industry Report." 2 June 2016. Satellite Industry Association and The Tauri Group 5 Sept. 2016: 11 <http://www.sia.org/wp-content/uploads/2016/06/SSIR16-Pdf-Copy-for-Website-Compressed.pdf>.

[237] Ibid. at 4.

Consumer Services
As mentioned above, consumer services are made up of satellite television, radio, and broadband services. The $104.3 billion in downstream consumer services revenue continued to surpass the combined revenues of upstream satellite industry segments (i.e. manufacturing, launch services, and ground equipment) in 2015, 93.8 % of that consumer revenue came from satellite television services (DBS/DTH). With about 230 million satellite television subscribers worldwide, it is a key driver in consumer services revenue, with increasing growth in emerging markets. In 2015, the U.S. attributed for about 42 % of global revenues from satellite television services. The overall rate of growth increased by 3.4 % in 2015, from 2.6 % in 2014, while satellite radio revenue grew by 9.5 % to $4.6 billion in 2015 from $4.2 billion in 2014 and satellite broadband revenue increased by 5.6 %, reaching $1.9 billion from $1.8 billion in 2014.[238] Satellite radio and broadband services amount to 6.2 % of the consumer services segment, with 29.6 million and 1.8 million subscribers respectively, each mainly coming from the U.S.[239]

Fixed Satellite Services
Fixed Satellite Services (FSS) refers to the use of spacecraft that utilise land terminals in fixed positions to broadcast. Whereas Consumer Services covers satellite broadband Internet, communications and network television and radio broadcasts, FSS relates to commercial signal agreements, such as transponder agreements and managed network services. The FSS segment saw unequal growth in 2015, with revenue for transponder agreements increasing by about 1 % from 4 % in 2014, whereas revenues from managed services, mainly from airborne services[240] grew by 15 % in 2015 from 4 % in 2014.

Mobile Satellite Services
Mobile Satellite Services (MSS) offer both mobile data service and mobile voice service (including satellite phones). MSS revenue grew by 3.0 % in 2015, earning $3.4 million, a modest increase from the 26.9 % growth in revenue in 2014. Here, mobile voice services grew by 9 % in 2015 from 19 % in 2014, whereas mobile data services grew by 4.3 % in 2015, from 27 % in 2014. Mobile voice services revenue reached $1.0 billion in 2015, from $900 million in 2014, while mobile data service revenue, used heavily in the aviation sector, grew to $2.4 billion in 2015, from $2.3 billion in 2014. This latter segment comprised 70.6 % of all mobile satellite services revenue.[241]

---

[238]Ibid. at 11.
[239]Ibid. at 12.
[240]Ibid. at 13.
[241]Ibid. at 11.

Earth Observation Services

Earth observation services refers to commercial companies that provide optical and radar images to the open market; however, demand for such services is mostly driven by government entities. Nevertheless, new entrants such as Terra Bella (formerly known as Skybox Imaging) and Planet Labs have continued to raise capital, and have begun to deploy initial constellations. Earth observation services revenue increased by 12.5 % in 2015, reaching $1.8 billion from $1.6 billion in 2014.[242]

Satellite Manufacturing

The total revenue of satellite manufacturers that built satellites both for governmental and commercial customers reached $16.6 billion in 2015, an increase of 4.4 % from the $15.9 billion generated in 2014 (Fig. 1.6). Whereas U.S. generated revenue of $10 billion for 2015 remained on par with 2014, non-U.S. revenue increased by 11.9 % to $6.6 billion in 2015 from $5.9 billion in 2014, marking steady growth over the past 3 years. The SIA reports that 42 % of the total revenues generated in satellite manufacturing came from communications satellites (i.e. 24 % from commercial communications, while 18 % were for civil/military communications). Next, military surveillance satellites accounted for 36 % of the revenue for the year, while navigation satellites were 9 % and Earth Observation satellites were 8 %. Meteorology satellites and scientific satellites each represented 2 % of the revenue for the year, while satellites developed for R&D purposes

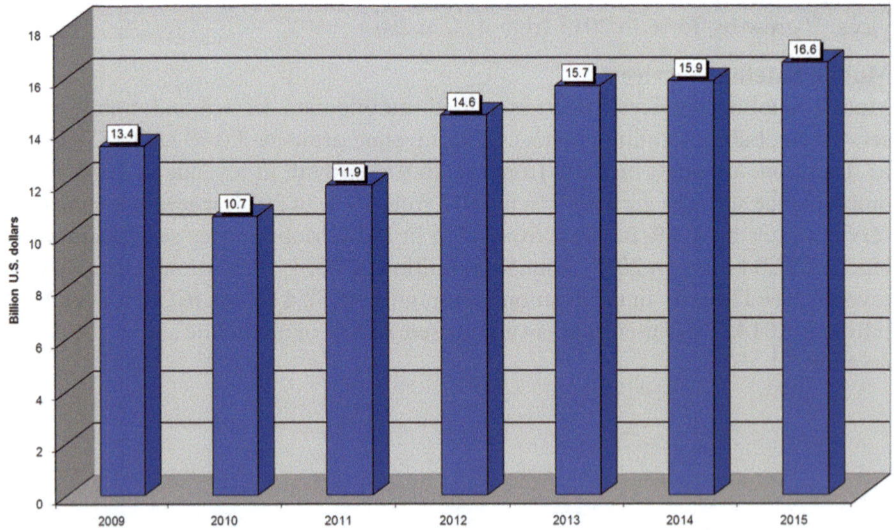

Fig. 1.6 World satellite manufacturing revenue (Source: SIA)

242Ibid. at 13.

amounted to 1 %. Cubesats continued to represent less than 1 % of the total revenue generated for the year.[243]

Commercial Space Launch

There were 22 commercial launches in 2015; with two failures including the Proton M carrying the MexSat-1 communications satellite for Mexico's government, and the failure of SpaceX's Dragon CRS-7 ISS resupply mission which carried ISS equipment and an allotment of eight Planet Labs cube satellites intended to be released from the ISS. The remaining 20 successful commercial launches carried 36 commercial services payloads into orbit (in addition to 29 commercial cube satellites intended to be released from the ISS). Commercial launches accounted for 25.3 % of the total 87 launches in 2015; and amounted to 23.8 % of the 265 payloads launched. Of the 265 payloads that were launched in 2015, 78 were cube satellites launched directly into orbit (including the failed maiden launch of the Super Strypi that resulted in the destruction of 13 cube satellites), while another 65 cube satellites were intended to be released into orbit from the ISS (8 cube satellites were destroyed, along with equipment intended for the ISS during the launch of the Dragon CRS-7). When not considering cube satellites, the percentage of commercial payloads launched amounted to 25.4 %, or 31 commercial payloads out of a total of 122 non-cube satellite payloads.

In 2015, launch providers in the U.S. conducted 8 commercial launches, out of a total of 20 launches; the U.S. share of total commercial launches decreased to 36.4 % for the year. Russia had the most launches in 2015, but only 5 out of its 26 launches were for commercial purposes; its share of total commercial launches increased to 22.7 % for 2015. China conducted 19 non-commercial launches, but did not have any commercial launch activity fin 2015. Next, Europe conducted 6 commercial launches out of a total of 12 in 2015; increasing its share of total commercial launches to 27.3 % for the year (not counting the actual number of payloads launched). And finally, two out of five launches by India were commercial, amounting to a 9.1 % share; while one of the four launches by Japan was commercial, amounting giving it a 4.5 % share. Iran's single launch for the year was non-commercial (Fig. 1.7).

The total estimated revenue from the 22 launches amounted to $2.15 billion, an 8.9 % decrease from the $2.36 billion earned in 2014. Europe generated the most commercial launch revenue for 2015, earning $1.066 billion—an increase of 15.9 % from the $920 million earned in 2014. The U.S. earned $617 million in 2015, a drop of 44.3 % from the $1.107 billion it earned in 2014. Russia held the third position earning $289 million, an increase of 32.6 % from the $218 million earned in 2014. Japan returned to the commercial launch scene in 2015, earning $113 million, while India earned an estimated $66 million in 2015, more than three times the revenue earned in 2014, marking a shift in reliance on the low-cost launch provider (Fig. 1.8).[244]

[243]Ibid. at 18.

[244]Federal Aviation Administration. The Annual Compendium of Commercial Space Transportation: 2016. Washington, DC: FAA, Jan. 2016: 40.

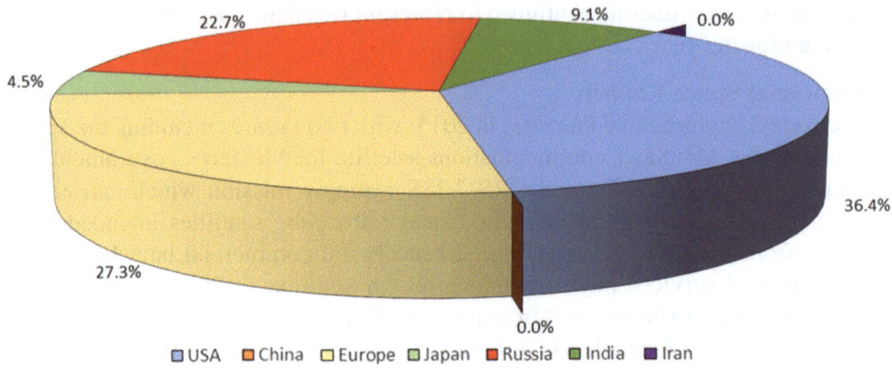

Fig. 1.7 Commercial Launch Activity by Country in 2015 (Source: FAA)

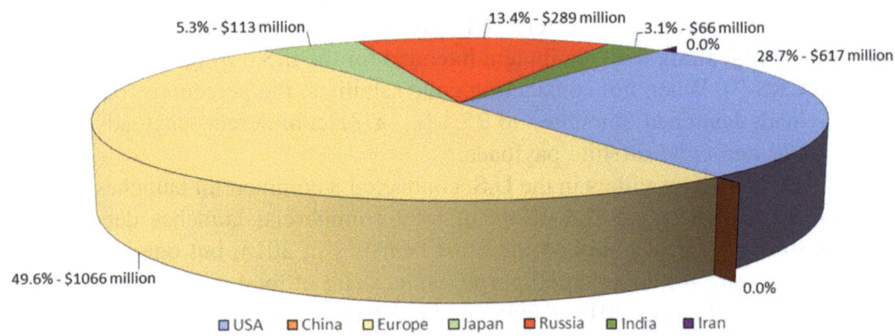

Fig. 1.8 Commercial Launch Revenues by Country in 2015 (Source: FAA)

In 2015, Arianespace conducted a total of 12 launches from French Guiana. Its Ariane 5 ECA launcher had six launches, which lifted nine commercial telecommunications satellites, the Franco-Italian SICRAL-2 military communications satellite, Eumetsat's MSG-4 meteorology satellite, and one civil government communications satellite into GEO orbit. The Europeanized Soyuz had three launches, which placed six Galileo navigation satellites (Galileo FOC-3 to -6, -8, and -9) to medium Earth orbit (MEO) for the European Commission. And three launches were conducted by the Vega launcher which lifted the IXV spacecraft and its AVUM VV04 upper stage on a suborbital trajectory; it also launched the Sentinel 2A into SSO and the LISA Pathfinder mission beyond Earth orbit for ESA.

Ground Equipment

Ground equipment revenue includes infrastructure elements, such as mobile terminals, gateways and control stations, and consumer equipment, such as very small aperture terminals (VSAT), ultra small aperture terminals (USAT), DTH broadcast dishes, satellite phones and digital audio radio satellite (DARS) equipment. Portable Navigation Devices (PND) form one of the sub-segments of end-user electronics that incorporate GPS chip sets.

Ground equipment revenues for 2015 increased only slightly from 2014 stemming from growth solely in network equipment and broadband equipment revenues, whereas consumer equipment for satellite navigation revenue remained flat reflecting continued migration from standalone devices to embedded chipsets for devices such as smartphones.[245] Network equipment revenue grew by 3.2 % in 2015, reaching $9.6 billion from $9.3 billion in 2014, while consumer broadband equipment revenue grew by 2.2 % in 2015, reaching $18.3 billion from $17.9 billion in 2014. In contrast, consumer equipment revenue continued to lower slightly, reducing by 0.3 % to $31.0 billion from $31.1 billion in 2014; even so, the segment represents more than half of the overall ground equipment revenue of 2015. Lastly, overall ground equipment revenues grew by 1.0 % to $58.9 billion in 2015 from $58.3 billion in 2014, constituting a 28.3 % share of the $208.3 billion world satellite industry revenue in 2015.

Garmin and TomTom, the two companies leading the PND market, seemed to switch roles in growth for 2015. Garmin experienced a 1.8 % decrease in revenue in 2015, amounting to $2.820 billion from $2.871 billion in 2014. The loss came mainly from a 15 % reduction in revenue from its automotive sector, whose impact would have been greater were it not offset by revenue gains of 16 % in its fitness segment, and 15 % in its marine segment.[246] In contrast, TomTom's revenue noted a 5.9 % uptick in revenue for 2015, amounting to 1.007 billion euros ($1.123 billion) from 950.292 million euros ($1.155 billion) in 2014. That growth mainly came from revenue gains of 27.3 % in its licensing segment, and 22.5 % in its telematics segment, while its consumer segment grew by just 0.7 %, and automotive decreased by 3.2 % in 2015 (Table 1.1).[247]

Insurance Sector

Several launch failures in 2015 appear likely to result in another losing year for insurance providers, which could mean that insurance premiums will increase in coming years. The 16 May 2015 failure of the Russian Proton-M, due partly to the collapse of a turbopump in the rocket's third stage engine, resulted in the total loss of the Mexican government's Centenario mobile communications satellite. Despite the launch failure, the Mexican government can expect to receive a $390 million insurance pay-out for the loss, after having made the atypical decision of purchasing full insurance coverage for the satellite, including $300 million for the satellite

[245]"2016 State of the Satellite Industry Report." 2 June 2016. Satellite Industry Association and The Tauri Group 5 Sept. 2016: 29 <http://www.sia.org/wp-content/uploads/2016/06/SSIR16-Pdf-Copy-for-Website-Compressed.pdf>.

[246]"Garmin Reports Q4 and Fiscal 2015 Results." 17 Feb. 2016. Garmin 3 July 2016 <http://newsroom.garmin.com/press-release/earnings/garmin-reports-q4-and-fiscal-2015-results>.

[247]"TomTom Annual Report 2015." 19 Feb. 2016. TomTom 3 June 2016: 33 <http://files.shareholder.com/downloads/TOMTOM/1941825123x0x876179/DC922427-C53C-4CE9-A5CC-43EB2855F9C7/TomTom_Annual_Report_2015.pdf>.

Table 1.1 Understanding TomTom & Garmin variables

		2015	2014
Total revenue			
TomTom		1.007 billion euros ($1.123 billion)	950.292 million euros ($1.155 billion)
Garmin		$2.820 billion	$2.871 billion
Geographical sales			
TomTom	Europe	771.491 million euros ($860.444 million)	718.767 million euros ($873.647 million)
	North America	186.115 million euros ($207.574 million)	163.461 million euros ($198.684 million)
	Rest of World	49.001 million euros ($54.651 million)	68.064 million euros ($82.730 million)
Garmin	Europe/Middle East/Africa	$1.013 billion	$1.054 billion
	Americas	$1.469 billion	$1.538 billion
	Asia Pacific region	$337.888 million	$278.092 million

and $90 million for the launch.[248] Moreover, the Tel Aviv, Israel-based Spacecom, which is vying for an increased share of the African satellite communications market, filed a total loss claim to its insurance underwriters amounting to $158 million (including $50 million for the satellite's propulsion system) for the loss of its Amos-5 which stopped communicating from its Geostationary orbit on 21 November 2015. While Amos-5's propulsion unit had experienced a glitch in 2013, causing Spacecom's insurers to place exclusions on claims based on any similar events in the future, this latest event was determined to be unrelated to its propulsion system, although its Russian manufacturer ISS Reshetnev has been unable to pinpoint another possible cause.[249]

The insurance sector seems to be evolving as operators seek to lower their insurance premiums while taking on higher risk. In late October 2015, the mobile satellite services provider Iridium announced an agreement with its creditors on a revised insurance regime for its second-generation Next constellation. In an aggregated approach, Iridium must secure insurance for the first 3 of 10 satellites to be launched for the 81 satellite Next constellation, and secure insurance for the rest of the block shortly after those initial launches. Here, insurance companies would not be required to cover the failure of up to two of those initial satellites, but the failure of a third satellite would require insurers to compensate Iridium for all three lost satellites in addition to a

[248] De Selding, Peter B. "Latest Proton Failure Leaves Customers, Insurers in a Bind." 29 May 2015. SpaceNews 13 Jan. 2016 <http://spacenews.com/proton-failure-leaves-customers-insurers-in-a-bind-2/>.

[249] De Selding, Peter B. "After Amos-5 Loss, Competitors Lure Spacecom Customers with Below-cost Pricing." 24 Nov. 2015. SpaceNews 14 Jan. 2016 <http://spacenews.com/after-amos-5-loss-competitors-lure-spacecom-customers-with-below-cost-pricing/>.

pro rata share of the launch costs.[250] Orbcom has a similar three satellite insurance deductible for its 17 small satellite constellation, however its insurance policy covers the full second-generation constellation and its launches.[251] According to an estimate by Eutelsat, insurance charges and other diverse expenses make up a 20 % share of the total cost of placing a typical GEOCOM satellite into orbit, while 30 % of the cost goes toward its launch, and 50 % goes toward its construction.[252]

On 6 August 2015, Orbital ATK, NASA and the State of Virginia agreed to split the $15 million cost of repairing Pad 0A of the Mid-Atlantic Regional Spaceport in Wallops Island, Virginia, following the 28 October 2014 explosion of Orbital ATK's Antares rocket NASA commercial resupply services (CRS) mission to the ISS. While at the time of the explosion, the state-owned launch pad had been uninsured, leading to dispute between the U.S. federal government, the state of Virginia, and Orbital ATK over responsibility for repairs and clean-up costs, under the terms of the new agreement, Orbital ATK will now secure launch insurance for Virginia assets, the launch pad and systems.[253]

1.3.3 Developments in the Space Industry

1.3.3.1 Industrial Developments in Europe

At the end of 2015, Arianespace commercial launch company revenue was announced to be 1.4 billion euros from the 1.399 billion euros earned in 2014. And while operating profit was not disclosed for the year, it was expected to break even when counting 100 million euros in price support from ESA. Arianespace conducted 12 launches in 2015, including 6 Ariane 5 ECA launches, 3 Europeanized Soyuz launches, and 3 launches for the Vega launcher.[254] In 2015, the launch provider won 14 contracts to lift satellites in GEO orbit using the Ariane 5 ECA launcher, 12 of which were commercial awards; in comparison, SpaceX won

---

[250]De Selding, Peter B. "Component Issue Delays Iridium Next Launches by Four Months." 29 Oct. 2015. SpaceNews 13 Jan. 2016 <http://spacenews.com/component-issue-delays-iridium-next-launches-by-four-months/>.

[251]De Selding, Peter B. "After Failure, Orbcomm Touts Constellation Resiliency." 10 Aug. 2015. SpaceNews 13 Jan. 2016 <http://spacenews.com/after-failure-orbcomm-touts-constellation-resiliency/>.

[252]De Selding, Peter B. "Eutelsat Does the Math on Reducing Future Satellite Costs." 3 Dec. 2015. SpaceNews 14 Jan. 2016 <http://spacenews.com/eutelsat-does-the-math-on-reducing-future-satellite-costs-by-at-least-20-percent/>.

[253]Leone, Dan. "Orbital ATK, Virginia Reach Accord on Pad Repair Bill, Insurance." 7 Aug. 2015. SpaceNews 13 Jan. 2016 <http://spacenews.com/orbital-atk-virginia-reach-accord-on-pad-repair-bill/>.

[254]De Selding, Peter B. "Arianespace Surpassed SpaceX in Commercial Launch Orders in 2015." 6 Jan. 2016. SpaceNews 17 June 2016 <http://spacenews.com/arianespace-surpassed-spacex-in-commercial-launch-orders-in-2015/>.

9 commercial geostationary launch awards, while International Launch Services (ILS) won a single commercial launch award.[255] Arianespace also booked its largest single launch contract on 25 June 2015, involving 21 Russian Soyuz rockets—some launching from Europe's spaceport in French Guiana—to lift OneWeb's 700 small satellite broadcasting constellation into LEO orbit in 32–36 satellite batches starting in 2017. The contract is valued at more than 1 billion euros, and is considered to be the largest commercial launch purchase in history, besting SpaceX's block contract with Iridium signed in 2010 valued at nearly $500 million.[256] Arianespace's total contract wins for 2015 amounted to 2.5 billion euros, with 875 million euros coming from Ariane 5 launch awards.[257]

Eutelsat earned a total revenue of 1.476 billion euros for the year-ending 30 June 2015, a 9.5 % increase from the 1.348 billion euros earned in the previous year. Its EBITDA margin remained at 76.7 % of revenue for the year, while its net profit grew by 17.2 % amounting to 355.2 million euros as at 30 June 2015 from 303.2 euros million in the previous year.[258] The share of revenue generated from European regions continued to decrease in the reporting year, reaching 59.0 % in 2015 from 64.0 % as of 30 June 2014, while revenue from the Americas jumped to 20.8 %, followed by the Middle East (12.1 %), Africa (5.6 %), and Asia (2.5 %).[259] Eutelsat's revenue in the 6 months ending 31 December 2015 reached 774.4 million euros, up 7.1 % from the 722.8 million euros earned in the previous year. Eutelsat ordered its first Eutelsat Quantum satellite from Airbus Defence and Space in July 2015, to be launched in 2019; another all-electric High Throughput Satellite was ordered from Thales Alenia Space in October 2015, also to be launched in 2019. And by the end of 2015, Eutelsat's order backlog had decreased to 5.8 billion euros from 6.2 billion euros as of mid-year.[260]

SES earned a total revenue of 2.014 billion euros in revenue for the year ending 31 December 2015, a 5.0 % increase from 1.919 billion euros earned in the previous year. Its EBITDA for 2015 was 1.494 billion euros (74.2 % of revenue), from 1.428 billion euros (74.4 % of revenue) in 2014. Moreover, its backlog for the year grew to 7.4 billion euros in 2015 from 7.3 billion euros in 2014. SES plans to launch

[255] De Selding, Peter B. "Three's Company – Commercial Launch Contracts." 18 Jan. 2016. SpaceNews Magazine 5 Sept. 2016 <http://www.spacenewsmag.com/feature/threes-company/>.

[256] Clark, Stephen. "OneWeb launch deal called largest commercial rocket buy in history." 1 July 2015. Spaceflight Now 17 June 2016 <https://spaceflightnow.com/2015/07/01/oneweb-launch-deal-called-largest-commercial-rocket-buy-in-history/>.

[257] De Selding, Peter B. "Arianespace Surpassed SpaceX in Commercial Launch Orders in 2015." 6 Jan. 2016. SpaceNews 17 June 2016 <http://spacenews.com/arianespace-surpassed-spacex-in-commercial-launch-orders-in-2015/>.

[258] "Full Year 2014–2015 Results." 30 July 2015. Eutelsat 9 June 2016 <http://news.eutelsat.com/pressreleases/full-year-2014-15-results-1197412>.

[259] Reference Document 2014–2015. 22 Oct. 2015. Eutelsat 10 June 2016: 140 <http://www.eutelsat.com/files/contributed/investors/pdf/AGM2015/ETL-DDR%202014-15-UK-web.pdf>.

[260] Press Release. "Eutelsat Communications First Half 2015–2016 Results." 17 Feb. 2016. Eutelsat 10 June 2016 <http://news.eutelsat.com/pressreleases/eutelsat-communications-first-half-2015-2016-results-1318802>.

seven new satellites between the end of 2015 through 2017, including SES-9 which will add 180 incremental wide beam transponders and 36 GHz of HTS capacity.[261]

Hispasat earned a total revenue of 219.6 million euros for 2015, an increase of 8.66 % from the 202.1 million euros earned in 2014. Its EBITDA remained above 80 % in 2015, growing by 1.4 points to 81.5 %, while its net profit grew by 37.3 % reaching 62.6 million euros in 2015 from 45.6 million euros in 2014. Hispasat's recent entrance into Morocco provided access to a new market and additional income. The revenue coming from space capacity rentals also grew by 9.7 %, reaching 216.4 million euros from 197.2 million euros in 2014; this year 65.4 % of the revenue came from clients in the Americas, while 34.6 % came from leasing space capacity to clients in Europe and North Africa.[262]

Telenor Satellite Broadcasting of Norway reported a 1.3 % decrease in revenue for the year ending 31 December 2015, with the decline attributable to its divestment of its subsidiary Conax, which had been deconsolidated in the first quarter of 2014. The Oslo-based satellite fleet operator reported revenue of 6.076 billion kroner (635.04 million euros) in 2015, from 6.155 billion kroner (679.50 million euros) in 2014. That revenue included Nordic DTH subscribers and households in SMATV networks, revenues from satellite services, revenues from terrestrial radio and TV transmission and sale of encryption and conditional access services for TV distribution. The launch of its new satellite, Thor 7, on 27 April 2015 was the main reason for the increase in the company's capital expenditure for 2015.[263]

Inmarsat's total revenue decreased by 0.9 % for the year ending 31 December 2015, generating $1.274 billion (1.17 billion euros) from $1.286 billion (1.058 billion euros) in 2014. Its EBITDA grew by 3.6 % to $726.0 million (664.41 million euros) from $701.0 million (576.66 million euros) in 2014, while its margin increased to 57.0 % from 54.5 % in 2014. Around 93.0 % of that revenue was generated from its Mobile Satellite Services (MSS) including Maritime, Government, Enterprise, Aviation, and other, amounting to $1.186 billion (1.09 billion euros); the remaining 7.0 % came from the LightSquared Cooperation Agreement which earned $88.1 million (80.63 million euros) in 2015. Its total revenue also included wholesale MSS revenues of $832.8 million (762.15 million euros), which increased by 5.2 % from $791.4 million (651.02 million euros) in 2014, wherein higher wholesale MSS revenue in Maritime and Aviation helped to offset its continued decline in government wholesale business.[264]

---

[261] "SES Annual report 2015 – New foundations." 7 Apr. 2016. SES 7 Sept. 2016 <http://www.ses.com/22199159/annual-report-15-en.pdf>.

[262] "HISPASAT increases total revenue by 8.7% in 2015." 15 Feb. 2016. Hispasat 13 June 2016 <http://www.hispasat.com/en/press-room/press-releases-2016/pagina-3/204/hispasat-increases-total-revenue--by-87-in-2015>.

[263] Annual Report 2015. Telenor Group 18 Apr. 2016: 88 <https://www.telenor.com/wp-content/uploads/2015/09/Godkjent-Annual-Report-2015_7y8erjhregj745.pdf>.

[264] "Global Reach Global Impact | Inmarsat PLC Annual Report and Accounts 2015." 6 Apr. 2016. Inmarsat 13 June 2016 <http://www.inmarsat.com/wp-content/uploads/2016/04/Inmarsat_AR15_Bookmarked.pdf>.

The Airbus Group's Airbus Defence and Space (Airbus D&S) division reported a 0.4 % increase in revenue for 2015, reaching 13.080 billion euros from 13.025 billion euros in 2014. The Airbus D&S Space Systems business line accounted for 29 % of that revenue, or around 3.79 billion euros. And the division's overall order intake increased by 18.1 %, reaching 14.440 billion euros from 12.225 billion euros in 2014, while its backlog lowered by 0.5 % to 42.861 billion euros from 43.075 billion euros in 2014. Airbus D&S booked orders for five telecommunications satellites in 2015, along with a contract for 900 small satellites to be built for OneWeb. Its ESA orders included a contract to build the Juice Orbiter, and a service module for the manned U.S. Orion space capsule. Airbus D&S also signed the first Quantum satellite contract with Eutelsat and ESA, which will be the first satellite capable of adapting its coverage, bandwidth, power and frequency, and its orbital position according to changing customer requirements.[265] On 21 December 2015, the Airbus Group signed a share-purchase agreement to sell its Vizada commercial mobile satellite division back to the company's previous owner Apax Partners, with the deal set to close in early 2016 subject to regulatory approvals.[266]

By 10 July 2015, the UK-based Surrey Satellite Technology Ltd. (SSTL) had delivered the DMC3 satellite constellation into orbit for its China-based customer Twenty-First Century Aerospace Technology Co. (21AT). SSTL was able to bypass U.S. ITAR export restrictions and deliver its three-satellite Earth observation constellation into orbit by launching the satellites aboard India's lower-cost PSLV launcher. SSTL, an arm-length subsidiary of Airbus D&S, entered into the contract with its anchor and sole customer 21AT (China's first non-government owned Earth observation imagery provider) in 2011, after receiving export consent by the governments of both the UK and China. The three small satellites were lifted into orbit on a single PSLV launcher that carried two additional payloads, which enabled 21AT's constellation to enter into full service soon afterward. By launching the three satellites on a single low cost PSLV launcher, SSTL's business model allowed it to charge a substantially lower price for the constellation than would normally be the case. Its contract with 21AT is valued at £110 million (149.23 million euros). SSTL used 21AT's purchase of all of the capacity of the DCM3 system for 7 years to pay for the system's construction, launch and insurance.[267] The system is spaced at 120 degrees apart at 647 km LEO orbit, and each satellite is capable of pointing up to 45 degrees nadir, allowing the system to revisit any point on Earth every 24 h. The system relies on xenon-electric propulsion to allow for around a 10-year operating life. Under SSTL's contract with 21AT, 21AT may send

[265]"Flying ahead | Annual Report 2015." 26 Apr. 2016. Airbus Group 17 June 2016: 24–28 <http://annualreport.airbusgroup.com/api/airbus_ar_2015.pdf>.

[266]Jasper, Christopher. "Airbus Agrees to Sell Commercial Satcoms Business Back to Apax." 21 Dec. 2015. Bloomberg 17 June 2016 <http://www.bloomberg.com/news/articles/2015-12-21/airbus-agrees-to-sell-commercial-satcoms-business-back-to-apax>.

[267]De Selding, Peter B. "SSTL Using Immediate Access as a Lure To Sell a 4th DMC-3." 16 July 2015. SpaceNews 13 Jan. 2016 <http://spacenews.com/sstl-using-immediate-access-as-a-lure-to-sell-a-4th-dmc-3/>.

image requests to the satellites directly when targets are on China's mainland, but its requests for targets outside of China would be subject to a right of review.[268]

SSTL hopes that the 21AT contract will attract other companies or nations to purchase an additional fourth DMC3 satellite, which would help to increase revisit rates and provide an in-orbit backup for the system, in addition to increasing its capacity.[269] The offer of immediate access to the full system is meant to be an additional lure for potential customers. Other competitors, including Airbus D&S, Deimos Imaging of Spain, and ImageSat of Israel are using the same business strategy. Meanwhile, another competitive challenge threatening to undo both SSTL and 21AT's business comes from China's decision to develop its own domestically-built constellation of 16 by the end of 2016, and 60 by 2020 which will offer a 30 min revisit capability across the globe; that constellation would increase to 138 satellites with a 10 min revisit capability by 2030.[270]

The Thales Group's Aerospace segment generated around 2.045 billion euros in sales in 2015 from 1.955 billion euros in 2014; the segment includes its Avionics business and revenue from its Space Global Business Unit e.g. Thales Alenia Space and Telespazio under its Space Alliance strategic partnership with Finmeccanica. In 2015, Thales Alenia Space entered into the C/D phase of the Neosat programme with ESA, and signed a contract with Eutelsat for a telecommunications satellite that will be the first commercial use of the all-electric Spacebus NEO satellite platform. Thales Alenia Space also signed contracts to build the Bangabandhu telecom satellite for Bangladesh, and to provide a payload for Argentina's ARSAT-3 satellite. Additionally, Thales Alenia Space has signed a contract to build eight additional satellites for the O3b Network, and it continues to work with LEOSAT on the feasibility and definition of its 80–120 broadband Internet satellite constellation. Some of Thales Alenia Space's institutional contracts in 2015 include its 65 % partnership agreement for the Comsat NG French military satellite communications programme, and its contract with ESA and the European Union to build the C and D models of the Sentinel 1 environmental monitoring satellite under the Copernicus programme.[271]

OHB AG of Germany increased its total revenue to 730.38 million euros for the fiscal year ending 31 December 2015; a 5.5 % decrease from the 772.95 million euros earned in 2014. Its EBITDA reached 52.13 million euros in 2015, decreasing

---

[268] Ibid.

[269] De Selding, Peter B. "India's PSLV Lofts UK-built Earth-observation Satellites Leased by Chinese Firm." 13 July 2015. SpaceNews 13 Jan. 2016 <http://spacenews.com/indias-pslv-lofts-uk-built-earth-observation-satellites-leased-by-chinese-firm/>.

[270] De Selding, Peter B. "China Launches High-resolution Commercial Imaging Satellite." 7 Oct. 2015. SpaceNews 13 Jan. 2016 <http://spacenews.com/china-launches-high-resolution-commercial-imaging-satellite/>.

[271] "Registration Document 2015 | Annual Financial Report." 22 Feb. 2016. Thales 17 June 2016: 116 <https://www.thalesgroup.com/sites/default/files/asset/document/2015_registration_document.pdf>.

by 2.4 % from 53.42 million euros in 2014.[272] In February 2015, OHB System completed the integration of its first GEOCOM satellite for the commercial telecommunications satellite operator Hispasat, while OHB Sweden and ÅAC Microtec received a contract from the Swedish National Space Board (SNSB) for the development of a highly capable "InnoSat" small satellite platform and for is first use on the Mesospheric Airglow/Aerosol Tomography and Spectroscopy (MATS) advanced science mission. The OHB-developed and built Galileo FOC satellites numbers 3 and 4 were successfully launched into orbit on a Europeanized Soyuz launcher on 28 March 2015, followed by the launch of numbers 5 and 6 on 11 September 2015, and the launch of numbers 8 and 9 on 17 December 2015. OHB System was also awarded a contract for the expansion of the German federal armed forces' large satellite ground station on 10 November 2015; and on 22 December 2015, OHB's strategic subsidiary Orbcomm launched its second mission of 11 OG2 satellites into orbit on a Falcon 9 launcher.[273] By the end of 2015, OHB's staff numbered 2056, a reduction of 50 personnel which was relatively stable by comparison to its previous reduction of 326 employees in 2014 due to the deconsolidation of its subsidiary Aerotech Peissenberg GmbH & Co. About 65.5 % of those employees worked in OHB's Space Systems business unit, while the remainder worked mainly in other Aerospace and Industrial Products. Moreover, 79.3 % of OHB's employees were based in companies in Germany, while 15.1 % were in other parts of Europe, and the remaining 5.6 % worked in the rest of the world.[274]

RUAG Space revenue decreased by 3.7 %, earning CHF 310 million (286.27 million euros) for the fiscal year ending 31 December 2015, from CHF 322 million (267.68 million euros) in 2014. Its EBITDA was CHF 45 million (41.56 million euros) in 2015, an increase of 4.7 % from the CHF 43 million (35.75 million euros) it had earned in the previous year.[275] Based in Switzerland, Sweden, Austria, and Finland, RUAG Space had 1204 employees in 2015, an increase of 41 staff from the previous year.[276] RUAG completed its acquisition of the space unit of Finnish company Patria on 7 May 2015. First announced on 17 December 2014, Patria Space's business operations and assets, along with its 29 employees were transferred to the newly founded company RUAG Space Finland.[277] Patria's space business has been active in various ESA satellite programmes including Sentinel-

[272]Annual Report 2015. 16 Mar. 2016. OHB 24 Apr. 2016: 41–43 <http://www.ohb.de/tl_files/ohb/pdf/finanzberichte_hauptversammlung/2015/ohb_gb_2015_e.pdf>.

[273]Ibid.

[274]Annual Report 2015. 16 Mar. 2016. OHB 24 Apr. 2016: 61 <http://www.ohb.de/tl_files/ohb/pdf/finanzberichte_hauptversammlung/2015/ohb_gb_2015_e.pdf>.

[275]RUAG Annual Report 2015. 17 Mar. 2016. RUAG 24 Apr. 2016: 25 <https://www.ruag.com/fileadmin/ruag/group/Annual_Report/2015/GB/RUAG_GB_2015_72dpi_E.pdf>.

[276]Ibid. at 18.

[277]Media release. "RUAG completes acquisition of Patria's space business." 12 May 2015. RUAG 6 July 2016 <http://www.ruag.com/space/media/media-releases/news/ruag-completes-acquisition-of-patrias-space-business/dea359f87ec7580485cade82cfb694d1/>.

2, Earthcare and Swarm and the Gaia space telescope, with key product areas in spacecraft control electronics, electrical power subsystems, electronic units, and related test equipment. RUAG also had cause to celebrate its successful launch of its 250th fairing on 22 June 2015, which encapsulated Europe's Sentinel-2a environmental satellite while transporting it into orbit.[278]

1.3.3.2 Industrial Developments in the United States

Boeing's Network & Space Systems segment under its Defence, Space & Security division reported a 3.1 % decrease in revenue, earning $7.751 billion for the year ending 31 December 2015 from $8.003 billion 2014, due to lower satellite volume in its fourth quarter. However its operating earnings increased by 4.0 % in 2015, reaching $726 million from $698 million in 2014. Moreover, its backlog at the end of 2015 was $7.4 billion, from $8.9 billion in 2014.[279]

Lockheed Martin's Space Systems segment revenue for in 2015 decreased by 1.1 % to $9.105 billion from $9.202 billion in adjusted revenue earned in 2014. Its operating earnings also decreased by 1.4 %, to $1.171 billion from $1.187 billion in 2014. During the fourth quarter of 2015, Lockheed Martin realigned certain programs among its business segments in connection with a strategic review of its government IT and technical services businesses, which included transferring space services programs from its Information Systems & Global Solutions (IS&GS) business segment to its Space Systems business segment.[280]

Orbital ATK's revenue for the year ending 31 December 2015 increased by 52.2 %, earning $1.137 billion from the combined revenue of $747 million earned in 2014—i.e. adjusted as of the 9 February 2015 merger of Orbital Sciences Corporation (Orbital) and Alliant Techsystems Inc. (ATK) that took place on 1 January 2014. Its operating earnings increased by 7.8 %, reaching $87.3 million from the adjusted $81.0 million in 2014. The company's backlog was $8.1 billion at the end of 2015.[281]

Intelsat's total revenue for 2015 was $2.352 billion, a 4.8 % decrease from the $2.472 billion earned in the previous year. Its adjusted EBITDA for 2015 was

---

[278] Media release. "Riding high: RUAG's 250th fairing launch." 19 June 2015. RUAG 6 July 2016 <http://www.ruag.com/space/media/media-releases/news/riding-high-ruag-s-250th-fairing-launch/064689ee1cecd0fcf08b4652ada0799d/>.

[279] "Boeing Reports Fourth-Quarter Results and Provides 2016 Guidance." 27 Jan. 2016. Boeing 7 July 2016 <http://investors.boeing.com/investors/investor-news/press-release-details/2016/Boeing-Reports-Fourth-Quarter-Results-and-Provides-2016-Guidance/default.aspx>.

[280] "Lockheed Martin Reports Fourth Quarter and Full Year 2015 Results." 26 Jan. 2016. Lockheed Martin 7 July 2016 <http://www.lockheedmartin.com/us/news/press-releases/2016/january/012516-lm-q4-fy2015-earnings.html>.

[281] "Orbital ATK Announces Fourth Quarter and Full Year 2015 Financial Results." 29 Feb. 2016. Business Wire 7 July 2016 <http://www.businesswire.com/news/home/20160229007038/en/Orbital-ATK-Announces-Fourth-Quarter-Full-Year>.

$1.854 billion (78.8 % of revenue), while its adjusted EBITDA in 2014 was $1.854 billion (79.2 % of revenue). Its backlog for 2015 was about four times its annual revenue for the year at $9.4 billion. The operator planned to launch its first Intelsat Epic next generation high throughput satellite, Intelsat 29e, into orbit near the beginning of 2016; other satellites it plans to launch in that year include its Intelsat 31, Intelsat 36, and Intelsat 33e.[282]

DigitalGlobe, the commercial high-resolution Earth observation satellite imagery provider, reported a 7.3 % revenue increase for 2015, earning $702.4 million from $654.6 million in 2014. Its adjusted EBITDA grew by 24.3 %, reaching $355.7 million, while its net income was $23.3 million for 2015. During 2015, DigitalGlobe announced an agreement with KACST and TAQNIA Space to develop and launch 6 or more sub-metre satellites, with DigitalGlobe providing ground infrastructure, operations expertise, production capabilities, and global sales and distribution capabilities, while KACST and TAQNIA Space will fully fund and build the satellites. DigitalGlobe also announced several early contractual agreements for direct access EO capacity from international defence and intelligence customers for its WorldView-4 that was slated to launch in September 2016.[283]

1.3.3.3 Industrial Developments in Russia

At the beginning of 2015, signs already indicated that the new Vostochny Cosmodrome would miss its aim of being ready before the end of the year, as construction was behind schedule by up to 3 months, and the 6000 workers at the site were less than half of what was needed. Despite a cash infusion of an additional 50 billion roubles ($1.3 billion) to finish construction on time,[284] the year 2015 saw similar delays, as more than 100 workers went on strike in March and April 2015 to protest unpaid wages, while a second manager was fired at the end of March 2015 because of delays and corruption scandals.[285] Ultimately, construction of the spaceport continued into 2016. Yet the importance of the Vostochny Cosmodrome for Russia should be stressed, as it will give Russia an alternative launch port to the

[282]"Intelsat Announces Preliminary Fourth Quarter and Full Year 2015 Results." 22 Feb. 2016. BusinessWire 7 Sept. 2016 <http://www.businesswire.com/news/home/20160222005419/en/Intelsat-Announces-Preliminary-Fourth-Quarter-Full-Year>.

[283]"DigitalGlobe Reports Full Year and Fourth Quarter 2015 Results." 25 Feb. 2016. RSS Feeds 7 July 2016 <http://investor.digitalglobe.com/phoenix.zhtml?c=70788&p=RssLanding&cat=news&id=2143469>.

[284]Bodner, Matthew. "Cash Infusion Reaffirms Putin's Commitment to New Launch Complex." 3 Sept. 2014. SpaceNews 16 July 2015 <http://spacenews.com/41747cash-infusion-reaffirms-putins-commitment-to-new-launch-complex/>.

[285]Foust, Jeff. "Russia Fires Launch Site Construction Manager Again." 30 Mar. 2015. SpaceNews 1 Jan. 2016 <http://spacenews.com/russia-fires-launch-site-construction-manager-again/>.

Baikonur Cosmodrome that has been leased from Kazakhstan since the Soviet Union's collapse in 1991.[286]

1.3.3.4 Industrial Developments in Japan

Mitsubishi Electric Co. (Melco) of Japan develops satellites within its Information and Communication Systems (ICS) business segment; however, as this segment does not separate satellite-related revenue from its telecommunication, information systems, and electronic systems business, it should only be seen as generating a small portion of the total revenue earned by this segment. In the year ending 31 March 2015, the ICS segment generated 11.2 % of Melco's total sales. The ICS segment increased net sales by 2.1 %, reaching ¥559.5 billion in 2015, from ¥548.2 billion in 2014. Operating income had a substantial increase in 2015, reaching ¥18.9 billion, following continued growth in recent years, due primarily to an increase in sales.[287] The revenue generated for the fiscal year grew mainly from sales due to progress in orders already received for projects in the defence systems business, which offset decreases in revenue from telecommunications equipment business, information systems and services business, and electronic systems business, which experienced respective decreases in demand for communications infrastructure products, in system integration business, and in orders due to decreases in large-scale projects in the defence system businesses.[288]

NEC Corporation of Japan is entering into satellite integration, building on its wealth of experience in developing communication subsystems in the last four decades, within its Public Business segment, which includes satellite-related revenue within its other Major Products and Services, such as systems integration, maintenance and support, outsourcing/cloud services, and system equipment, Hence, its satellite-related business should only be seen as generating a small portion of the total revenue earned by this segment.[289] In the year ending 31 March 2015, NEC's Public Business segment generated 28 % of its total sales. The segment increased net sales by 11.3 %, reaching ¥821.9 billion in 2015, from ¥738.4 billion in 2014. Operating income increased by 1.6 %, reaching ¥62.0

[286]Soldatkin, Vladimir. "Putin orders building hastened at new Russian spaceport." 27 Aug. 2014. Reuters 16 July 2015 <http://www.reuters.com/article/2014/09/02/us-russia-space-idUSKBN0GX1AV20140902>.

[287]"2015 Annual Report | For the year ended March 31, 2015." 15 July 2015. Mitsubishi Electric 13 June 2016: 10 <http://www.mitsubishielectric.com/company/ir/library/annual_report/pdf/ar2015.pdf>.

[288]"Investor Relations – Results by Business Segment." 3 July 2015. Mitsubishi Electric 13 June 2016 <http://www.mitsubishielectric.com/company/ir/highlights/segment/index.html>.

[289]"Annual Report 2015 – Year ended March 31, 2015." 22 June 2015. NEC 9 July 2016 <http://www.nec.com/en/global/ir/pdf/annual/2015/ar2015-e.pdf>.

billion, from ¥58.6 billion in 2014.[290] The Japanese government launched its Advanced Satellite with New System Architecture for Observation (ASNARO-1) on 6 November 2014. The ASNARO-1 is the first of a new generation of low-cost Earth observation satellite platforms built by NEC Corp. Rather than develop a more powerful bus for Western customers, its focus is on emerging Asian and South American countries that do not yet require powerful bus systems. NEC plans to offer its low cost satellites in package deals that include communications, ground support and, in the case of Earth observation satellites, geospatial information products and services to enable full use of the systems.[291]

1.3.3.5 Industrial Developments in China

AsiaSat of Hong Kong had a 4.0 % decrease in revenue in 2015, earning HK$1.311 billion from HK$1.365 billion in 2014. Its operating profit decreased by 16.2 %, reaching HK$608.7 million from HK$726.3 million.[292] AsiaSat underwent a comprehensive rebranding exercise following the Carlyle Group's acquisition of General Electric Company's 50 % voting interest and 49.5 % economic stake in AsiaSat's majority shareholder Bowenvale Ltd. at the beginning of 2015; Bowenvale Ltd., a joint venture with China's government-backed CITIC Group, has about a 74 % stake in AsiaSat.[293] AsiaSat successfully regained access to the video market in China, and expected to generate additional revenues from a new AsiaSat 6 customer from January 2016. And lastly, the construction of AsiaSat 4's replacement, AsiaSat 9, remained on track for completion in late 2016 or early 2017.[294]

Three DFH-4 telecommunication satellites were launched into orbit in 2015, including TJS-1 on 12 September, Apstar-9 on 16 October, and LaoSat-1 on 20 November. While the first two satellites are operated by Chinese customers, LaoSat-1 was sold by CASC's China Great Wall Industry Corporation (CGWIC) commercial arm to Laos in a contract signed in March 2010.[295] Of the 15 DFH-4

[290]"Annual Report 2015 (Financials) – Year ended March 31, 2015." 28 Aug. 2015. NEC 9 July 2016 <http://www.nec.com/en/global/ir/pdf/annual/2015/ar2015_fin.pdf>.

[291]Kallender-Umezu, Paul. "Japan's NEC Looks To Expand Commercial Market Footprint." 24 Nov. 2014. SpaceNews 11 Feb. 2015 <http://spacenews.com/42644japans-nec-looks-to-expand-commercial-market-footprint/>.

[292]"Teamwork, together we reach new heights – 2015 Annual Report." 16 Mar. 2016. AsiaSat 13 June 2016: 63 <http://103.15.135.80/sites/default/files/E1135_AR_0421_0854.pdf>.

[293]Fellman, Joshua. "Carlyle Fund to Buy GE's Stake in AsiaSat, Make Buyout Offer." 24 Dec. 2014. Bloomberg 13 June 2016 <http://www.bloomberg.com/news/articles/2014-12-23/carlyle-fund-to-buy-ge-s-stake-in-asiasat-and-make-buyout-offer>.

[294]Media Release. "AsiaSat Reports 2015 Annual Results." 16 Mar. 2016. AsiaSat 13 June 2016 <http://103.15.135.80/sites/default/files/2015%20AsiaSat%20Annual%20Results_English_final_0.pdf>.

[295]"LaoSat 1." 17 Apr. 2016. Gunter's Space Page 7 Sept. 2016 <http://space.skyrocket.de/doc_sdat/laosat-1.htm>.

commercial satellites that have been launched since the platform's development in 2006, CGWIC has sold 8 to international customers; it plans 12 more DFH-4 spacecraft in the coming years.

1.3.3.6 Industrial Developments in India

ISRO's Antrix commercial arm reported a 17.0 % increase in revenue earning ₹1860.71 crore for the year ending 31 March 2015 from ₹1590.52 crore in 2014. Its profit increased by 2.5 %, reaching ₹205.50 crore from ₹200.56 crore in 2014. During the year, Antrix signed eight PSLV launch services agreements to carry 16 international customer satellites as co-passengers; the company has a total of 33 satellite launch agreements.[296]

1.3.3.7 World

Canada's Space Systems/Loral (SS/L), a subsidiary of MDA Corp. based in the U.S., once again won the most commercial satellite contracts in 2015, outbidding the usual top prime contractors for 5 of the roughly 26 orders that were open for competitive bidding. Europe's Thales Alenia Space and Airbus Defense & Space each had four orders, while Boeing had three, and Lockheed Martin and Orbital ATK each won two orders. China's CGWIC commercial arm won three commercial orders, while India's Antrix commercial arm won two, and Russia's ISS Reshetnev won one.[297] The revenue earned by SS/L's parent company, MDA Corp., for the fiscal year-ended 31 December 2015 increased by 0.9 % to C $2.117 billion, from C$2.099 billion in 2014.[298]

Com Dev International is Canada's second largest space company after MDA Corp, providing space hardware subsystems and components. At the beginning of 2015, fewer tax incentives, a drought of U.S. institutional orders caused by stresses in the U.S. space budget in 2014, and a loosening of U.S. ITAR restrictions on how satellite components are treated, pushed Com Dev to close its facilities in California, and bid on U.S. military contracts while working in lower-cost Canada.[299] At that time Com Dev still hoped to acquire a U.S. company that did not rely on

[296]"Annual Report 2014–15." 30 Nov. 2015. Antrix 9 July 2016 <http://www.antrix.gov.in/sites/default/files/article-attachments/ANNUAL%20REPORT%202014%20-2015%28ENGLISH%29.pdf>.

[297]"Recently awarded GEO-Sat Contracts." 6 May 2016. Gunter's Space Page 12 May 2016 <http://space.skyrocket.de/doc_sat/sat-contracts.htm>.

[298]"MDA reports fourth quarter and full year 2015 results." 24 Feb. 2016. MDA Corp. 7 July 2016 <http://mdacorporation.com/news/pr/pr2016022401.html>.

[299]De Selding, Peter B. "Com Dev Closing California Plant But Expects To Find Another U.S. Foothold." 16 Jan. 2015. SpaceNews 12 Jan. 2016 <http://spacenews.com/com-dev-closing-california-plant-but-expects-to-find-another-u-s-foothold/>.

government orders. Com Dev's focus broadened to tap the UK's high-growth space market, with its purchase of MESL Holdings and MESL Microwave of Scotland. The purchase was valued at C$23 million. Com Dev hoped its MESL acquisitions would generate an additional C$10 million for the 2015 year. Yet on 5 November 2015, Com Dev International announced its acquisition by Canadian company Honeywell and the spinout of Com Dev's exactEarth subsidiary. Honeywell's acquisition of Com Dev is valued at approximately C$455 million.[300] In the year prior to its acquisition, Com Dev had reported total revenue of C$208.2 million for the fiscal year-ended 31 October 2014, down 3.4 % from C$215.5 million in 2013.[301]

Com Dev's subsidiary exactEarth Ltd. provides maritime traffic information to coastal authorities from the signals that they transmit to exactEarth's LEO constellation. In the 9-months ended 31 July 2015, exactEarth earned C$19.1 million, up 68 % from the 9 months ended 1 August 2014. Its EBITDA over that same period reached C$3.6 million by 31 July 2015, from C$1.2 million in the 9 months ended 1 August 2014.[302] Prior to its spinoff from Com Dev following the parent company's acquisition by Honeywell on 5 November 2015, on 9 June 2015, exactEarth and Harris Corp. of the U.S. announced a strategic partnership, wherein Harris Corp will mount exactEarth developed payloads on 58 next-generation Iridium mobile communication satellites. Under the arrangement, Harris Corp. will gain immediate access to exactEarth's constellation of eight small LEO Automatic Identification Satellites (AIS). Under the new partnership, exactEarth expects to increase its AIS constellation to 11 satellites by 2017, in addition to increasing sales to U.S. government customers.[303] While Harris Corp. will assume the cost of integrating exactEarth technology in its Iridium Next payloads, exactEarth has agreed to pay $10 million in commitment fees through to June 2016.[304] Under exactEarth's partnership with Harris Corp., exactEarth will pay Harris Corp. $3 million per year after the full deployment of Harris' payloads hosted on 72 Iridium satellites. Those Iridium satellites were to begin to launch starting in early 2016, wherein during this initial period, Harris will pay exactEarth

[300] "COM DEV announces acquisition by Honeywell and spinout of exactEarth." 6 Nov. 2015. Com Dev International 16 June 2016 <https://www.comdev.ca/images/financial-reports/CDV_Honeywell_release_2015-11-05.pdf>.

[301] "COM DEV Announces Fourth Quarter and Year-End Fiscal 2014 Results." 15 Jan. 2015. COM DEV International 10 Feb. 2015 <http://www.comdev.ca/images/financial-reports/CDV_Q4_14_Financial_Release2.pdf>.

[302] "COM DEV announces acquisition by Honeywell and spinout of exactEarth." 6 Nov. 2015. Com Dev International 16 June 2016 <https://www.comdev.ca/images/financial-reports/CDV_Honeywell_release_2015-11-05.pdf>.

[303] De Selding, Peter B. "Harris, exactEarth To Place AIS Gear on Iridium Craft." 9 June 2015. SpaceNews 13 Jan. 2016 <http://spacenews.com/harris-exactearth-to-place-ais-gear-on-iridium-craft/>.

[304] De Selding, Peter B. "Harris, exactEarth Aim To Ride Iridium Next to Growth in AIS." 15 June 2015. SpaceNews 13 Jan. 2016 <http://spacenews.com/harris-exactearth-aim-to-ride-iridium-next-to-growth-in-ais/>.

15 % of its institutional revenues until it reaches $339,000 in revenue or by March 2016; from there, its payment to exactEarth increases to 50 % of government revenue up to a ceiling of $40 million. After reaching that ceiling, exactEarth will then receive 33 % of the generated revenues.[305] By 13 July 2015, the ownership configuration of exactEarth had exhibited signs of change, as its parent Com Dev and minority owner Hisdesat (73 % Com Dev, 27 % Hisdesat) announced their decision to make available 40–45 % of their ownership in exactEarth in an initial stock offering expected to generate C$88 million.[306]

Thailand's Thaicom satellite operator earned 12.453 billion baht for the year ended 31 December 2015, an increase of 4.7 % over its restated revenue of 11.893 billion baht earned in 2014. Its profits from its operations continued to grow, this time by 52.2 %, to 2.305 billion baht in 2015, from 1.514 billion baht in 2014.[307] Its satellite and related services amounted to 73.8 % of the revenue generated in 2015, an incremental increase from the 71.4 % share in 2014, and 69.6 % share in 2013. Its sales and services revenue by geographic area came from Thailand (63.2 %), Australia (10.2 %), Japan (5.2 %), Myanmar (5.0 %), India (4.4 %), China (2.6 %), and Others (9.4 %).[308] Among the significant developments for Thaicom over 2015, on 17 August 2015, the company and Thailand's Ministry of Defence signed a memorandum of understanding for collaboration in satellite communications and the possibility of co-developing a satellite project between the two parties. Other developments near the end of the year, included the Thaicom 7 satellite reaching a 100 % booking of transponders on 19 November 2015, followed by the company recognising the 454 million baht impairment loss on investment on one of its Australian broadband satellite service providers.[309]

On 16 April 2015, NewSat Ltd. of Australia filed for bankruptcy protection with the U.S. Bankruptcy Court of Delaware (applying Australian administration proceedings).[310] The start-up satellite fleet operator, which reported a 17 % drop in revenue for the 6 months ending 31 December 2014, totalling 13.7 million Australian dollars, and a net loss of 39.7 million Australian dollars,[311] had been in financial constraints in recent years after defaulting on around $300.5 million in loans from the U.S. Export-Import (Ex-Im) Bank, and a $115 million loan

[305] De Selding, Peter B. "Harris, exactEarth Aim To Ride Iridium Next to Growth in AIS." 15 June 2015. SpaceNews 13 Jan. 2016 <http://spacenews.com/harris-exactearth-aim-to-ride-iridium-next-to-growth-in-ais/>.

[306] De Selding, Peter B. "exactEarth's IPO Plans Coming into Focus." 17 July 2015. SpaceNews 13 Jan. 2016 <http://spacenews.com/exactearths-ipo-plans-coming-into-focus/>.

[307] "The Next Phase 2016 – Annual Report 2015." 3 Mar. 2016. Thaicom Public Company Limited 15 June 2016: 155° E, 156° W <http://thcom.listedcompany.com/ar.html>.

[308] Ibid. at 061° E.

[309] Ibid. at 051° E – 053° E.

[310] De Selding, Peter B. "NewSat Seeks Bankruptcy Protection." 17 Apr. 2015. SpaceNews 13 Jan. 2016 <http://spacenews.com/newsat-seeks-bankruptcy-protection/>.

[311] De Selding, Peter B. "NewSat's Losses Deepen as Creditors Circle." 27 Feb. 2015. SpaceNews 12 Jan. 2016 <http://spacenews.com/newsats-losses-deepen-as-creditors-circle/>.

guarantee provided by France's Coface to launch its Jabiru-1 satellite aboard an Ariane 5 launcher.[312] NewSat's lenders had grown unwilling to supply additional funds for the development of the Jabiru-1 after it missed a 30 November 2014 deadline to raise an additional $40 million in cash or equity.[313] By that time, NewSat had used mainly Ex-Im Bank funds to invest $193 million in Jabiru-1, while around $78 million was still needed to complete the satellite.[314] A bankruptcy filing appeared unavoidable as NewSat was beset by breach of contract notices from its two principal suppliers (i.e. for the non-payment of $21 million to Lockheed Martin and $42.4 million to Arianespace) early in 2015, and refusals by its lenders Coface and Ex-Im Bank to approve loan waivers that would have permitted funding to resume.[315]

On 22 May 2015, the bankruptcy court ruled that the contract between NewSat and Lockheed Martin was no longer in force because NewSat, Ex-Im Bank and Lockheed Martin had failed to resolve their difference by a mutually agreed upon deadline. As NewSat was unable to pay past-due bills to Lockheed Martin, Lockheed was granted permission to take possession of the mostly completed Jibaru-1.[316] However, the cancellation of the contract potentially left the Ex-Im Bank with a loss of over $100 million, as Ex-Im Bank's security claim was placed on NewSat that no longer had a claim over the Jabiru-1.[317] On 29 May 2015, the court ordered a standstill agreement between NewSat and Arianespace, allowing Arianespace to stop further work on the contract while NewSat was given until 1 August 2015 to finalize the sale of its slot on the Ariane 5 launcher to satellite fleet operator Measat of Malaysia.[318]

---

[312] De Selding, Peter B. "NewSat Seeks Bankruptcy Protection." 17 Apr. 2015. SpaceNews 13 Jan. 2016 <http://spacenews.com/newsat-seeks-bankruptcy-protection/>.

[313] De Selding, Peter B. "Jabiru-1 Launch Slips Further as NewSat, Creditors Haggle over New Financing." 9 Feb. 2015. SpaceNews 12 Jan. 2016 <http://spacenews.com/jabiru-1-launch-slips-further-as-newsat-creditors-haggle-over-new-financing/>.

[314] De Selding, Peter B. "Aug. 1 Deadline Set for Selling NewSat's Ariane 5 Launch Slot to Measat." 29 July 2015. SpaceNews 13 Jan. 2016 <http://spacenews.com/aug-1-deadline-set-for-selling-newsats-ariane-5-launch-slot-to-measat/>.

[315] C.f. De Selding, Peter B. "NewSat in Credit Bind after Coface Rebuffs Waiver Request." 8 Apr. 2015. SpaceNews 13 Jan. 2016 <http://spacenews.com/coface-refusal-puts-newsat-in-tight-spot-with-creditors/>; and De Selding, Peter B. "Ex-Im Hasn't Given Up on Recovering $100 Million NewSat Loss." 9 June 2015. SpaceNews 13 Jan. 2016 <http://spacenews.com/ex-im-hasnt-given-up-on-recovering-100-million-newsat-launch/>.

[316] SpaceNews Editor. "Editorial | NewSat's Failure and the Ex-Im Bank." 15 June 2015. SpaceNews 13 Jan. 2016 <http://spacenews.com/editorial-newsats-failure-and-the-ex-im-bank/>.

[317] De Selding, Peter B. "Aug. 1 Deadline Set for Selling NewSat's Ariane 5 Launch Slot to Measat." 29 July 2015. SpaceNews 13 Jan. 2016 <http://spacenews.com/aug-1-deadline-set-for-selling-newsats-ariane-5-launch-slot-to-measat/>.

[318] De Selding, Peter B. "Aug. 1 Deadline Set for Selling NewSat's Ariane 5 Launch Slot to Measat." 29 July 2015. SpaceNews 13 Jan. 2016 <http://spacenews.com/aug-1-deadline-set-for-selling-newsats-ariane-5-launch-slot-to-measat/>.

1.3.4 Industrial Overview

1.3.4.1 Launch Sector

The launch sector is an enabler rather than a primary economic activity. Yet, with the growth in low-cost launch services, the marginal revenue the launch sector generates is becoming a more important factor to watch.

Launch activity decreased in 2015, with a total of 87 launches conducted by launch providers from Russia, the United States, China, Europe, India, Japan, and Iran. Amid the positive developments to be presented in this section, there were also some notable setbacks that affected the launch sector in 2015. There were four launch failures in 2015, resulting in the loss of two ISS resupply missions, and up to 25 payloads of various sizes. The first failure occurred on 28 April 2015, with the Progress M-27M ISS resupply mission, which reached orbit, but lost telemetry just before separating from the Soyuz 2.1a upper stage. The next ISS resupply mission, SpaceX's Dragon CRS-7 valued at $61.2 million, also failed on 28 June 2015 due to the presence of too much pressure in the Falcon 9's upper stage liquid oxygen tank about 139 seconds into the flight. The failure of the third stage of the Proton M Briz-M (Ph.3) on 16 May 2015 resulted in the loss of the MexSat 1 communications satellite intended to secure communications for Mexico's national security needs. There was also the failed maiden launch of the three-stage Super Strypi small satellite launch vehicle on 4 November 2015, which malfunctioned during its stage one burn.

When looking into the launches of specific countries (Table 1.2), Russia was once again the world leader in the number of launches for 2015, conducting 30.2 % of total launches. The United States remained in second position with a 23.3 % share, followed closely by China which had a 22.1 % share of the total. Europe stayed in fourth position with a 12.8 % share, ahead of India with a 5.8 %, Japan with a 4.3 % share, and Iran which accounted for 1.2 % of the total launch figure (see Fig. 1.9).[319]

Russia conducted 26 launches using 11 different launch system configurations.[320] The United States also used 11 different launch system configurations for its 20 launches.[321] China used 7 different launch configurations for its 19 launches.[322] Europe relied on its workhorse Ariane 5 ECA launcher, in addition to its Vega launcher and Europeanized Soyuz 2.1b and Soyuz STB Fregat-MT to have 12 launches (6 Ariane 5 ECA, 3 Vega, 1 Soyuz 2.1b, and 2 Soyuz STB Fregat-MT). India used three launcher configurations (i.e. three PSLV XL, one PSLV XL,

[319]Federal Aviation Administration. The Annual Compendium of Commercial Space Transportation: 2016. Washington, DC: FAA, Jan. 2016: 39.

[320]i.e. Dnepr, Proton M Blok DM 03, Proton-M Briz-M (Ph.1 mod. 2), Proton M Briz-M (Ph.3), Rokot KM, Soyuz 2.1a, Soyuz 2.1b, Soyuz 2.1v Volga, Soyuz FG, Soyuz U, and Zenit 3F.

[321]i.e. Atlas 5 (401), Atlas 5 (421), Atlas 5 (501), Atlas 5 (551), Delta 4 Medium+ (4,2), Delta 4 Medium+ (5,4) (upgrade), Delta II 7320-10C, Falcon 9 v1.1, Falcon 9 v1.1(R), Falcon 9FT, and Super Strypi.

[322]i.e. Long March 11, Long March 2D, Long March 3B, Long March 3C, Long March 4B, Long March 4C, and Long March 6.

Table 1.2 Worldwide launches in 2015 per country, number of launched systems, and commercial status (Source: FAA)

Launchers	Number of launch systems active in 2015	Total number of launches	Commercial launches	Non-commercial launches
Russia	11	26	5	21
United States	11	20	8	12
China	7	19	0	19
Europe	4	12	6	6
India	3	5	2	3
Japan	3	4	1	3
Iran	1	1	0	1
Total	40	87	22	65

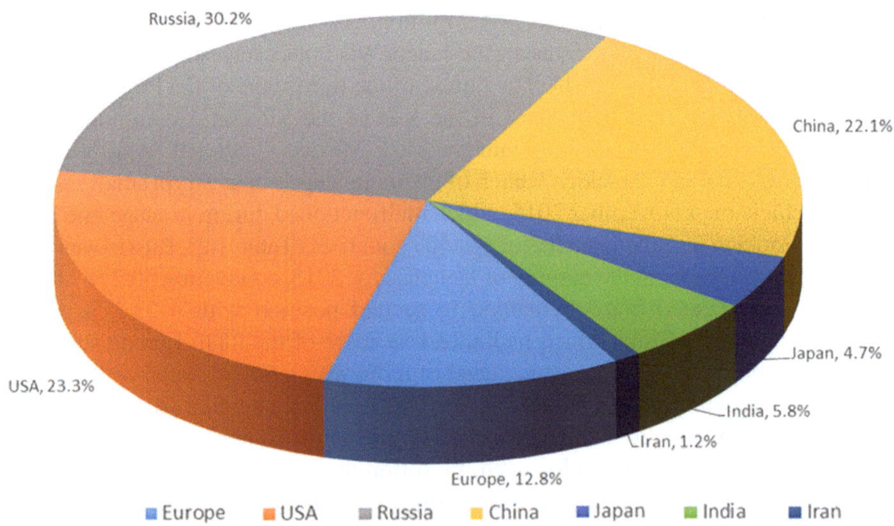

Fig. 1.9 Worldwide launches by country in 2015 (Source: FAA)

and one GSLV MK2) for its five launches; while Japan used three launcher configurations (i.e. two H-IIA 202, one H-IIA-204, and one H-IIB 304) for its four launches. And Iran's single launch was conducted using its Safir 2 launcher.

Launch system utilization moved back to 40 active launch systems in 2015 from 36 used in 2014. Russia, the U.S., and China accounted for 75.9 % of the number of launches for 2015 while launch activity in Europe, India, and Japan steadily developed. Moreover, what is not reflected when looking at the number of launches is the fact that certain launchers have dual-launch capabilities such as Europe's Ariane 5 ECA, which can lift two standard-size payloads to geostationary orbit. Hence the number of launches does not reflect the number of payloads brought to orbit.

While Europe had the second highest number of commercial launches in 2015, its Ariane 5 dual payload capability helped it to earn the greatest amount of commercial launch revenue for the year. Inversely, the U.S. conducted the most commercial launches, but was in second position in terms of launch revenue in 2015. Russia increased its number of commercial launches to five, placing it at third position in terms of commercial revenue generated. And Japan's single commercial launch placed it in fourth position in terms of commercial launch revenue, while India's two commercial launches placed it at the end of the pack of commercial launch providers. When considering non-commercial launches, Russia had a substantial lead, conducting 32.3 % of the launches, China was in second position with 29.2 %, followed by the U.S. with an 18.5 % share. The remaining 20 % of non-commercial launches was split by Europe, India, Japan, and Iran.

The ratio of commercial launches to non-commercial launches was unchanged in 2015, with an equilibrium of one commercial launch for every three non-commercial launches. Yet, in 2015 the number of payloads launched decreased to 265 from 295 in 2014, which can be attributed to the reduced number of cube satellites intended to be released from the International Space Station.

In terms of the global share of payloads launched in 2015 (Fig. 1.10), the U.S. was first in the number of payloads it placed in orbit, lifting 123 payloads, which represents 46.4 % of the total. China moved to second place, launching 45 payloads (nearly doubling its number in 2014) resulting in a 17.0 % share. Russia was in third position, launching 32 payloads, which amounted to a 12.1 % share. Europe and Japan came in fourth position, each with 22 payloads (8.3 %), followed closely by India with 20 payloads (7.5 %), and Iran with 1 payload

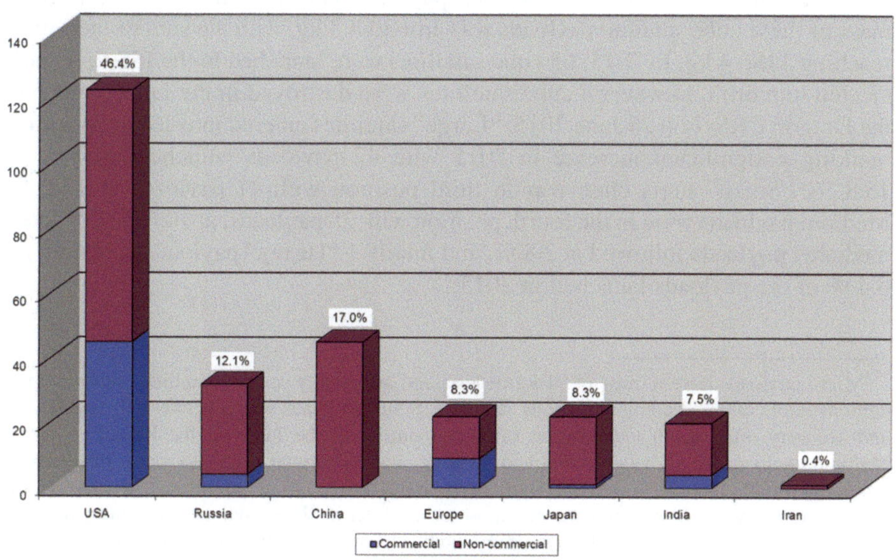

Fig. 1.10 Total payloads launched in 2015 by country, share and commercial status (Source: FAA)

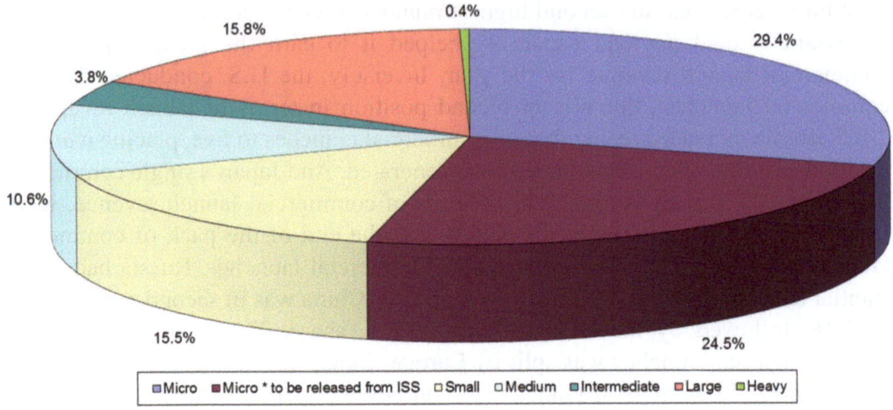

Fig. 1.11 Distribution of the payloads launched in 2015 by mass class (Source: FAA)

(0.4 %). The global share of payloads launched changes considerably when excluding the total 143 cube satellite payloads from the assessment. In this case, the U.S. remained in first position with a decreased 29.5 % share, followed by Russia whose share increases to 24.6 % of the total. Moreover, Europe and China tie in third position, each with an 18.0 % share, while India's share drops to fourth position with 6.6 %, followed by Japan with a 3.3 % share; Iran's cube satellite is not considered in this scenario.

There were also some changes in the distribution of payload sizes in 2015 (Fig. 1.11 and Table 1.3). The number of "Micro" sized payloads reduced to 143, accounting for 54.0 % of the total payloads launched in 2015. The average mass of these cube satellite payloads was around 8.3 kg, with the sum of their mass reaching 1186.4 kg. In 2015, 65 cube satellites were launched to the ISS to be later ejected into orbit, however 8 cube satellites were destroyed in the failed launch of the Dragon CRS-7 on 28 June 2015. "Large" satellites entered into second position, marking a significant increase in 2015 with 42 payloads launched; a share of 15.8 %. "Small" mass class was in third position with 41 payloads at 15.5 %. Medium payloads were in the fourth position with 28 payloads, at 10.6 %; 10 "Intermediate" payloads followed at 3.8 %, and finally 1 "Heavy" payload accounted for 0.4 % of the payloads launched in 2015.[323]

[323]Micro payloads have a mass of 91 kg or less, and are mainly science satellites, technological demonstrators or small communications satellites. Small payloads weigh between 92 and 907 kg and are very often Earth Observation satellites, similar to the Jason or the RapidEye series. Medium payloads weigh between 908 and 2268 kg, and feature the most diverse set of satellites, including small satcoms in geostationary orbit, Earth Observation satellites, and most of the Russian military satellites from the Kosmos series. Intermediate payloads, weighing between 2269 and 4536 kg, comprise medium satcoms and big scientific satellites. Large payloads, between 4537 and 9072 kg, refer to big satcoms, as well as to the Soyuz and Progress spacecraft flying to the ISS. Finally, Heavy payloads, exceeding 9072 kg, are linked to ISS activity, such as the cargo spacecraft, ATV, HTV, etc. *See* Commercial Space Transportation: 2011 Year in Review, 32.

Table 1.3 Distribution of the payloads launched in 2015 by mass class (Source: FAA and Gunter's Space Page)

Payloads by mass class	2015	Percentage	Average mass (kg)	Mass sum (kg)
Micro	78	29.4	10.1	791.4
Micro * to be released from ISS	65	24.5	6.1	395.0
Small	41	15.5	373.4	15,308.0
Medium	28	10.6	1533.4	42,936.0
Intermediate	10	3.8	3143.1	31,431.0
Large	42	15.8	6107.4	256,510.0
Heavy	1	0.4	16,500.0	16,500.0
	265	100.00	1373.1	363,871.4

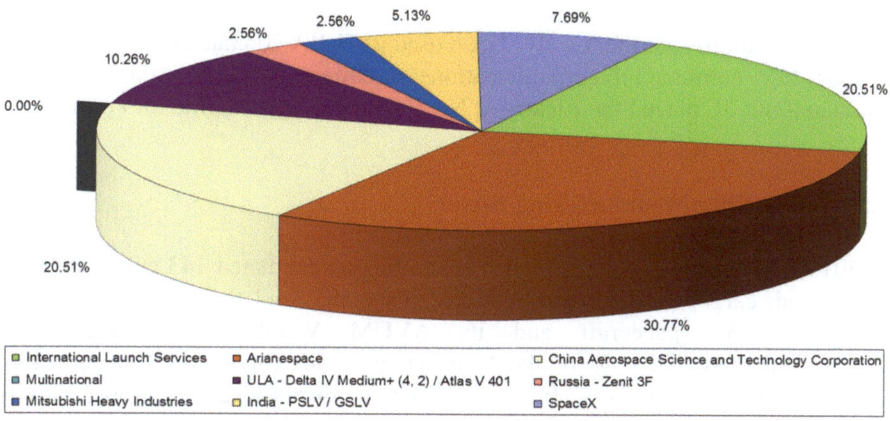

Fig. 1.12 Share of launch contracts for GEO satellites in 2015 by launch service provider

In 2015, Arianespace conducted the most launches in GEO, with a 30.8 % share, followed by the China Aerospace Science and Technology Corporation (CASC), and the Russian International Launch Services (ILS), each with 20.5 % (Fig. 1.12 and Table 1.4). Arianespace placed 10 communications satellites, along with the Italian/French Sicral 2 military communications satellite, and Eumetsat's MSG-4 meteorological satellite into GEO orbit using six Ariane 5 ECA launchers. China's CASC launched four communication satellites, two Beidou navigation satellites, its Gaofen 4 remote sensing satellite, and its TJSSW 1 early warning satellite to GEO orbit mainly on Long March 3B launchers, and one 3C launcher. The Russian ILS launched eight GEO communication satellites on eight Proton M launchers; including a failed launch of the MexSat 1 satellite. The U.S. ULA had four launches to lift four communications satellites (MUOS 3, MUOS 4, WGS 7, and MexSat 3) into GEO orbit. And the SpaceX Falcon 9 was used twice to launch three commercial communications satellites (ABS 3A, Eutelsat 115, and TürkmenÄlem 52E/ MonacoSat-1) into GEO orbit. India's PSLV XL launcher lifted the IRNSS 1D navigation satellite to GEO orbit, while its GSLV Mk.2 launched the GSAT

Table 1.4 Share of launch contracts for GEO satellites in 2015 by launch service provider (Source: FAA)

Launch services provider	Launches	Payloads	Percentage
Arianespace	6	12	30.77
China Aerospace Science and Technology Corporation (CASC)	8	8	20.51
International Launch Services (ILS)	8	8	20.51
United Launch Alliance (ULA)	4	4	10.26
SpaceX	2	3	7.69
India	2	2	5.13
Mitsubishi Heavy Industries	1	1	2.56
Russia	1	1	2.56
	32	39	100.00

6 communications satellite. Lastly, Japan used its H-IIA 204 launcher to launch the Telstar 12V commercial communications satellite into GEO orbit, while the Russian Zenit 3F placed the Elektro-L No. 2 meteorology satellite into orbit.

1.3.4.2 Satellite Manufacturing Sector

In 2015, 265 payloads were launched (including an estimated 143 cube satellites, 15 crewed, cargo, or hardware missions to the ISS, and the test launch of the Europe's IXV spacecraft and its AVUM VV04 upper stage). The U.S. manufactured 52.1 % of the launched payloads (including 104 cubesats), while China accounted for 17.0 %, and Russia produced 10.2 %. Europe, with its 24 satellites mostly built for navigation and communication purposes, accounted for 9.1 % of the payloads launched, while Japan and India each accounted for 1.5 %. The remaining 8.7 % of payloads were from various parts of Asia, the Middle East, and the Americas.[324]

Of the 248 satellites launched in 2015,[325] 166 were non-commercial. Among the large satellite integrators, China's CAST was once again in first position in manufacturing 16 non-commercial satellites, followed by the Reshetnev Company which produced 10 non-commercial satellites, while the regions of Asia, North America, Europe, and Russia also showed a strong presence in this market. About 16 of the 24 European manufactured satellites were non-commercial. Boeing produced seven non-commercial satellites, followed by OHB Systems with six non-commercial satellites. ISRO produced four non-commercial satellites, while

[324]Federal Aviation Administration. The Annual Compendium of Commercial Space Transportation: 2016. Washington, DC: FAA, Jan. 2016: 65–69. Payloads are assigned to the nation that commissioned them, not according to the nationality of the manufacturer.

[325]i.e. not counting the 15 crewed, cargo, or hardware missions to the ISS, and the test launch of the Europe's IXV spacecraft and its AVUM VV04 upper stage.

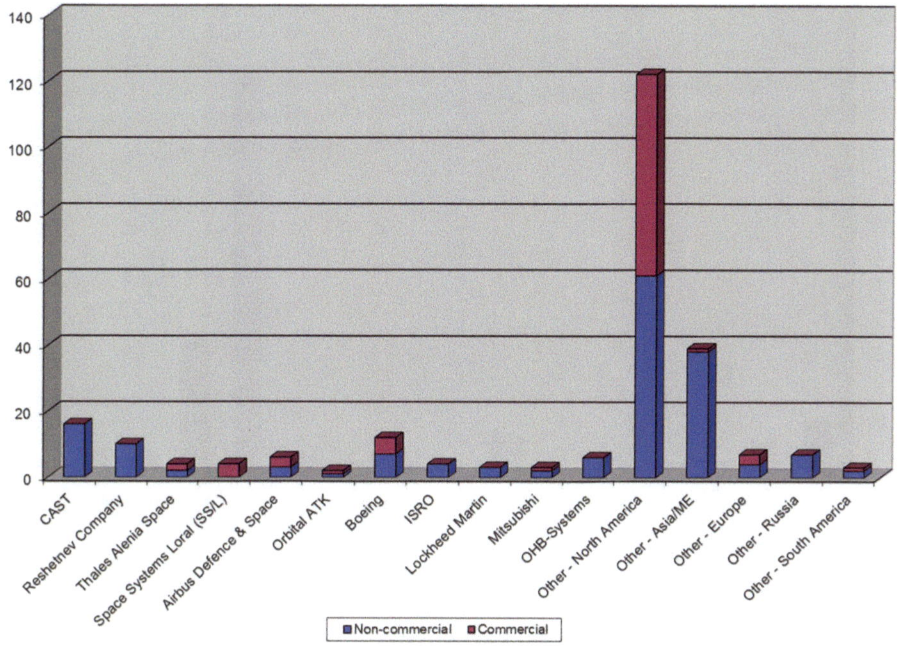

Fig. 1.13 Satellites launched in 2015 by manufacturer and commercial status (Source: FAA)

three were developed each by Airbus Defence & Space and Lockheed Martin, along with two developed each by Thales Alenia Space and Mitsubishi, and one non-commercial satellite developed by Orbital ATK (Fig. 1.13).[326]

In 2015, Boeing took the top position among the commercial prime contractors with the launch of five of its commercial satellites. Canada's Space Systems Loral (SS/L) took the second position manufacturing four commercial satellites, while Airbus Defense & Space was third in terms of its three commercial satellites. Thales Alenia Space built two commercial satellites, while Mitsubishi, and Orbital ATK each developed one commercial satellite (Fig. 1.13).[327]

In 2015, 15.7 % of the 248 satellites launched were geostationary satellites (Fig. 1.14). In this field, 35.9 % of the 39 GEO satellites came from North America (including 7 by Boeing, 4 by SS/L, 2 by Lockheed Martin, and 1 by Orbital ATK). Europe provided 20.5 % of the GEO satellites (four by Airbus Defence & Space and four by Thales Alenia Space, built mainly for telecommunications purposes). China's CAST also provided 20.5 % of the GEO satellites launched in 2015, launching a total of eight satellites to GEO orbit following a hiatus in 2014. Russia's Reshetnev Company had a 7.7 % share of GEO satellites launched into

[326]Federal Aviation Administration. The Annual Compendium of Commercial Space Transportation: 2016. Washington, DC: FAA, Jan. 2016: 65–69.
[327]Ibid.

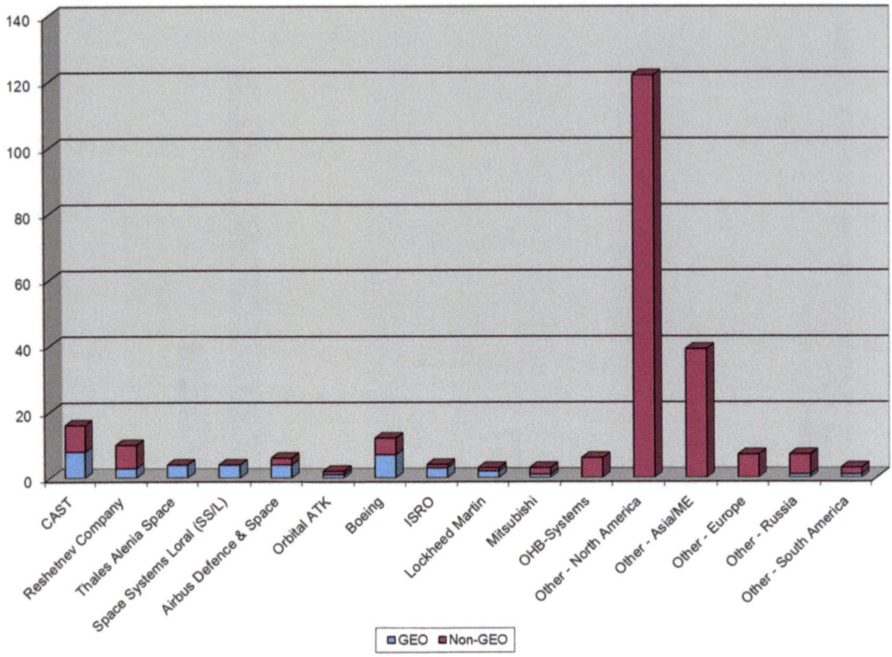

Fig. 1.14 Satellites launched in 2015 by manufacturer and orbit type (Source: FAA)

orbit, as did India's ISRO, each launching of three satellites. Japan's Mitsubishi had a 2.6 % share, launching one satellite to GEO orbit, in addition to one satellite by Other—Russia, and one by Other—South America.[328]

Among the 209 non-GEO orbiting satellites, North America's share grew to 61.7 % (5 by Boeing, 1 by Lockheed Martin, 1 by Orbital ATK, and another 122—mostly cube satellites—from other parts of North America). Manufacturers in Asia held a 23.9 % share with 50 non-GEO satellites (8 by China's CAST, 2 by Japan's Mitsubishi, 1 by India's ISRO, and another 39 developed by in Other—Asia/ME). Europe's share was 7.2 % with 15 non-GEO satellites (6 by OHB Systems, 2 by Airbus Defence & Space, and the remainder from other European makers). Russia's share was 6.2 % with 13 non-GEO satellites (7 by the Reshetnev Company, and the remaining by Other Russian manufacturers). The remaining two non-GEO satellites were developed in South America, and accounted for a 1.0 % share (Fig. 1.15).[329]

[328]Federal Aviation Administration. The Annual Compendium of Commercial Space Transportation: 2016. Washington, DC: FAA, Jan. 2016: 40.

[329]Federal Aviation Administration. The Annual Compendium of Commercial Space Transportation: 2016. Washington, DC: FAA, Jan. 2016: 65–69.

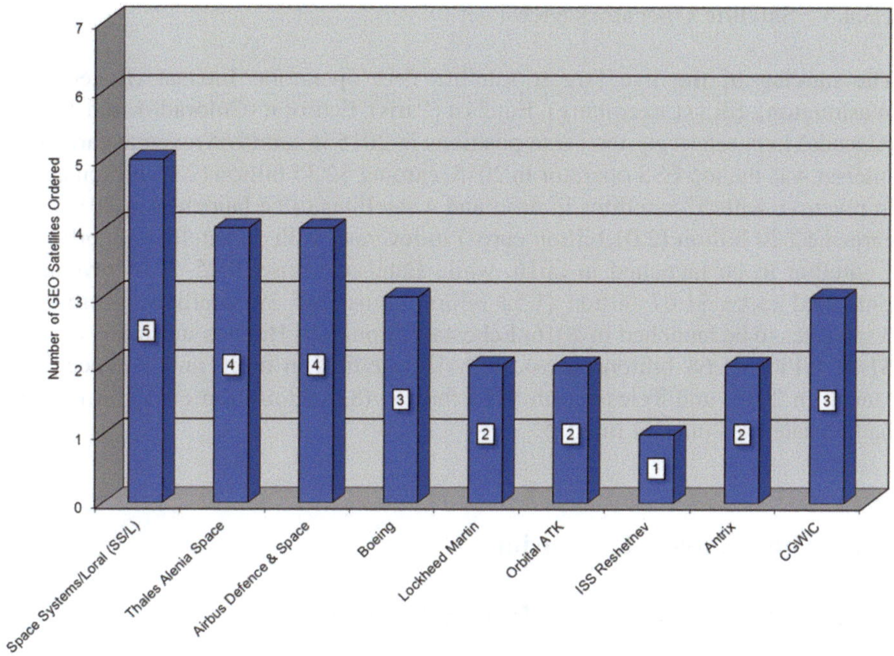

Fig. 1.15 GEO satellite orders in 2015 by manufacturer

In 2015, North American prime spacecraft manufacturers had a combined lead in orders for commercial GEO communications satellites, accounting for 46.1 % of the 26 contracts awarded, whereas European contracts decreased to 30.8 % of the available awards (Fig. 1.15). SS/L won the most commercial orders, five contracts (including the Azerspace 2, BSat 4a, Telkom 4, Telstar 18V, and Telstar 19V). Europe's Airbus Defence & Space with four orders (Inmarsat-6 F1, Inmarsat-6 F2, SES 14, and Eutelsat Quantum), along with Thales Alenia Space, which also had four orders (Bangabandhu 1, Eutelsat African Broadband Satellite, Comsat-NG 1, and Comsat-NG 2), came next. U.S. prime contractors' Boeing had three contracts (ABS 8, SES 15, and Silkwave 1), while Lockheed Martin won two commercial contracts for the year (ArabSat 6A and HellasSat 4) along with Orbital ATK (SES 16/GovSat and an unknown satellite). China's CGWIC had an 11.5 % share by winning three contracts (APStar 6C, ChinaSat 6C, and ChinaSat 18), while India's Antrix had a 7.7 % share winning two contracts (GSat 17 and GSat 18). Lastly, Russia's ISS Reshetnev won a contract to build the Russian Yamal 601 communication satellite resulting in a 3.8 % share.[330]

[330]"Recently awarded GEO-Sat Contracts." 6 May 2016. Gunter's Space Page 12 May 2016 <http://space.skyrocket.de/doc_sat/sat-contracts.htm>.

1.3.4.3 Satellite Operators Sector

The ranking of the five largest satellite fleet operators: Intelsat (Luxemburg, Washington), SES (Luxemburg), Eutelsat (Paris), Echostar (Colorado), and Telesat (Canada) kept relatively the same positions in 2015 in terms of revenue earned.[331] Intelsat was the top FSS operator in 2015, earning $2.35 billion (2.15 billion euros) in revenue, with 52 satellites in orbit and 4 satellites to be launched in 2016. SES earned $2.20 billion (2.01 billion euros) in revenue, with 53 satellites in orbit and 1 satellite to be launched in 2016, while Eutelsat whose 2015–2016 revenue is estimated to be $1.67 billion (1.53 billion euros) has 39 satellites in orbit and 3 satellites to be launched in 2016. Echostar (through its Hughes subsidiary) earned $1.84 billion (1.68 billion euros), with 18 satellites in orbit, and 5 expected to launch in 2016, and Telesat with $955 million (873.98 million euros) in revenue had 15 satellites in orbit in 2015.

1.4 The Security Dimension

1.4.1 The Global Space Military Context

The Space Report 2016 listed space-related military spending in 2015 as having reached $34.151 billion, decreasing by 3.3 % from $35.325 billion due more to unfavourable dollar exchange rates rather than decreased national spending.[332] As is typical with the nature of dual-use technology in space activity, there is a risk that certain military activities have been already included in larger budgets, which can result in double counting. Moreover, while missions, often listed as civil programmes, may also serve dual-purpose military objectives, their expenditure is not included in this section.

According to The Space Report 2016, the U.S. generated about 69.0 % ($23.572 billion) of global military space spending in 2015, a notable increase from the 63.6 % share estimated for 2014. Non-U.S. global military spending decreased to 31.0 % ($10.579 billion) from 36.4 % in 2014.[333] However, due to fluctuating exchange rates, variations in purchasing power, and different employment costs, a direct comparison of the budgets of these countries in fixed dollar values does not present a clear picture of their relative space defence efforts.

---

[331] "World Teleport Association Publishes Top Operator Rankings for 2015." 16 Dec. 2015. WTA 5 Sept. 2016 <http://www.worldteleport.org/news/266131/World-Teleport-Association-Publishes-Top-Operator-Rankings-for-2015.htm>.

[332] The Space Report 2016. Colorado Springs: The Space Foundation, 2014: 38.

[333] Ibid.

1.4.2 Europe

According to estimates by Eurospace, total funding for European military space programmes was about $753.051 million (675.2 million euros) in 2015. France had the highest military budget at $362.696 million (325.2 million euros), while the United Kingdom budgeted $317.860 million (285.0 million euros), with Germany at $55.765 million (50.0 million euros), and Italy at $16.730 million (15.0 million euros). Yet, there is a discrepancy between estimates provided in the previous 2014 reporting period and the current figures, due once again to currency exchange rates, but also to the use of Eurospace figures as an authority in lieu of Euroconsult figures. It should also be noted that Eurospace figures do not account for expenditures in services.

Financial problems prevented Italy from participating in France's latest Comsat NG dedicated military communications system. Nevertheless, as the two countries previously collaborated in other milsat space programmes such as Cosmo-SkyMed and the Sicral 2 UHF and SH-band satellite, their collaboration sets an example that the other three closely allied European countries owning military satellite systems, i.e. Britain, Germany, and Spain, may benefit from adoption. The five countries have often indicated a desire to do so, however coordinating cooperative opportunities has proven difficult. In addition to financial constraints, as in the case of Italy, another barrier to collaboration was the desire to use existing capacity before investing in new capacity. Coordinating investment on a shared milsat programme will be a challenge, as the different stage life cycles of the military systems in orbit mean that countries may be unwilling to procure additional capacity while their own systems are expected to operate for several more years. Yet collaborating on a shared milsat programme should lower costs by an estimated 2 billion euros according to the European Defence Agency (EDA), and will substantially save costs over the long term while still making a bounty of capacity available.[334] Collaboration may also help to quell the unequal approach by the EC to milsatcom procurement of services from operators.

In a move to bolster its military communications space infrastructure and provide encrypted jam-resistant capacity to NATO-allied governments, Airbus D&S has moved its Skynet 5A military X- and UHF-band satellite to the Asia-Pacific region (covering 178 degrees West to 165 degrees East). The move will provide near-global telecommunications coverage for military users including the UK government, Skynet's anchor customer, along with NATO and individual NATO governments, which lease capacity from the system.[335] While Airbus D&S owns the Skynet satellite system, it has been in a £3.6 billion (4.62 billion

---

[334]SpaceNews. "Editorial | Despite Compelling Logic, Cost Sharing Remains Elusive for European Milspace." 20 Jan. 2015. SpaceNews 8 Jan. 2016 <http://spacenews.com/cost-sharing-elusive-for-european-milspace/>.

[335]De Selding, Peter B. "Britain's Skynet 5A Reassigned to Asia-Pacific." 16 Mar. 2015. SpaceNews 12 Jan. 2016 <http://spacenews.com/britains-skynet-5a-reassigned-to-asia-pacific/>.

euros) contract with the British Defence Ministry since 2003. The current system management contract will end by August 2022, at which point the four Skynet 5 satellites and the four older Skynet 4 satellites will revert to British military control. The British government is now considering whether to continue the contract with Airbus or enter into a similar contract with another company. It is also considering the advantages of purchasing two satellites of its own, one in 2019 and another in the following year or two, and simply extending the life of the Skynet 5 system under British military control into the 2020s.[336]

1.4.3 The United States

The U.S. Department of Defense (DoD) space budget slightly increased by 4.8 % in 2015, to $23.572 billion from $22.483 billion in 2014, while the budget for the U.S. Missile Defense Agency (U.S. MDA) continued to decrease in 2015, reducing by another 1.9 % to around $7.46 billion in 2015 from $7.6 billion in 2014; however, it requested $8.127 billion in FY 2016 to improve and expand the development of its defences.[337] Near mid-2015, members of the U.S. Congress' House Armed Services Committee showed renewed interest in a 2009 concept to put several miniaturized kill vehicles on an interceptor missile to overcome the U.S. missile defense system's inability to reliably distinguish between missile warheads and relatively low-tech decoys. The U.S. MDA would make the multi-object kill vehicle (MOKV) a long-term technology, which would come after the completion of its redesigned kill vehicle (RKV) expected to be ready around 2020.[338] In August, Boeing, Lockheed Martin, and Raytheon each won study contracts worth approximately $9.7 million to develop MOKV concepts that are expected to be completed by May 2016. The three companies will also develop specific subsystems for the RKV, to modularize and simplify the interceptor's production; the RKV's design review is expected at the beginning of 2016, while bidding for the production contract is expected in 2018.[339]

---

[336]De Selding, Peter B. "Britain Backpedals on Privatized Milcom Satellites." 5 Nov. 2015. SpaceNews 13 Jan. 2016 <http://spacenews.com/britain-backpedals-on-privatized-milcom-satellites/>.

[337]C.f. "Missile Defense Agency Fiscal Year (FY) 2015 Budget Estimates | Overview." 20 Feb. 2014. MDA 10 July 2016 <http://www.mda.mil/global/documents/pdf/budget_overview.pdf>.

"Missile Defense Agency Fiscal Year (FY) 2016 Budget Estimates | Overview." 26 Jan. 2015 MDA 10 July 2016 <http://mda.mil/global/documents/pdf/budgetfy16.pdf>.

[338]Gruss, Mike. "House Bill Would Revive Dormant Missile Defense Kill Vehicle Project." 11 May 2015. SpaceNews 22 Dec. 2015 <http://spacenews.com/house-bill-would-revive-dormant-missile-defense-kill-vehicle-project/>.

[339]Gruss, Mike. "MDA Pursues Kill Vehicles with Cost-cutting Mandate." 24 Aug. 2015. SpaceNews 24 Dec. 2015 <http://spacenews.com/mda-pursues-kill-vehicles-with-cost-cutting-mandate/>.

Following the 31 March 2015 launch of Russia's Cosmos 2504 spacecraft; its resulting manoeuvres, including at least one case where the upper stage of its launcher appeared to be nudged to a higher orbit, motivated the U.S. to early deployment of its once-classified Geosynchronous Space Situational Awareness Program (GSSAP) satellites.[340] Two GSSAP satellites, which had been in checkout procedures and operational testing since their launch in July 2014, were taken out of test mode twice by 16 September 2015 to make observations of specific objects in geosynchronous orbit. The satellites performed as expected, with initial operational capability for the satellites expected in the following weeks. Another two GSSAP satellites could launch in 2016. Since the GSSAP programme's declassification in February 2014, U.S. defence officials have acknowledged that the satellites will perform their own rendezvous and proximity manoeuvres to allow close-up looks at spacecraft in GEO orbits.[341]

1.4.4 Russia

On 3 August 2015, the Russian government merged its Air Force and its recently-formed Aerospace Defence Forces (VKO) under one unified command structure, representing an evolution in Russian military thinking from an era where its air and space forces existed as separate branches with little overlap in command authority to one where air and space will be treated more as a seamless theatre of war.[342] By 17 November 2015, Russia seemed to have taken a page from the U.S. playbook during the Gulf War, as it involved 10 imagery and electronic warfare reconnaissance satellites, including civilian-use spacecraft, to provide support for its operation against ISIS forces in Syria. Russia's Aerospace Defence Force's strikes in Syria intensified after 31 October 2015, following the ISIS downing of a Russian aircraft that killed 224 people.[343]

Russia launched a second mysterious and potentially threatening space object known as Cosmos 2504 on 31 March 2015—less than a year after the launch of the 2014-28E spacecraft which began exhibiting unusual behaviour in November 2014 that might be consistent with—but not indicative of—anti-satellite weapons tested

[340] Gruss, Mike. "Maneuvering Russian Satellite Has Everyone's Attention." 17 July 2015. SpaceNews 20 Mar. 2016 <http://spacenews.com/maneuvering-russian-satellite-has-everyones-attention/>.

[341] Gruss, Mike. "Space Surveillance Sats Pressed into Early Service." 18 Sept. 2015. SpaceNews 11 Jan. 2016 <http://spacenews.com/space-surveillance-sats-pressed-into-early-service/>.

[342] Bodner, Matthew. "Russian Military Merges Air Force and Space Command." 3 Aug. 2015. The Moscow Times 10 July 2016 <http://www.themoscowtimes.com/business/article/russian-military-merges-air-force-and-space-command/526672.html>.

[343] "Russia involves 10 reconnaissance satellites in Syria operation — General Staff." 17 Nov. 2015. TASS 20 Mar. 2016 <http://tass.ru/en/science/837273>.

by the Soviet Union during the Cold War.[344] By early July 2015, Cosmos 2504 had made at least 11 close approaches to the upper stage that had launched it into orbit, which led to the early deployment of two U.S. GSSAP satellites to observe the activity.[345] Russia's Olimp K satellite (alternatively known as Luch), operated by Russia's Aerospace Defence Forces (ADF), which launched on 27 September 2014, also raised some concerns when it parked into a GEO orbit between two of Intelsat's commercial GEOCOM satellites, Intelsat 7 and Intelsat 901, converging at a distance of 10 km from both spacecraft over a period of 5 months, before relocating near the Intelsat 905 satellite in September 2015. While on-orbit inspection technology is not unusual, the close proximity of the lurking satellite has raised concerns from Intelsat, and likely the U.S. government, as Intelsat has no way of knowing which way to manoeuvre its own satellite to maintain a safe distance without information about Olimp K's planned manoeuvres.[346]

1.4.5 Japan

Concern over China's 2007 direct-ascent anti-satellite weapon test, and subsequent experiments focusing on jamming and laser-blinding satellites, is said to have motivated Japan's shift toward placing security at the forefront of its national space policy.[347] Concern also exists over North Korea which has continued development of long-range ballistic missiles, one of which provocatively overflew Japan dropping its first stage very near Japanese territory in a test conducted in 1998.[348] Japan continued increasing its ties with the U.S. throughout 2015. For instance, in April 2015 the two countries updated their joint defence guidelines for the promotion of "seamless" cooperation, while also calling for an increase in Japan's use of space in support of the U.S.-Japan Security Alliance. And by July 2015, Japan had

[344] Ferster, Warren. "General: Russian, Chinese Launches Demonstrate Growing Space Threat." 15 Apr. 2015. SpaceNews 22 Dec. 2015 <http://spacenews.com/u-s-general-says-russian-chinese-launches-demonstrate-growing-space-threat/>.

[345] Gruss, Mike. "Maneuvering Russian Satellite Has Everyone's Attention." 17 July 2015. SpaceNews 20 Mar. 2016 <http://spacenews.com/maneuvering-russian-satellite-has-everyones-attention/>.

[346] Gruss, Mike. "Russian Luch Satellite Relocates — Next to Another Intelsat Craft." 16 Oct. 2015. SpaceNews 11 Jan. 2016 <http://spacenews.com/russian-luch-satellite-relocates-next-to-another-intelsat-craft/>.

[347] Kallender-Umezu, Paul. "Japan Boosts Space Spending In Support of Security Focus." 2 Feb. 2015. SpaceNews 10 Dec. 2015 <http://spacenews.com/japan-boosts-space-spending-in-support-of-security-focus/>.

[348] Rose, Frank A. "Commentary | Strategic Stability in U.S.-China Relations." 29 Jan. 2015. SpaceNews 10 Dec. 2015 <http://spacenews.com/commentary-strategic-stability-in-u-s-china-relations/>.

ended its ban on collective self-defence, allowing it to more readily come to the aid of U.S. in the event of a conflict in East Asian waters.[349]

By 8 December 2015, the U.S. Missile Defense Agency and Japan's Ministry of Defence had conducted a second live-fire test of the new Standard Missile-3 (SM-3) Block IIA missile under the Aegis Ballistic Missile Defense program. The test evaluated the missile's performance based on telemetry and other data from the test, and validated the new sensor seeker developed for the redesigned kill vehicle. Its first test took place on 6 June 2015.[350] According to the U.S. Missile Defense Agency, the U.S. has spent $2 billion on the SM-3 IIA programme, along with a contribution of around $1 billion from Japan. In the coming years, the new missile will be deployed on U.S. Aegis destroyers and Japan's Kongo ships.[351]

Spending in defence areas continued to drive increased spending through 2015, amounting to ¥324.5 billion (2.51 billion euros) for the fiscal year 2015 (beginning on 1 April 2015 and ending 30 March 2016).[352] About 61 % of Japan's Cabinet Office budget of ¥22.3 billion (172.41 million euros) has been earmarked for the expansion of the QZSS regional navigation system to an enhanced seven satellite constellation.[353] JAXA also received funding to begin the new development of a laser-optical data relay satellite, an advanced Earth observation satellite carrying a ballistic missile early warning sensor as a hosted payload for Japan's Ministry of Defence, and a new line of multipurpose small satellites capable of rapid production and deployment for a range of missions.[354]

---

349 Kallender-Umezu, Paul. "What's Behind Japan's Sudden Thirst for More Spy Satellites." 13 Nov. 2015. SpaceNews 7 Jan. 2016 <http://spacenews.com/whats-behind-japan-sudden-thirst-for-more-spy-satellites/>.

350 "US, Japan Test Advanced SM-3 Missile for Aegis System – US Navy." 10 Dec. 2015. Sputnik News 21 Jan. 2016 <http://sputniknews.com/military/20151210/1031509097/us-japan-aegis.html>.

351 Shalal, Andrea. "U.S.-Japan funded Raytheon SM-3 missile aces second test: Pentagon." 8 Dec. 2015. Reuters 21 Jan. 2016 <http://www.reuters.com/article/us-usa-japan-missile-defense-idUSKBN0TR2V320151209>.

352 Kallender-Umezu, Paul. "Japan Boosts Space Spending In Support of Security Focus." 2 Feb. 2015. SpaceNews 10 Dec. 2015 <http://spacenews.com/japan-boosts-space-spending-in-support-of-security-focus/>.

353 Rajagopalan, Rajeswari Pillai. "Op-ed | Japan's Space Policy Shift Reflects New Asian Realities." 23 Feb. 2015. SpaceNews 10 Dec. 2015 <http://spacenews.com/op-ed-japans-space-policy-shift-reflects-new-asian-realities/>.

354 Kallender-Umezu, Paul. "Japan Boosts Space Spending In Support of Security Focus." 2 Feb. 2015. SpaceNews 10 Dec. 2015 <http://spacenews.com/japan-boosts-space-spending-in-support-of-security-focus/>.

1.4.6 China

China's space defence spending almost doubled between the years 2009 and 2014, growing from an estimated $977 million to $1.91 billion respectively.[355] Yet, a similar observation can be made about its civil space spending, which is estimated to have grown from $1.28 billion to $2.66 billion over that same period.[356] China's overall defence budget was expected to reach about $145 billion in 2015, while one U.S. DoD estimate suggested that its actual defence spending in 2015 could be greater than $175 billion.[357]

China's military space capability has long attracted heightened speculation, as China's space sector has long been intimately connected to the People's Liberation Army (PLA). Yet, China's recent Middle and Long Term Development Plan for State Civil Space Infrastructure (2015–2025) suggests a potential departure from that union, appearing more focussed on enhancing its space capabilities for domestic purposes—particularly in establishing its state civil space infrastructure system, and enhancing its competitiveness.[358]

Nonetheless, some concerns remain about China's military space capabilities following the publication of the U.S. DoD's annual report to Congress on China's military and security developments, and China's publication of its white paper on military strategy shortly afterward. The U.S. Pentagon published its 'Military and Security Developments Involving the People's Republic of China 2015' on 8 May 2015.[359] Among other observations of the development of China's space and counter-space capabilities, including directed-energy weapons and satellite jammers, the report described a launch event conducted on 23 July 2014 that had a similar profile to its January 2007 ASAT test which intercepted a no-longer functioning Chinese weather satellite and created a myriad of space debris. While the 2014 launch did not result in the destruction of a satellite or creation of space debris, it does follow on the heels of another disturbing launch that was conducted on 13 May 2013. The ballistic trajectory of that earlier launch had its apex at an altitude above 30,000 km, i.e. near GEO orbit where numerous communication and Earth observation satellites are stationed; yet the rocket was

[355]Profiles of Government Space Programs. Paris: Euroconsult, 2015: 16.

[356]Profiles of Government Space Programs. Paris: Euroconsult, 2015: 14.

[357]Fisher, Richard D. Jr. "China's Rising Defense Budgets Stoke Regional Concerns." 8 Jan. 2015. Aviation Week & Space Technology 12 June 2015 <http://aviationweek.com/defense/china-s-rising-defense-budgets-stoke-regional-concerns>.

[358]Google Translated. "关于印发国家民用空间基础设施中长期发展规划(2015–2025年)的通知." 26 Oct. 2015. National Development and Reform Commission 2 Feb. 2016 <http://www.sdpc.gov.cn/zcfb/zcfbghwb/201510/t20151029_756376.html>.

[359]"Office of the Secretary of Defense. Annual Report to Congress – Military and Security Developments Involving the People's Republic of China 2015." 8 May 2015. Defense.gov 2 Feb 2016 <http://www.defense.gov/Portals/1/Documents/pubs/2015_China_Military_Power_Report.pdf>.

not on the appropriate trajectory to place objects in orbit and no new satellites were released.[360]

In rebuttal, China's Ministry of National Defence (MOD) released a white paper on China's Military Strategy on 26 May 2015.[361] In the narrative of its national security situation, an initial concern points to the provocative actions of the U.S. and other offshore neighbours, which in China's view have illegally reinforced their military presence on its reefs and islands in the South China Sea, in addition to maintaining constant close-in air and sea surveillance and reconnaissance against China.[362] In response to that perceived encroachment, the white paper outlines China's strategic guideline of active defence, along with modernizing its military to adapt to new changes in its maritime security environment.[363] And in terms of military space and counter-space activities, aside from a paragraph indicating that China will strengthen its capabilities for strategic deterrence and nuclear counterattack, and medium- and long-range precision strikes, it later goes on to say that "China will keep abreast of the dynamics of outer space, deal with security threats and challenges in that domain, and secure its space assets to serve its national economic and social development, and maintain outer space security".[364]

1.4.7 India

India is developing its own space military programme; however, the majority of the activities of the Indian Space Research Organization (ISRO) still focus on civil applications.

[360]"Office of the Secretary of Defense. Annual Report to Congress – Military and Security Developments Involving the People's Republic of China 2015." 8 May 2015. Defense.gov 2 Feb 2016: 14 <http://www.defense.gov/Portals/1/Documents/pubs/2015_China_Military_Power_Report.pdf>.

[361]Minnick, Wendell. "White Paper Outlines China's Ambitions." 27 May 2015. Defense News 22 Jan. 2016 <http://www.defensenews.com/story/defense/policy-budget/warfare/2015/05/26/china-us-pentagon-taiwan-report-south-east-sea-islands-reefs-s400-su35-missiles-satellite-space-deterrence/27957131/>; *see also* "China's Military Strategy." 26 May 2015. Ministry of National Defense – The People's Republic of China 3 Feb. 2016 <http://eng.mod.gov.cn/Database/WhitePapers/index.htm>.

[362]"I. National Security Situation." 26 May 2015 Ministry of National Defense – The People's Republic of China 3 Feb. 2016 <http://eng.mod.gov.cn/Database/WhitePapers/2015-05/26/content_4586688.htm>.

[363]"III. Strategic Guideline of Active Defense." 26 May 2015 Ministry of National Defense – The People's Republic of China 3 Feb. 2016 <http://eng.mod.gov.cn/Database/WhitePapers/2015-05/26/content_4586711.htm>.

[364]"IV. Building and Development of China's Armed Forces." 26 May 2015 Ministry of National Defense – The People's Republic of China 3 Feb. 2016 <http://eng.mod.gov.cn/Database/WhitePapers/2015-05/26/content_4586713.htm>.

On 22 November 2015, India's Defence Research Development Organization (DRDO) conducted a validation test of its Advanced Air Defence (AAD) missile. Having undergone trials since 2007, this latest test focussed on the kill effect of the AAD in flight mode by analysing data from multiple tracking sources. The AAD, along with the Prithvi Air Defense (PAD) missile, which had its first test launch on 27 April 2014, are both part of the first phase of India's Ballistic Missile Defence (BMD) Initiative. While the AAD system is optimised for surface-to-air strikes against aircraft and UAVs, the PAD provides exoatmospheric defence; however, a full-scale BMD system would incorporate technology from both the PAD and AAD systems.[365]

India's PAD can also be seen as a further step toward developing its own anti-satellite capabilities. In this pursuit, the DRDO is looking at the feasibility of developing such an anti-satellite vehicle by integrating its Angi-3 missile with its PAD. If it succeeds, the anti-satellite missile would have an effective range of about 1400–1500 km, and would advance India's missile capabilities to be on a par with U.S. and China.[366]

[365]Panda, Ankit. "India Tests Supersonic Advanced Air Defense Missile." 23 Nov. 2015. The Diplomat 10 July 2016 <http://thediplomat.com/2015/11/india-tests-supersonic-advanced-air-defense-missile/>.

[366]"India Contemplates Anti-Satellite Vehicle Integration with Agni-III Ballistic Missile." 15 Oct. 2013. Missile Threat 12 May 2014 <http://missilethreat.com/india-contemplates-anti-satellite-vehicle-integration-with-agni-iii-ballistic-missile/>.

Chapter 2
Developments in Space Policies, Programmes and Technologies Throughout the World and in Europe

Cenan Al-Ekabi

2.1 Space Policies and Programmes

The major space policy developments worldwide were presented in Chap. 1, above, to identify the principal space faring nations' strategies in 2015. In the section below, there will be a brief discussion of developments in technology related areas, including policies and access to space technologies. The aim of this section is to clarify how these strategies interact with and influence specific space programmes, and related research and development projects.

2.2 Space Transportation

2.2.1 Europe

As Europe's launch sector begins its substantial reorganisation to make it more competitive in the global market, the Arianespace launch consortium aims to increase the launch rate of its Vega launcher to benefit from economies of scale and increased demand. The per-launch cost of the Vega launcher is expected to decrease as its prime contractor Avio SpA ramps up production to enable three launches per year. Yet as favourable EU/U.S. exchange rates are likely to have helped grow the Vega's backlog to 11 launches for a total of 13 satellites as of June 2015, the demand for Vega has also allowed Arianespace to increase its launch price. Whereas in 2013, Vega's backlog of four satellite launches was valued at

C. Al-Ekabi (✉)
European Space Policy Institute, Vienna, Austria
e-mail: cenan.al-ekabi@espi.or.at

C. Al-Ekabi et al. (eds.), *Yearbook on Space Policy 2015*, Yearbook on Space Policy, DOI 10.1007/978-3-7091-4860-0_2

130 million euros, or 32.5 million euros per launch, the value of this new backlog has grown to 400 million euros, or 36.4 million euros per launch.[1]

On 12 August 2015, ESA and CNES signed a contract to develop the Ariane 6 launch pad and horizontal launcher integration facilities in Europe's Guiana Space Centre in French Guiana. Valued at 600 million euros, construction of the new launch installation will need to be completed before the intended demonstration launch of the Ariane 6 in 2020. French industry can expect to receive 52 % of the value of the contract in proportion to its government's investment in the ESA Ariane 6 launcher programme. Likewise, Germany's industry will receive a 23 % share of the contract, while the remaining 25 % will be spread among Italy, and other Ariane 6 participating member states under ESA's geographic return policy.[2]

Until now an Ariane 5 launcher has sold for 150 million euros; however, its launch cost is 170 million euros, requiring institutional support of Arianespace of around 100 million euros annually.[3] With orders for the Ariane 5 launcher booked into 2017, Arianespace and ASL aim to decrease the cost of the Ariane 5 by 5–6 % in the next batch of launch orders starting in 2019 through improvements in production and operation. They also plan to raise the price for heavier satellites intended for the Ariane 5's upper-berth, as demand for that slot has increased following the recent failed launches by competitors ILS and Sea Launch. With a total of 69 consecutive successful launches as at the end of 2015, the Ariane 5's reliability is a strong motivator for commercial satellite operators. Moreover, the price for the Ariane 5's lower berth, targeted for small satellites and telecom satellites using electric propulsion to reach geostationary transfer orbit, will be reduced to make it more competitive with SpaceX prices. Another advantage exists in the current EU-U.S. exchange rate that will allow customers to purchase an Ariane 5 launch with a lower value Euro.[4] Yet, once the Ariane 6 becomes operational in 2020, the Ariane 5 will undergo a 3-year phase out.[5]

ASL's first priority in developing the Ariane 6 launcher is to reduce the cost of launching a heavy telecomm satellite to geostationary transfer orbit by 50 % in

---

[1]De Selding, Peter B. "Vega Launches Sentinel-2A Observation Satellite." 23 June 2015. SpaceNews 5 Jan. 2016 <http://spacenews.com/vega-launches-sentinel-2a-observation-satellite/>.

[2]De Selding, Peter B. "ESA Inks $3.8 Billion in Contracts for Ariane 6, Vega-C and Spaceport Upgrades." 12 Aug. 2015. SpaceNews 13 Jan. 2016 <http://spacenews.com/esa-inks-3-8-billion-in-contracts-for-ariane-6-vega-c-and-spaceport-upgrades/>.

[3]De Selding, Peter B. "With Revenue Looking Up, Arianespace Seeks To Bring Ariane 5 Costs Down." 21 Oct. 2015. SpaceNews 13 Jan. 2016 <http://spacenews.com/with-revenue-looking-up-arianespace-seeks-to-bring-ariane-5-costs-down/>.

[4]De Selding, Peter B. "Arianespace Assures French Parliament it Can Outcompete SpaceX." 13 May 2015. SpaceNews 5 Jan. 2016 <http://spacenews.com/arianespace-assures-french-parliament-it-can-outcompete-spacex/>.

[5]De Selding, Peter B. "With Revenue Looking Up, Arianespace Seeks To Bring Ariane 5 Costs Down." 21 Oct. 2015. SpaceNews 13 Jan. 2016 <http://spacenews.com/with-revenue-looking-up-arianespace-seeks-to-bring-ariane-5-costs-down/>.

comparison to the Ariane 5.[6] In addition to its modernized and simpler design, and the reshaped role of industry, expenditures on the Ariane 6 will be reduced by integrating the launcher horizontally; a notable departure from the vertical integration of previous Ariane launchers.[7] Capable of lifting two telecom satellites with a combined mass of 9500 kg to geostationary transfer orbit at a total cost of around 90 million euros, the 64 configuration of the Ariane 6 with four boosters will be sold for 96.34 million euros per launch (or 48.17 million euros per customer).[8] The 62 configuration of the Ariane 6 with two boosters for small satellites will cost around 75 million euros.[9]

ESA's ties with China have grown throughout the year, with the ESA actively working on eventually placing a European astronaut on China's space station by its completion in 2022. While no specific plans have been made for an ESA astronaut mission, ESA astronauts have visited China's astronaut facilities, while some have begun learning to speak Chinese in collaboration with China's Manned Space Flight Office.[10] As the year progressed, it was increasingly realised by European stakeholders that China's inclusion in the global space exploration endeavour would be an asset.[11] Chinese officials echoed those sentiments, inviting international participation in the form of foreign modules on its space station, along with foreign cargo and crewed visits, at the 66th IAC held on 12–16 October 2015 in Jerusalem, Israel. Europe and Russia have each already signed initial space station cooperation agreements with China.[12] Yet, an open question remains as to whether China's space station will use the same international docking system standard used by Europe and the U.S. on the ISS, or whether a docking adapter will be needed.

---

[6]De Selding, Peter B. "ESA Approval Paves Way for Ariane 6, Vega-Contracts." 17 July 2015. SpaceNews 13 Jan. 2016 <http://spacenews.com/esa-approval-paves-way-for-ariane-6-vega-contracts/>.

[7]De Selding, Peter B. "ESA Inks $3.8 Billion in Contracts for Ariane 6, Vega-C and Spaceport Upgrades." 12 Aug. 2015. SpaceNews 5 Jan. 2016 <http://spacenews.com/esa-inks-3-8-billion-in-contracts-for-ariane-6-vega-c-and-spaceport-upgrades/>.

[8]De Selding, Peter B. "With Revenue Looking Up, Arianespace Seeks To Bring Ariane 5 Costs Down." 21 Oct. 2015. SpaceNews 13 Jan. 2016 <http://spacenews.com/with-revenue-looking-up-arianespace-seeks-to-bring-ariane-5-costs-down/>.

[9]Gallois, Dominique. "Ariane 6, un chantier européen pour rester dans la course spatiale." 1 Dec. 2014. Le Monde 9 Mar. 2016 <http://www.lemonde.fr/economie/article/2014/12/01/les-europeens-s-appretent-a-mettre-ariane-6-en-chantier_4532259_3234.html>.

[10]De Selding, Peter B. "ESA Hikes Budget, Takes Steps To Send Astronaut to Chinese Space Station." 16 Jan. 2015. SpaceNews 8 Jan. 2016 <http://spacenews.com/esa-hikes-budget-takes-steps-to-send-astronaut-to-chinese-space-station/>.

[11]C.f. De Selding, Peter B. "Dordain Says ESA Cannot Go it Alone in Debris Mitigation." 21 May 2015. SpaceNews 22 Dec. 2015 <http://spacenews.com/dordain-says-esa-cannot-go-it-alone-in-debris-mitigation/>.

[12]De Selding, Peter B. "China's Space Station Planners Put out Welcome Mat." 13 Oct. 2015. SpaceNews 11 Jan. 2016 <http://spacenews.com/chinas-space-station-planners-put-out-welcome-mat/>.

2.2.2 *United States*

Despite the success of NASA's Orion Multi-Purpose Crew Vehicle (MPCV) Exploration Flight Test 1 mission on the Space Launch System (SLS) on 5 December 2014, the ongoing White House and Congress debate on funding levels and prioritisation resulted in missed opportunities, with the debut of the SLS delayed by a year. NASA estimates given in November 2015 place both the SLS and its ground systems on schedule to support a first launch in June 2018, while Orion could be launched as soon as September 2018. Moreover, while NASA announced in September 2015 that the first crewed Orion launch on SLS could be delayed to as late as April 2023, there was still a hope that it can be advanced to August 2021.[13]

On 16 December 2015, the U.S. Congress approved the National Defense Authorization Act for 2016 (NDAA-16) which lifted the ban on the Russian-built RD-180 engine that was enacted in the previous year's NDAA. Amid continuing tensions between the U.S. and Russia following Russia's 2014 incursion in Ukraine, the act prohibited U.S. companies from contracting with Russian suppliers of rocket engines or renewing current contracts for space launch activities.[14] The act directly affected United Launch Alliance (ULA) that has had the U.S. government launch services market to itself since its formation in 2006 but now faces increasing competition from SpaceX. Since ULA uses RD-180 engines for the first stage of its Atlas 5 launcher, the restriction limited its use before ULA's follow-on Vulcan launcher—powered by Blue Origin's BE-4 engine—is ready in 2020. While ULA was granted an exemption of 5 RD-180 engines that were already on order at the time of the law's enactment, in addition to a batch of 29 RD-180 engines that were ordered while the NDAA-15 was still pending,[15] it needed access to at least 13 more RD-180 engines in order to remain competitive with SpaceX.[16] By 23 December 2015, ULA had ordered 20 more RD-180 engines, in addition to the 29 RD-180 engines it had ordered before Russia's annexation of Crimea from Ukraine.[17] The reversal had mixed responses in the Senate Armed Services Committee, with some considering whether to place a complete and indefinite restriction on RD-180

[13]Foust, Jeff. "NASA Counting on Budget Increase for SLS and Orion." 5 Nov. 2015. SpaceNews 10 July 2016 <http://spacenews.com/nasa-counting-on-budget-increase-for-sls-and-orion/>.

[14]Section 1608. Carl Leven and Howard P. "Buck" McKeon National Defense Authorization Act for Fiscal Year 2015, Pub. L. no 113-291 (2014) <http://www.gpo.gov/fdsys/pkg/CPRT-113HPRT92738/pdf/CPRT-113HPRT92738.pdf>.

[15]Gruss, Mike. "House-Senate Conference Measure To End Pentagon Use of RD-180." 5 Dec. 2014. SpaceNews 22 July 2015 <http://spacenews.com/42701house-senate-conference-measure-to-end-pentagon-use-of-rd-180/>.

[16]Ferster, Warren. "Defense Bill Curbs ULA Use of Russian Engines but Draws Veto Threat." 30 Sept. 2015. SpaceNews 24 Dec. 2015 <http://spacenews.com/defense-bill-limits-ula-to-9-more-russian-built-engines/>.

[17]Shalal, Andrea. "ULA Orders 20 More RD-180 Rocket Engines." 23 Dec. 2015. SpaceNews 14 Jan. 2016 <http://spacenews.com/ula-orders-20-more-rd-180-rocket-engines/>.

engines in the NDAA for 2017. The NDAA-16 also provided $143 million in addition to the $84 million requested by the US Air Force to accelerate the development of a U.S. made replacement engine expected to be ready to fly by 2019.[18]

2.2.3 Russia

Development of the Vostochny Cosmodrome, located in the Far Eastern region of Russia, is continuing to ensure the completion of the spaceport's Soyuz-2 and Angara launch pads by 2016; which should also coincide with the completion of its Angara launcher.[19] The final construction stage of the Vostochny Cosmodrome, taking place between 2016 and 2018, will develop facilities for a super-heavy launch vehicle that will be capable of delivering 120–150 tons into low Earth orbit by 2020.[20]

Russia's Proton launcher, manufactured by Moscow-based Khrunichev State Research and Production Space Centre, and marketed in the U.S. by its commercial-arm International Launch Services (ILS), has suffered a string of failures since 2012. Uncertainty over the Proton M's reliability has led to more commercial contracts seeping to SpaceX and to Arianespace. Both launch providers split evenly the commercial launch contracts in 2014, each winning nine contracts for geostationary satellites,[21] while also ahead in 2015. To overcome its recent failure record and resulting unfavourable insurance rates, Khrunichev aims to introduce its Angara rocket family to replace its Proton launcher. In the near future, as the SpaceX has a large backlog and Arianespace's backlog for its Ariane 5 is booked into 2017, ILS might want to consider charging its consumers a lower price for its Proton launcher in light of the lower-valued rouble to offset the increased price customers must pay for insurance.[22]

---

[18]Gruss, Mike. "Spending Bill Lifts RD-180 Ban, Puts ULA Back in Competitive Game." 16 Dec. 2015. SpaceNews 7 Jan. 2016 <http://spacenews.com/spending-bill-lifts-rd-180-ban-puts-ula-back-in-competitive-game/>.

[19]Nowakowski, Tomasz. "Putin pledges $1.5 billion for completion of Vostochny Cosmodrome, OKs plans for creating super-heavy rockets." 9 Sept. 2014. Spaceflight Insider 16 July 2015 <http://www.spaceflightinsider.com/organizations/roscosmos/putin-pledges-1-5-billion-completion-vostochny-cosmodrome-oks-plans-creating-super-heavy-rockets/>.

[20]Bodner, Matthew. "Putin Pledges $1 Billion for Completion of New Cosmodrome." 2 Sept. 2014. The Moscow Times 16 July 2015 <http://www.themoscowtimes.com/business/article/putin-pledges-1-billion-for-completion-of-new-cosmodrome/506321.html>.

[21]De Selding, Peter B. "Arianespace Assures French Parliament it Can Outcompete SpaceX." 13 May 2015. SpaceNews 5 Jan. 2016 <http://spacenews.com/arianespace-assures-french-parliament-it-can-outcompete-spacex/>.

[22]De Selding, Peter B. "ILS Mission Assurance VP Replaces Slack as President." 11 Sept. 2015. SpaceNews 13 Jan. 2016 <http://spacenews.com/ils-mission-assurance-vp-replaces-slack-as-president/>.

Until SpaceX's commercial crew capsule begins service in late 2017, Russia's Soyuz launcher is currently the sole vehicle capable of sending astronauts and cosmonauts to the ISS. By August 2015, the cost of sending six astronauts to the ISS had grown to $490 million, or $82 million per seat, up from $70.7 million per seat in 2013.[23] In contrast, the average cost of sending an astronaut to the ISS on a U.S. crewed vehicle is expected to cost on average $58 million per seat.[24]

2.2.4 Japan

JAXA has increased its funding toward the development of a lower cost H-3 successor to Japan's workhorse H-2A launcher. Of the ¥182 billion accorded to JAXA's 2015 fiscal year budget ending 30 March 2016, funding for the new development of the launcher increased by ¥5.5 billion in 2015, amounting to a total of ¥12.5 billion for 2015. The H-3 is estimated to need $1.9 billion (¥194.46 billion) for full development by 2020. The two-stage H-3 should be able to lift up to 6.5 metric ton payloads to GTO at a cost ranging between $50 million (¥5.12 billion) and $70 million (¥7.16 billion) per launch.[25] The H-3 launcher is expected to feature a liquid hydrogen/liquid oxygen core stage with up to six solid-fuel strap-on boosters to offer a wide range of payload-to-orbit capabilities.[26]

As Japan starts developing its H-3 launcher, the status of its recent three-stage solid-fuel Epsilon launcher remains unclear. Likewise developed to cut launcher costs, the Epsilon launcher uses the same solid-fuel strap-on booster used by the H-2A.[27] The first launch of the Epsilon-1 was conducted on 14 September 2013, yet Japan has sought to perfect the launch capability of the launcher in the following years. In 2015, JAXA's development of an Enhanced Epsilon Launch Vehicle continued smoothly, which included the completion of a static firing test of the sub-sized upper-stage motor in March 2015, followed by a pyro-shock test of the shocks and behaviour due to the payload separation impact conducted in June

[23]"NASA: Seats on Russian rockets will cost U.S. $490 million." 6 Aug. 2015. CBS 7 Sept. 2016 <http://www.cbsnews.com/news/nasa-seats-on-russian-rockets-will-cost-u-s-490-million/>.

[24]Kermer, Ken. "These Astronauts Will be the First to Launch With SpaceX and Boeing." 10 July 2015. Io9 7 Sept. 2016 <http://io9.gizmodo.com/these-astronauts-will-be-the-first-to-launch-with-space-1717036938>.

[25]Onuki, Misuzi. "Japan Approves $1.9B for H-3 Rocket." 13 Jan. 2014. SpaceNews 30 May 2014 <http://www.spacenews.com/article/civil-space/39069japan-approves-19b-for-h-3-rocket>.

[26]Onuki, Misuzu. "MHI Formally Selected as H-X Prime Contractor, Operator." 25 Mar. 2014. SpaceNews 30 May 2014 <http://www.spacenews.com/article/financial-report/39971mhi-formally-selected-as-h-x-prime-contractor-operator>.

[27]Matsuda, Shogo. "Japan's Epsilon rocket shoved aside?" 16 Jan. 2015. Nikkei Asian Review 21 Jan. 2016 <http://asia.nikkei.com/Tech-Science/Tech/Japan-s-Epsilon-rocket-shoved-aside>.

2015.[28] Another combustion test of the enhanced engine was conducted on 21 December 2015.[29] The Epsilon launcher is designed to be capable of placing a 1.2 tons payload in low Earth orbit (LEO), and 700 kg payload into sun-synchronous orbit (SSO).[30]

2.2.5 China

China conducted the maiden launch of its three-stage Long March 6 on 19 September 2015 from the northern Taiyuan Satellite Launch Centre in China. With a launch capacity of 1500 kg into a low earth orbit (and 1080 kg into a 700 km sun synchronous orbit), the launcher lifted 20 small to cube-sized satellites to an sun-synchronous orbit of around 524 km.[31] Long March 6 is part of a new family of more environmentally friendly launchers being developed in China, including the Long March 5 (with a 14 tons to GEO lift capacity) and Long March 7 (5.5 tons to LEO). The launcher's first and second stages are fuelled by a liquid oxygen and kerosene mixture which is less toxic than the hypergolic propellants used in the earlier generation of its launchers, while its restartable third stage uses unsymmetrical dimethylhydrazine in combination with dinitrogen tetroxide as an oxidizer. In addition to the Taiyuan launch site, China plans to launch the Long March 6 from the newly constructed Wenchang Satellite Launch Centre on the north-eastern coast of Hainan Island.[32]

2.2.6 India

By the end of 2015, India's workhorse Polar Satellite Launch Vehicle (PSLV) had conducted its 31st successful launch out of 32 launches in total.[33] Capable of lifting 3700 kg to Low Earth Orbit and 800 kg to Geosynchronous Transfer Orbits (GTO),

[28] "Development progress for Enhanced Epsilon." 6 Aug. 2015. JAXA 21 Jan. 2016 <http://global.jaxa.jp/projects/rockets/epsilon/>.

[29] "JAXA tests improved Epsilon rocket engine." 21 Dec. 2015. NHK World 21 Jan. 2016 <http://www3.nhk.or.jp/nhkworld/english/news/20151221_28.html>.

[30] Federal Aviation Administration. The Annual Compendium of Commercial Space Transportation: 2014. Washington, DC: FAA, Feb. 2015: 10.

[31] "China's Long March 6 Rocket successfully completes Maiden Flight." 23 Sept. 2015. Spaceflight101.com 28 Jan. 2016 <http://spaceflight101.com/chinas-long-march-6-rocket-successfully-completes-maiden-flight/>.

[32] "CZ-6 (Chang Zheng-6)." 21 Dec. 2015. Gunter's Space Page 28 Jan. 2016 <http://space.skyrocket.de/doc_lau/cz-6.htm>.

[33] "PSLV." 6 Feb. 2016. Gunter's Space Page 15 Feb. 2016 <http://space.skyrocket.de/doc_lau/pslv.htm>.

the PSLV has launched a total of 51 satellites for customers in 20 countries. In addition to its successful track record, having one initial failure in 1993, and a partial success in 1997, another draw card of the launcher is its low cost; its latest 16 December 2015 launch of six satellites cost Singapore's government 26 million euros ($30 million).[34] Incidentally, India's low cost launch capacity, and India's refusal to sign a commercial launch accord with the U.S. government to ensure fair market-based pricing, had been the reason U.S. companies have been restricted from using the PSLV in the past decade.[35]

India is also nearing the completion of its Geosynchronous Satellite Launch Vehicle (GSLV)-Mark III. Following the GSLV-Mark III's first sub-orbital test on 18 December 2014, ISRO tested the launcher's indigenous restartable upper-stage engine on 28 April 2015.[36] On 27 August 2015, ISRO launched the GSLV-Mark II variant to place the GSAT-6 satellite in orbit. This third launch of the GSLV-Mark II enabled ISRO to test its upgraded indigenous cryogenic upper stage engine.[37] The engine used in the GSLV-Mark II's previous launch in 2014 was an older version of its indigenous cryogenic upper stage, while earlier GSLV launcher variants used a cryogenic system provided by Russia. The new GSLV Mark III rocket will be able to launch 4000 kg into GTO, while the GSLV-Mark II can lift 2500 kg into GTO. Aside from the GSLV-Mark III's restartable liquid fuel engine, it shares a number of components with its predecessor GSLV-Mark I and II launchers.[38] The next GSLV-Mark II launcher is expected to be developed and launched by July 2016, while the GSLV-Mark III should be ready for institutional and commercial launches by 2017.[39]

By 28 November 2015, ISRO announced that it was ready to conduct the initial test flight of its Reusable Launch Vehicle (RLV) Technology Demonstration Program. First conceived by ISRO in 2009, the long delayed first flight test of the scramjet propulsion system was to occur sometime in January or February 2016 from the Sriharikota spaceport in India.[40] Sitting aboard two RH 560 sounding rockets in a two-stage to orbit configuration, the 1500 kg technology demonstrator

[34]Jayaraman, K.S. "ISRO Launches Six Satellites for Singapore." 16 Dec. 2015. SpaceNews 6 Jan. 2016 <http://spacenews.com/isro-launches-six-satellites-for-singapore/>.

[35]SpaceNews Editor. "Getting the Cubesat Revolution Out of Low Gear." 24 Nov. 2015. SpaceNews 6 Jan. 2016 <http://spacenews.com/getting-the-cubesat-revolution-out-of-low-gear/>.

[36]SpaceNews Staff. "India Test Fires New Engine for GSLV." 29 Apr. 2015. SpaceNews 5 Jan. 2016 <http://spacenews.com/india-test-fires-new-engine-for-gslv/>.

[37]De Selding, Peter B. "Successful Indian GSLV Launch Features Domestic Upper Stage." 27 Aug. 2015. SpaceNews 5 Jan. 2016 <http://spacenews.com/successful-indian-gslv-launch-features-homemade-upper-stage/>.

[38]"GSLV MARK III." ISRO 19 Apr. 2013 <http://www.isro.org/Launchvehicles/GSLVMARKIII/mark3.aspx>.

[39]Ibid.

[40]"'Scramjet Propulsion' Technology Test in Jan." 28 Nov. 2015. The New Indian Express 11 Feb. 2016 <http://www.newindianexpress.com/cities/thiruvananthapuram/Scramjet-Propulsion-Technology-Test-in-Jan/2015/11/28/article3150688.ece>.

will ascend to an altitude of 70 km, and will autonomously test the RLV's hypersonic aerodynamic properties, avionics, thermal protection and control systems, and its mission management. This initial mission is the first of four flights of the technology demonstrator, which will be followed by a landing experiment, a launch to orbit and return mission, and a scramjet propulsion demonstration.[41]

2.3 Space Science and Exploration

In this section, space science is understood to mean using mainly remote observation to make discoveries on the origin, evolution and future of the Universe, its galaxies, our Solar System, and other celestial bodies e.g. stars, exoplanets, comets, and asteroids. Space exploration, on the other hand, involves human and robotic spaceflight missions. While traditional governmental space agencies dominate in both these fields, progress in the latter category can be seen with the development of exploration involving commercial players, and with new space powers demonstrating the technology needed to carry out such missions.

2.3.1 Human Spaceflight Activities

Human spaceflight was focused in Low Earth Orbit (LEO), with the International Space Station (ISS) at centre stage, following its formal extension to 2024. Following the retirement of NASA's Space Shuttle, Roscosmos is the sole launch provider relied upon to transport crew regularly to the ISS and, using Progress and Soyuz, it also provided ISS cargo resupply services, along with U.S. commercial resupply missions provided by SpaceX and Orbital ATK, Europe's Automated Transfer Vehicle (ATV) and Japan's H-II Transfer Vehicle (HTV) providing auxiliary support.

Among ESA's newest batch of astronauts selected on 22 November 2010, Samantha Cristoforetti, Andreas Mogensen, and Timothy Peake, journeyed to the ISS in 2015; ESA astronauts Luca Parmitano and Alexander Gerst participated on the station in Expeditions 36/37 and 40/41 respectively in 2013 and 2014, while ESA astronaut Thomas Pesquet was scheduled to join the ISS for Expedition 50/51 in November 2016.

ESA astronaut Samantha Cristoforetti began a 5-month Expedition 42/43 mission on the ISS under the mission banner 'Futura' on 24 November 2014.[42] While

---

[41] Jayaraman, K.S. "India To Fly RLV Tech Demo by June." 5 Mar. 2015. SpaceNews 1 Jan. 2016 <http://spacenews.com/india-to-fly-rlv-tech-demo-by-june/>.

[42] "ESA astronaut Samantha Cristoforetti arrives at Space Station." 24 Nov. 2014. ESA 26 May 2015 <http://www.esa.int/Our_Activities/Human_Spaceflight/Futura/ESA_astronaut_Samantha_Cristoforetti_arrives_at_Space_Station>.

Samantha expected to return to Earth by mid-May 2015, the Futura mission was extended until 11 June 2015 due to the failure of the Russian Progress M-27M cargo resupply mission on 28 April 2015, which burned up in the atmosphere in an uncontrolled re-entry; the incident also placed further launches to the ISS on hold pending the results of an investigation into the failure.[43] On January 2015, ESA astronaut Samantha Cristoforetti assisted NASA astronaut Butch Wilmore in capturing the Dragon CRS spacecraft, which brought an assortment of experiments involving fruit flies, flatworms, human, and shellfish immune cells.[44] In March 2015, Cristoforetti and NASA astronaut Terry Virts participated in their final session of the Airway Monitoring experiment which measured the amount of nitric oxide expelled by the two astronauts while they were inside the station's cylindrical Quest airlock as air was being pumped out to reduce air pressure by 30 %—equivalent to being on a mountain at 3000 m altitude. The experiment was meant to test the use of nitric oxide as a tool for monitoring lung inflammation as well as charting lung health in astronauts as their lungs are at risk of harm due to ever-floating dust on the station; dust might be an even greater concern when astronauts return to the Moon and probably Mars, as it will likely stick to astronauts through static electricity and have sharp edges.[45] Other experiments performed by ESA astronaut Samantha Cristoforetti involved studying gene changes over generations of fruit flies, investigating small particles behaviour in liquids without gravity's interference, looking at microscopic worms that appear to thrive in weightlessness, and observing how plants grow in weightlessness compared to Earth. Samantha was the prime operator for the undocking of ESA's final Automated Transfer Vehicle (ATV) 'George Lemaître' on 14 February 2015,[46] and also captured a second Dragon CRS spacecraft to the station with the assistance of NASA astronaut Terry Virts on 17 April 2015.[47]

ESA astronaut Andreas Mogensen began a short-duration visit to the ISS, for an ESA science mission named 'IrISS', on 4 September 2015.[48] Andreas Mogensen, Europe's first Danish astronaut, and Kazakh cosmonaut Aidyn Aimbetov, spent

[43]"Samantha's longer stay on Space Station." 12 May 2015. ESA 27 May 2016 <http://m.esa.int/Our_Activities/Human_Spaceflight/Futura/Samantha_s_longer_stay_on_Space_Station>.

[44]"Fresh supplies and experiments for Samantha." 13 Jan. 2015. ESA 27 May 2016 <http://m.esa.int/Our_Activities/Human_Spaceflight/Futura/Fresh_supplies_and_experiments_for_Samantha>.

[45]"Testing astronauts' lungs in Space Station airlock." 9 Mar. 2015. ESA 27 May 2016 <http://m.esa.int/Our_Activities/Human_Spaceflight/Futura/Testing_astronauts_lungs_in_Space_Station_airlock>.

[46]"Futura Mission In Brief." 30 Sept. 2015. ESA 27 May 2016 <http://www.esa.int/Our_Activities/Human_Spaceflight/Futura/Futura_mission_in_brief>.

[47]"Second Dragon, fruit flies and fresh coffee for Samantha." 22 Apr. 2015. ESA 27 May 2016 <http://m.esa.int/Our_Activities/Human_Spaceflight/Futura/Second_Dragon_fruit_flies_and_fresh_coffee_for_Samantha>.

[48]"ESA astronaut Andreas Mogensen begins busy International Space Station tour." 4 Sept. 2015. ESA 27 May 2016 <http://www.esa.int/Our_Activities/Human_Spaceflight/iriss/ESA_astronaut_Andreas_Mogensen_begins_busy_International_Space_Station_tour>.

10 days in space delivering a new Soyuz spacecraft intended to return NASA astronaut Scott Kelly and Roscosmos cosmonauts Mikhail Kornienko and Sergei Volkov to Earth in March 2016. Kelly and Kornienko are the 1-year mission crew members, while Volkov is the Soyuz commander who flew to the ISS with Mogensen and Aimbetov.[49] While on the ISS, ESA astronaut Andreas Mogensen tested ESA's 'SkinSuit' for 2 days to test its effectiveness in preventing astronauts' spines stretching in weightlessness.[50] Andreas Mogensen also worked on around 20 European experiments, which focussed on Mogensen's blood vessels and muscles, bones and brain to see how they fared in space. He also helped to test new ways of interacting with ESA mission control to improve operations by relying on just-in-time-training from 3D software and wearing a headset to stream live video to mission control to conduct tasks, instead of training on the ground. Mogensen also participated in the Meteron project, which aims to control robotic systems from space, by operating a rover in the Netherlands from the ISS using a feedback joystick to move its twin on Earth which allowed him to 'feel' objects remotely.[51] Mogensena and Aimbetov returned to Earth with Russian cosmonaut Gennady Padalka aboard the Soyuz TMA-16M spacecraft on 12 September 2015.[52]

ESA astronaut Tim Peake began his 6-month Expedition 46/47 mission on the station under the mission banner 'Principia' on 15 December 2015. Peake is the first British astronaut to travel to the ISS; while serving as Flight Engineer, he plans to conduct dozens of experiments for researchers on Earth including growing crystals and blood vessels in space, simulating atomic structures, and charting areas in the brain as they adapt to stressful situations.[53] ESA astronaut Tim Peake is scheduled to return from the station on 5 June 2016.[54]

ESA's final Automated Transfer Vehicle (ATV) mission, ATV-5 George Lemaître, ended on 15 February 2015, following its separation from the ISS a day earlier.[55] Prior

---

[49]"Three Soyuz Crew Members Wrap Up Mission on Space Station." 11 Sept. 2015. NASA 30 May 2016 <https://blogs.nasa.gov/spacestation/2015/09/11/three-soyuz-crew-members-wrap-up-mission-on-space-station/>.

[50]*See further* "Suit up for Skinsuit." 10 Jan. 2014. ESA 30 May 2016 <http://m.esa.int/Our_Activities/Human_Spaceflight/Astronauts/Suit_up_for_Skinsuit>.

[51]"100 days to Andreas Mogensen's mission." 25 May 2015. Phys.Org 30 May 2016 <http://phys.org/news/2015-05-days-andreas-mogensen-mission.html>.

[52]"Andreas Mogensen lands after a busy mission on Space Station." 12 Sept. 2015. ESA 27 May 2016 <http://www.esa.int/Our_Activities/Human_Spaceflight/iriss/Andreas_Mogensen_lands_after_a_busy_mission_on_Space_Station>.

[53]"ESA astronaut Tim Peake begins six-month stay on Space Station." 15 Dec. 2015. ESA 27 May 2016 <http://m.esa.int/Our_Activities/Human_Spaceflight/Principia/ESA_astronaut_Tim_Peake_begins_six-month_stay_on_Space_Station>.

[54]"Tim Peake: British astronaut's space mission at a glance." 15 Dec. 2015. The Guardian 27 May 2016 <https://www.theguardian.com/science/2015/dec/15/tim-peake-british-astronauts-space-mission-at-a-glance>.

[55]"Last ATV reentry leaves legacy for future space exploration." 15 Feb. 2015. ESA 30 May 2016 <http://m.esa.int/Our_Activities/Human_Spaceflight/ATV/Last_ATV_reentry_leaves_legacy_for_future_space_exploration>.

to undocking, ESA astronaut Samantha Cristoforetti had installed the Break-Up Camera prototype within the capsule's cargo area and trained it on the forward hatch to record the ATV's death throes in the infrared as it plummeted through the atmosphere. During the fall, the camera recorded nearly 6000 pictures at a rate of 10 frames per second; it also took accelerometer and magnetometer readings, details of the sphere rotation and a temperature reading, and successfully transferred those results to the SatCom heat shield-protected sphere.[56] The SatCom survived the break-up of the capsule, and managed to broadcast at least one message to the Iridium telecom satellite constellation as it fell to Earth, indicating that the internal temperature of the sphere remained moderate and there were no signs of any thermal issues. However, the SatCom team was unable to retrieve all the data, and they are investigating why further data packets didn't make it through the scorching plasma—known for blocking radio signals.[57] ESA is using the knowledge gained from the ATV programme to build its European Service Module for NASA's Orion spacecraft that will fly astronauts to the Moon and beyond.[58]

Japan launched its fifth H-2 Transfer Vehicle (HTV) to the ISS on 19 August 2015; while it was supposed to launch to the station in July 2014, it incurred numerous and multifaceted external delays that pushed its launch to the later date. The HTV-5, also known as Kounotori 5, was launched on a H-IIB rocket from the Tanegashima Space Centre in Southern Japan, carrying 5.7 tons of cargo, including water and food supplies, crew commodities, system components, and science experiments.[59] Its cargo also included the CALorimetric Electron Telescope (CALET), which will be placed on the exposed facility attached to Japan's Kibo module for a period of 2–5 years, to search for signatures of dark matter and provide the highest energy direct measurements of cosmic rays and high energy gamma rays.[60] Other equipment included a new Fluids Control and Pump Assembly for the Urine Processor Assembly, Water Processing Assembly Multifiltration Beds, a new SAFER (Simplified Aid For EVA Rescue) unit, and an assortment of satellites to be released from the station. The HTV5 was released from the ISS on 28 September 2015; it burned up in Earth's atmosphere carrying the station's

[56]"Camera to record doomed ATV's disintegration – from inside." 6 Feb. 2015. ESA 30 May 2016 <http://m.esa.int/Our_Activities/Space_Engineering_Technology/Camera_to_record_doomed_ATV_s_disintegration_from_inside>.

[57]"ATV's internal camera delivered data, but not images." 20 Feb. 2015. ESA 30 May 2016 <http://m.esa.int/Our_Activities/Space_Engineering_Technology/ATV_s_internal_camera_delivered_data_but_not_images>.

[58]"Last ATV reentry leaves legacy for future space exploration." 15 Feb. 2015. ESA 30 May 2016 <http://m.esa.int/Our_Activities/Human_Spaceflight/ATV/Last_ATV_reentry_leaves_legacy_for_future_space_exploration>.

[59]Gebhardt, Chris, and Chris Bergin. "HTV-5 Kounotori sets sail for the ISS." 18 Aug. 2015. NASASpaceFlight.com 28 Jan. 2016 <http://www.nasaspaceflight.com/2015/08/htv-5-kounotori-launch-space-station/>.

[60]"CALorimetric Electron Telescope (CALET)." 24 Apr. 2014. JAXA 28 Jan. 2015 <http://iss.jaxa.jp/en/kiboexp/ef/calet/>.

refuse.[61] The next HTV6 mission is expected to launch near the end of 2016, along with four additional HTV missions that are intended in subsequent years.[62]

In 2015, Russia launched four expeditions to the ISS on its Soyuz launcher: TMA-16M on 27 March 2015 with the Expedition 43/44 crew of Gennady Padalka, Mikhail Kornienko, and Scott Kelly[63]; TMA-17M on 22 July 2015 with the Expedition 44/45 crew of Oleg Kononenko, Kimiya Yui, and Kjell Lindgren[64]; TMA-18M on 02 September 2015 with the Expedition 45/46 crew of Sergei Volkov, Andreas Mogensen, and Aidyn Aimbetov[65]; and TMA-19M on 15 December 2015 with the Expedition 46/47 crew of Yuri Malenchenko, Timothy Kopra, and Timothy Peake.[66] Russian cosmonaut Gennady Padalka became the first four-time ISS commander, and now holds the record for having spent the most time in space; adding 168 mission days to increase his total number of days in space to 879 over five trips.[67] Cosmonaut Mikhail Kornienko and NASA astronaut Scott Kelly are both participating in the first 1-year crew mission to the ISS, covering Expeditions 43–46.[68] Cosmonaut Oleg Kononenko, along with JAXA astronaut Kimiya Yui and NASA astronaut Kjell Lindgren spent 141 days aboard the space station, and conducted two landing simulations to refresh their knowledge of necessary procedures in their final week in orbit.[69] Russian cosmonaut Sergei Volkov stayed aboard the ISS for Expeditions 45/46, supporting various scientific activities and maintenance aboard the Russian segment, while crewmates ESA astronaut Andreas Mogensen and Kazakh Space Agency (KSA) cosmonaut Aidyn Aimbetov completed a 10-day mission returning to Earth with Gennady Padalka on 12 September 2015.[70] Prior to reaching the ISS on

[61]Bergin, Chris. “HTV-5 departs ISS – following slight glitch – ahead of fiery demise.” 28 Sept. 2015. NASASpaceFlight.com 28 Jan. 2016 <http://www.nasaspaceflight.com/2015/09/htv-5-departs-iss-fiery-demise/>.

[62]Gebhardt, Chris, and Chris Bergin. “HTV-5 Kounotori sets sail for the ISS.” 18 Aug. 2015. NASASpaceFlight.com 28 Jan. 2016 <http://www.nasaspaceflight.com/2015/08/htv-5-kounotori-launch-space-station/>.

[63]“Soyuz TMA-16M: Starting “Year in Space” mission.” 29 Feb. 2016 Russian Space Web 25 May 2016 <http://www.russianspaceweb.com/iss_soyuz_tma16m.html>.

[64]“Soyuz TMA-17M mission.” 12 Dec. 2015. Russian Space Web 25 May 2016 <http://www.russianspaceweb.com/iss_soyuz_tma17m.html>.

[65]“Soyuz TMA-18M mission.” 3 Mar. 2016. Russian Space Web 25 May 2016 <http://www.russianspaceweb.com/iss-soyuz-tma18m.html>.

[66]“Soyuz crew in close-call docking with ISS.” 29 Apr. 2016. Russian Space Web 25 May 2016 <http://www.russianspaceweb.com/iss-soyuz-tma19m.html>.

[67]“Russian cosmonaut record-breaker Padalka returns to Earth.” 12 Sept. 2015. BBC News 25 May 2016 <http://www.bbc.com/news/science-environment-34231700>.

[68]Kramer, Miriam. “One-Year Crew Begins Epic Trip on International Space Station.” 28 Mar. 2015. Space.com 25 May 2016 <http://www.space.com/28960-one-year-space-station-trip-begins.html>.

[69]“Soyuz TMA-17M mission.” 12 Dec. 2015. Russian Space Web 25 May 2016 <http://www.russianspaceweb.com/iss_soyuz_tma17m.html>.

[70]“Sergei Volkov.” 25 Aug. 2015. Spaceflight 101 27 May 2016 <http://spaceflight101.com/iss-expedition-45/sergei-volkov/>.

15 December 2015 for Expedition 46/47, following the failure of the Soyuz capsule's Kurs automated rendezvous system, Russian cosmonaut Yuri Malenchenko had to manually dock the spacecraft to the station, succeeding on his second attempt just 3 min from entering Earth's shadow, and another 15 min from going out of range of communications with Russia's mission control centre in Korolev.[71] Russia also conducted five resupply missions (including one failure) to the ISS with its Progress cargo transfer vehicles: M-26M on 17 February 2015, M-27M on 28 April 2015 (which failed to reach the ISS due to an anomaly experienced at orbit insertion—it re-entered the atmosphere out of control), M-28M on 3 July 2015, M-29M on 1 October 2015, and MS on 21 December 2015.[72]

An example of the softening U.S. hard-line position toward China can be found in the recent arrangement to have a Chinese experiment ferried to the ISS. Brokered by NanoRacks, which facilitates the use of the ISS for businesses, the commercial arrangement was seen to fall outside the ambit of Congress' 2011 restriction that continues to prohibit NASA and the White House Office of Science and Technology Policy (OSTP) from bilaterally cooperating with China. The experiment will investigate how the space environment affects DNA, with particular focus on whether space radiation and microgravity can cause mutations to gene encoding antibodies. The experiment, conducted by the Beijing Institute of Technology, will be lifted to the ISS on a SpaceX cargo supply mission in 2016, and will be housed in Japan's Kibo module for a 15-day duration.[73] The experiment will be performed by astronauts aboard the ISS, while China will receive only the data and experiment samples via NanoRacks.[74] It is hoped that the experiment can serve as a bridge for greater cooperation between the space powers in the future.

After nearly 2 years of inactivity for China's human spaceflight programme, China plans to launch its Tiangong 2 space lab in 2016 aboard a Long March 2F/G rocket. The space lab will use the same basic module as the Tiangong 1 space lab, but will include a new regenerative life support system and will be resupplied by China's first cargo spacecraft, Tianzhou.[75] As China's second space habitat, with the first space lab to be eventually de-orbited,[76] the Tiangong 2 will receive a single

---

[71]"Soyuz crew in close-call docking with ISS." 29 Apr. 2016. Russian Space Web 25 May 2016 <http://www.russianspaceweb.com/iss-soyuz-tma19m.html>.

[72]"A complete chronology of ISS missions." 10 Apr. 2016. Russian Space Web 27 May 2016 <http://www.russianspaceweb.com/iss_chronology_flights.html>.

[73]David, Leonard. "US-China Space Freeze May Thaw with Historic New Experiment." 21 Aug. 2015. Space.com 22 Jan. 2016 <http://www.space.com/30337-chinese-experiment-international-space-station.html>.

[74]Cowing, Keith. "Commercial Payload From China to Fly on ISS." 3 Aug. 2015. NASA Watch 22 Jan. 2016 <http://nasawatch.com/archives/2015/08/commercial-payl.html>.

[75]Jones, Morris. "Progress for Tiangong 2." 1 Sept. 2015. SpaceDaily.com 31 May 2016 <http://www.spacedaily.com/reports/Progress_for_Tiangong_2_999.html>.

[76]Wenz, John. "China to Launch a New Space Station in 2016." 10 Mar. 2015. Popular Mechanics 31 May 2016 <http://www.popularmechanics.com/space/a14480/china-space-station-2016-tiangong-2/>.

crew of Chinese astronauts on board the Shenzhou 11 for up to 1 month. China will attempt an automatic docking of its Tianzhou cargo spacecraft to the Tiangong-2 shortly after the launch of the Shenzhou 11.[77] China plans to launch another experimental core module of the future space station around 2018, while China's full space station is to be completed by 2022.[78]

2.3.2 *Lunar Science*

Interest in the Moon is ongoing for both its science and exploration value, in addition to being the finish line for several private space companies competing to win the Google Lunar X Prize. This year, the U.S. and China continued to progress toward a robotic and human lunar presence, however budget constraints still have the potential to delay well-intentioned initiatives. Moreover, Google Lunar X-Prize competitors have begun to partner up to better their odds of winning the symbolic award.

NASA's Lunar Reconnaissance Orbiter (LRO), launched in June 2009, is scouting the Moon in preparation for future lunar exploration, including finding landing sites, locating resources such as water, ice and hydrogen, and investigating the long-term effects of the lunar environment. The LRO is in the midst of its second 2-year extended science mission, which began on 15 September 2014 and is slated for completion in September 2016.[79] By 4 February 2015, a research team using the LRO's Lunar Exploration Neutron Detector (LEND) instrument, which detects hydrogen by counting the number of neutrons flying off the lunar surface, discovered that there is a greater abundance of hydrogen on the Moon's Pole-Facing Slopes (PFS) than on its Equator-Facing Slopes (EFS), by an average of about 23 parts-per-million-by-weight (ppmw). There is also a greater abundance of hydrogen in the Moon's southern hemisphere, beginning at between 50 and 60 degrees south latitude, wherein slopes closer to the South Pole show a larger hydrogen concentration difference. The presence of hydrogen could help reduce the need to transport it from Earth to the Moon's surface, it being an important component for fuel and drinking water.[80] On 4 May 2015, NASA completed a manoeuvre, consisting of two station keeping burns to lower the LRO's orbit to

[77]Whittington, Mark. "China preparing to send crewed Shenzhou 11 to Tiangong 2 space station in 2016." 2 Sept. 2015. Examiner.com 31 May 2016 <http://www.examiner.com/article/china-preparing-to-send-crewed-shenzhou-11-to-tiangong-2-space-station-2016-1>.

[78]Jones, Morris. "The Last Tiangong." 20 Oct. 2015. SpaceDaily.com 31 May 2016 <http://www.spacedaily.com/reports/The_Last_Tiangong_999.html>.

[79]Keller, J. "The Lunar Reconnaissance Orbiter and the New Moon: Mission Highlights and Two More Years of Science From Lunar Orbit!" 22 Nov. 2014. LRO – Lunar Planetary Institute 16 Mar. 2015 <http://www.lpi.usra.edu/meetings/leag2014/presentations/keller.pdf>.

[80]"NASA's LRO Discovers Lunar Hydrogen More Abundant on Moon's Pole-Facing Slopes." 4 Feb. 2015. NASA 24 Apr. 2016 <http://www.nasa.gov/content/goddard/lro-lunar-hydrogen>.

within 20 km of the Moon's South Pole and 165 km from its North Pole, to enable improved measurements of the LRO's Lunar Orbiter Laser Altimeter (LOLA) and its high resolution Diviner instrument over the South Pole. By magnifying the sensitivity of the LRO instruments, researchers will be able to better understand the mechanisms by which water or other volatiles might be trapped there.[81]

China's Chang'e 5-TI test capsule returned to lunar orbit in the week of 12 January 2015.[82] Launched on 23 October 2014, the prototype sample-return capsule reached the Moon within a day, circling it before returning to eject its sample capsule at a higher than average velocity into Earth's atmosphere. Following the successful release of the capsule on 1 November 2015, Chang'e 5-TI began making its way to the Earth-Moon Lagrange (L2) point on the opposite side of the Moon. Reaching L2 by late November 2015, the service module then completed three circles around that point prior to returning to lunar orbit.[83] In addition to testing critical breaking manoeuvres, the Chang'e 5-TI carries a camera system that will help to identify future landing sites for the Chang'e 5 robotic sample-return mission planned for launch in 2017 or 2018. This later Chang'e 5 mission will involve a soft landing on the Moon and the collecting of 200 g of samples prior to bringing them to Earth.[84]

China is also planning to launch a second lunar lander mission, Chang'e 4, after the Chang'e 5 mission. Following the success of the Chang'e 3 lunar lander and rover mission conducted at the end of 2013 and into 2014, China wants to send a similar upgraded version of the mission to the Moon in 2020. This second mission will use a more powerful launcher, allowing for the use of a heavier spacecraft, and will land in a different region of the Moon.[85] China is also developing plans for a manned lunar landing sometime between 2025 and 2030.[86]

The Google Lunar X Prize is a competition for a grand prize of $20 million and a second prize of $5 million for the first two privately funded teams to safely land on the Moon, travel at least 500 m across its surface, and send high-definition video, images, and data back to the Earth. In late 2013, the X Prize Foundation and Google announced a series of interim 'milestone' prizes available to assist the competing teams in accessing financing at a critical point in their mission timeline and to raise public excitement and support for the teams. By 26 January 2015, the X Prize

[81]"NASA's LRO Moves Closer to the Lunar Surface." 5 May 2015. NASA 24 Apr. 2016 <http://www.nasa.gov/feature/goddard/nasas-lro-moves-closer-to-the-lunar-surface>.

[82]David, Leonard. "Chinese Spacecraft Enters Orbit around the Moon." 20 Jan. 2015. SpaceNews 8 Jan. 2016 <http://spacenews.com/chinese-spacecraft-enters-orbit-around-the-moon/>.

[83]"Service Module of Chinese Probe Successfully Enters Lunar Orbit." 11 Jan. 2015. Sputnik News 22 Jan. 2016 <http://sputniknews.com/science/20150111/1016776437.html>.

[84]Foust, Jeff. "China's Mars Exploration Program Facing Delays." 1 Apr. 2015. SpaceNews 8 Jan. 2016 <http://spacenews.com/chinas-mars-exploration-program-facing-delays/>.

[85]Foust, Jeff. "China's Mars Exploration Program Facing Delays." 1 Apr. 2015. SpaceNews 8 Jan. 2016 <http://spacenews.com/chinas-mars-exploration-program-facing-delays/>.

[86]"China considering manned lunar landing in 2025–2030." 24 May 2009. China View 21 Apr. 2013 <http://news.xinhuanet.com/english/2009-05/24/content_11425131.htm>.

Foundation had awarded a total of $5.25 million in awards to 5 of the 18 teams in the competition as rewards for demonstrating progress in landing system technologies, imaging systems, and mobility systems.[87] The awards can also be seen as a litmus test to see where competitors are in achieving their self-defined milestone goals; i.e. here Astrobiotic Technology won $1.75 million for its progress in all three areas, while Moon Express won $1.25 million for its progress in landing system technologies and imaging systems. Team Indus won $1 million for its landing system technologies, while Part-Time Scientists won $750,000 for demonstrating their mobility and imaging systems; and Hakuto won $500,000 for its mobility system.[88]

In the follow-up to those awards, Astrobiotic Technology and Hakuto decided to partner up to share risks and rewards, in an arrangement where both teams will share the costs of the launch of the overall spacecraft, and then share the winning proceeds if one of their teams wins the grand prize. Upon reaching the Moon's surface, Astrobiotic Technology's lander will release its one rover along with the two rovers developed by Hakuto; the three rovers will then race to achieve the 500 m requirement, with the winning team splitting the proceeds according to an undisclosed ratio.[89] As Astrobiotic Technology plans to launch its lander on the SpaceX Falcon 9 rocket, it is expected to have an estimated 1250 kg of the 2500 kg launch capacity available for other customers. By 27 October 2015, Astrobiotic Technology had signed on another Google Lunar X Prize competitor, Team AngelicvM of Chile, along with two non-competing payloads (one from the Mexican Space Agency, and one from the British crowd-sourced Lunar Mission One).[90]

The Google Lunar X Prize competition was extended to the end of 2017, after two teams had met the X Prize Foundation's 16 December 2014 stipulation requiring at least one team to have made launch arrangements by the end of 2015.[91] An Israeli team, SpaceIL was the first to be confirmed by the foundation on 7 October 2015, after purchasing a spot on SpaceX's Falcon 9 launcher.[92] On

---

[87]The five teams considered are the following: Astrobiotic of Pittsburgh, USA; Moon Express (Mountain View, USA); Team Indus (Bengaluru, India); Hakuto (Tokyo, Japan); Part-Time Scientists (Multinational).

[88]Foust, Jeff. "5 Lunar X Prize Teams Land Payday; Only 2 Landed Hardware." 30 Jan. 2015. SpaceNews 8 Jan. 2016 <http://spacenews.com/5-lunar-x-prize-teams-land-payday-only-2-landed-hardware/>.

[89]Foust, Jeff. "Google Lunar X Prize Teams Partner To Share Risks and Rewards." 24 Feb. 2015. SpaceNews 8 Jan. 2016 <http://spacenews.com/google-lunar-x-prize-teams-partner-to-share-risks-and-rewards/>.

[90]Foust, Jeff. "Astrobotic Adds Another Google Lunar X Prize Team to Its Lander." 27 Oct. 2015. SpaceNews 11 Jan. 2016 <http://spacenews.com/astrobotic-adds-another-google-lunar-x-prize-team-to-its-lander/>.

[91]Foust, Jeff. "Google Lunar X Prize Extends Competition Deadline." 17 Dec. 2014. SpaceNews 23 July 2015 <http://spacenews.com/google-lunar-x-prize-extends-competition-deadline/>.

[92]Foust, Jeff. "Israeli X Prize Team Announces Launch Contract for Lunar Mission." 7 Oct. 2015. SpaceNews 5 Jan. 2016 <http://spacenews.com/israeli-x-prize-team-announces-launch-contract-for-lunar-mission/>.

8 December 2015, the X Prize Foundation confirmed a second launch agreement by Moon Express, which has purchased three launches on Rocket Lab's still in development small Electron launcher. If Moon Express wins the competition with one of its two launches planned 2017, the team plans to sell the remaining capacity on later missions to scientists or space agencies.[93] Astrobiotic Technology is waiting to finish signing up launch customers before finalizing its launch contract with SpaceX for a late 2017 mission. By the end of 2015, only 16 teams remained in the competition; these teams now have until the end of 2016 to submit their own launch documentation to remain in the competition.[94]

2.3.3 Mars Science

The focus for Mars science has for decades remained the investigation of the planet's habitability, in a search for the presence of water. The collected data continues to suggest that Mars was once partially covered by large oceans, and that life could have been possible in many locations on the planet's surface.

ESA's Mars Express orbiter, launched in June 2003, continued its mission imaging the entire surface of the planet at high resolution, including maps of its mineral composition and atmosphere, and determining the structure of the sub-surface to a depth of a few kilometres, the effect of the atmosphere on the surface, and the interaction of the atmosphere with the solar winds. On 15 January 2015, images taken by the MRO helped to reveal the location of the UK-led Beagle-2 Mars lander, which hitched a ride on ESA's Mars Express mission, but went radio silent after successfully touching down on Mars' surface on 25 December 2003.[95] Images taken by the high-resolution stereo camera on Mars Express on 12 July 2015 just to the south of the mouth of Minio Vallis, near volcanic giants such as Olympus Mons in the south-western portion of the Tharsis bulge, helped to show how volcanic activity may have triggered the melting of subsurface ice, and consequently the formation of the water-carved channels in the region.[96] Additionally, the course of the Mars Express mission has revealed that the planet has its own ultraviolet aurora despite not having a global magnetic field. Likely caused by weak residual magnetism in the crust, such as in the highlands of the southern hemisphere, the ultraviolet auroras last only a few seconds and tend to be rare and

---

[93]Foust, Jeff. "Moon Express Buys Rocket Lab Launches for Lunar Missions." 1 Oct. 2015. SpaceNews 5 Jan. 2016 <http://spacenews.com/moon-express-buys-rocket-lab-launches-for-lunar-missions/>.

[94]Foust, Jeff. "X Prize Verifies Moon Express Launch Contract." 8 Dec. 2015. SpaceNews 6 Jan. 2016 <http://spacenews.com/x-prize-verifies-moon-express-launch-contract/>.

[95]"Beagle-2 lander found on Mars." 16 Jan. 2015. ESA 25 Apr. 2016 <http://www.esa.int/Our_Activities/Space_Science/Mars_Express/Beagle-2_lander_found_on_Mars>.

[96]"Flash floods in Mangala Valles." 15 Oct. 2015. ESA 25 Apr. 2016 <http://www.esa.int/Our_Activities/Space_Science/Mars_Express/Flash_floods_in_Mangala_Valles>.

transient. A total of 16 detections were made from following 113 night side orbits since the beginning of the mission in 2003.[97]

The ExoMars mission is a joint endeavour between ESA and Russia that addresses whether life has ever existed on Mars by drilling the surface of the planet and analyzing the samples in situ. ESA is providing the Trace Gas Orbiter (TGO) and the Entry, Descent and Landing Demonstrator Module (EDM) 'Schiaparelli' in 2016,[98] and the carrier and rover in 2018; while Roscosmos is responsible for the 2018 descent module and surface platform, and will provide launchers for both missions. Near the end of 2015, ExoMars' orbiter and entry, descent and landing module mission, slated for launch in January 2016, was delayed by 2 months due to the presence of leaky pressure transducers. The same lot of transducers also forced delays in six other ESA missions currently in development, including ESA's Solar Orbiter and Cheops. While the transducers were meant to monitor pressure in the entry, descent and landing module's hydrazine fuel tank and its helium pressurization tank, ESA chose to remove them without replacement, as their role was non-critical to the lander's functioning, and only increased the risk to the mission's descent stage. ESA was made aware of the defective components by their manufacturer Moog Inc. of Aurora New York in the summer of 2015, after the company had found that a welding machine defect had made cracks in the transducers, making them at risk of leakage. ExoMars new launch date is expected to take place sometime between 14 and 25 March 2016.[99] The second part of the ExoMars mission will be launched in 2018, and will involve increased Russian participation. That mission is still missing over 175 million euros for development, yet the mission partners might be willing to pay more to avoid schedule delays.[100]

NASA's Mars Odyssey mission, launched on 7 April 2001, is the longest-operating spacecraft to be sent to Mars. The satellite orbited the planet's poles at about 5 AM/PM local Mars time for the first 6 years of its mission, providing optimal temperature for its Gamma Ray Spectrometer search for evidence of water near the Martian surface (e.g. how widely water ice and other elements are distributed on Mars). In the next 3 years, a 4 AM/PM orbit allowed the orbiter's Thermal Emission Imaging System (THEMIS) to more easily identify the infrared signatures of minerals. By 2012, after providing radio-relay support for the landing of the Curiosity Mars rover, the orbit was manoeuvred on a slow drift to later times of the day to aid in preserving battery life. A study of Mars' largest known carbonate deposit revealed merely twice the amount of carbon as in the current Mars

[97]"Shining a light on the aurora of Mars."

[98]"ExoMars Lander Module Named Schiaparelli." 8 Nov. 2013. ESA 28 Mar. 2014 <http://exploration.esa.int/mars/53145-exomars-lander-module-named-schiaparelli/>.

[99]De Selding, Peter B. "Faulty Component that Delayed ExoMars Affects Other ESA Programs." 22 Sept. 2015. SpaceNews 11 Jan. 2016 <http://spacenews.com/faulty-component-that-delayed-exomars-affects-other-esa-programs/>.

[100]De Selding, Peter B. "ExoMars Work at Frenzied Pace To Make 2016 Launch Date." 25 Nov. 2015. SpaceNews 11 Jan. 2016 <http://spacenews.com/exomars-work-at-frenzied-pace-to-make-2016-launch-date/>.

atmosphere, and that the combination of all known carbon reservoirs would not have been enough for Mars to have had a thick atmosphere at the time when rivers flowed on the planet. Rather than being a warm and wet planet at the time of its valley network formation, Mars might have been cold and wet with an atmosphere that had already thinned. There is also uncertainty about how much of that loss of carbon from the atmosphere occurred before the period of valley formation, which NASA hopes its MAVEN orbiter will help to uncover.[101]

NASA's Mars Reconnaissance Orbiter (MRO) continued to provide valuable data for the purpose of determining whether or not life has existed on Mars, characterising the climate and geology, and preparing for future human exploration. By 7 February 2015, the MRO had completed 40,000 orbits of Mars, while in its fourth extended mission, and ninth year of returning data; it reached its tenth year on 12 August 2015.[102] The MRO is now investigating seasonal and longer-term changes in Mars' atmosphere, surface and subsurface, including some warm-season flows that are the strongest evidence so far of liquid water on Mars today.[103] From 7 to 21 June 2015, the Mars solar conjunction occurred, wherein the Sun was directly between the Earth and the Red Planet, and blocked any signals from being sent to the MRO and other Mars' spacecraft. As the Sun disrupts radio communications between the planets during the conjunction period, commands to these spacecraft during that time were reduced, though data continued to be collected.[104] The MRO helped to reveal the presence of glass deposits in impact craters on Mars, which could preserve bio signatures and other evidence about whether Mars ever had life at the time of impact; on Earth, similar research has shown organic molecules and plant matter entombed in glass formed by an impact that occurred millions of years ago in Argentina.[105] And on 28 September 2015, data from the MRO's imaging spectrometer provided the strongest evidence yet for NASA to confirm that a briny liquid water flows intermittently on present-day Mars.[106]

[101]"What Happened to Early Mars' Atmosphere? New Study Eliminates One Theory." 2 Sept. 2015. NASA 25 Apr. 2016 <http://mars.jpl.nasa.gov/odyssey/news/whatsnew/index.cfm?FuseAction=ShowNews&NewsID=1852>.

[102]"One Decade after Launch, Mars Orbiter Still Going Strong." 10 Aug. 2015. NASA Jet Propulsion Laboratory 16 May 2016 <http://mars.nasa.gov/mro/news/whatsnew/index.cfm?FuseAction=ShowNews&NewsID=1849>.

[103]"NASA Spacecraft Completes 40,000 Mars Orbits." 9 Feb. 2015. NASA Jet Propulsion Laboratory 16 May 2016 <http://mars.nasa.gov/mro/news/whatsnew/index.cfm?FuseAction=ShowNews&NewsID=1778>.

[104]"Mars Conjunction." 3 June 2015. NASA Jet Propulsion Laboratory 16 May 2016 <http://mars.nasa.gov/mro/news/whatsnew/index.cfm?FuseAction=ShowNews&NewsID=1824>.

[105]"NASA Spacecraft Detects Impact Glass on Surface of Mars." 8 June 2015. NASA Jet Propulsion Laboratory 16 May 2016 <http://mars.nasa.gov/mro/news/whatsnew/index.cfm?FuseAction=ShowNews&NewsID=1827>.

[106]"NASA Confirms Evidence That Liquid Water Flows on Today's Mars." 28 Sept. 2015. NASA Jet Propulsion Laboratory 16 May 2016 <http://mars.nasa.gov/mro/news/whatsnew/index.cfm?FuseAction=ShowNews&NewsID=1858>.

NASA's Mars Science Laboratory (MSL) rover, Curiosity, has eight scientific objectives, i.e. determining the nature and inventory of organic carbon compounds; conducting an inventory of the chemical building blocks of life; identifying features that may represent the effects of biological processes; investigating the chemical, isotopic, and mineralogical composition of Martian geological materials; interpreting the processes that have formed and modified rocks and soils; assessing 4-billion-year timescale atmospheric evolution processes; determining the present state, distribution, and cycling of water and carbon dioxide; and characterizing the broad spectrum of surface radiation, including galactic cosmic radiation, solar proton events, and secondary neutrons.[107] By 5 February 2015, Curiosity had begun to analyse its second drilled rock sample at Mount Sharp, after using a new low-percussion-level drilling technique more suited for fragile rock. Using its Chemistry and Mineralogy (CheMin) instrument to analyse sample powder from the "Mojave 2" rock target, preliminary results showed a significant amount of jarosite, an oxidized mineral containing iron and sulphur that forms in acidic environments. The powder was also delivered to the internal Sample Analysis at Mars (SAM) suite of instruments for chemical analysis.[108] By 25 February 2015, Curiosity had collected another sample at its third drilling site at the base of Mount Sharp called "Telegraph Peak", finding higher ratios of silica to magnesium and silica to aluminium, which would be expected if there has been some acidic leaching.[109] By 24 March 2015, Curiosity's SAM instrument suite had made its first detection of nitrogen, in the form of nitric oxide, on Mars' surface.[110] By the end of 2015, the abundance of silica found in samples during the year was a contrast to what had been seen in the first 2 years of Curiosity's mission, highlighting considerable variability in minerals within very short distances.[111]

NASA's Mars Atmosphere and Volatile EvolutioN (MAVEN) mission aims to explore the planet's upper atmosphere, ionosphere and interactions with the Sun and solar wind, which will be used to determine the role that the loss to space of volatile compounds from the Mars atmosphere has played in the history of Mars'

---

[107]Mars Science Laboratory (MSL). 14 May. 2012. NASA NSSDC 14 Jan. 2013 <http://nssdc.gsfc.nasa.gov/nmc/spacecraftDisplay.do?id=2011-070A>.

[108]"NASA's Curiosity Analyzing Sample of Martian Mountain." 5 Feb. 2015. NASA Jet Propulsion Laboratory 16 May 2016 <http://mars.nasa.gov/msl/news/whatsnew/index.cfm?FuseAction=ShowNews&NewsID=1777>.

[109]"NASA's Curiosity Mars Rover Drills at 'Telegraph Peak'." 25 Feb. 2015. NASA Jet Propulsion Laboratory 16 May 2016 <http://mars.nasa.gov/msl/news/whatsnew/index.cfm?FuseAction=ShowNews&NewsID=1782>.

[110]"Curiosity Rover Finds Biologically Useful Nitrogen on Mars." 24 Mar. 2015. NASA Jet Propulsion Laboratory 16 May 2016 <http://mars.nasa.gov/msl/news/whatsnew/index.cfm?FuseAction=ShowNews&NewsID=1791>.

[111]"Rocks Rich in Silica Present Puzzles for Mars Rover Team." 17 Dec. 2015. NASA Jet Propulsion Laboratory 16 May 2016 <http://mars.nasa.gov/msl/news/whatsnew/index.cfm?FuseAction=ShowNews&NewsID=1879>.

habitability.[112] In other words, by studying the planet's upper atmosphere and measuring current rates of atmospheric loss, MAVEN scientists hope to understand how Mars transitioned from a warm, wet planet to its current dry desert state.[113] From 10 to 18 February 2015, MAVEN conducted its first of five deep-dip manoeuvres designed to gather measurements closer to the lower end of the Martian upper atmosphere. The campaign placed MAVEN at an altitude of 125 km, allowing scientists to make measurements at atmospheric densities more than ten times greater than recorded from its normal altitude of 150 km.[114] From the beginning of 2015, MAVEN had detected an unexplained high-altitude dust cloud using its Langmuir Probe and Waves (LPW) instrument, which has been present throughout MAVEN's operation. MAVEN also detected an aurora spanning Mars' northern hemisphere using its Imaging Ultraviolet Spectrograph (IUVS) instrument, which surprisingly occurred much deeper than on Earth or elsewhere on Mars.[115] By 21 September 2015, MAVEN had been in orbit around Mars for an entire year, carrying out 10 months of observations during MAVEN's primary mission, along with four deep-dip campaigns.[116] And by 5 November 2015, NASA announced that MAVEN measurements indicate that the solar wind strips away gas at a rate of about 100 g (equivalent to roughly 1/4 pound) every second; atmospheric erosion also increases significantly during solar storms, as observed following a series of dramatic solar storms affecting Mars' atmosphere in March 2015.[117]

ISRO's Mars Orbiter Mission (MOM) lifted into space on 5 November 2013, entering Mars' orbit on 24 September 2014.[118] Shortly thereafter, the spacecraft began its science activities fully, activating its five indigenous scientific payloads consisting of the Mars Colour Camera, the Methane Sensor For Mars, the Thermal

---

[112] "MAVEN." University of Colorado Boulder 14 Jan. 2013. <http://lasp.colorado.edu/home/maven/>.

[113] "NASA Launches Mission to Study Upper Atmosphere of Mars." 18 Nov. 2013. NASA 31 Mar. 2014 <http://www.nasa.gov/press/2013/november/nasa-launches-mission-to-study-upper-atmosphere-of-mars/#.Uzl3uvmSwj5>.

[114] "NASA's MAVEN Spacecraft Completes First Deep Dip Campaign." 19 Feb. 2015. NASA Mars Exploration 17 May 2016 <http://mars.nasa.gov/news/whatsnew/index.cfm?FuseAction=ShowNews&NewsID=1780>.

[115] "NASA Spacecraft Detects Aurora and Mysterious Dust Cloud around Mars." 18 Mar. 2015. NASA Mars Exploration 17 May 2016 <http://mars.nasa.gov/news/whatsnew/index.cfm?FuseAction=ShowNews&NewsID=1789>.

[116] "NASA's MAVEN Celebrates One Year at Mars." 21 Sept. 2015. NASA Mars Exploration 17 May 2016 <http://mars.nasa.gov/news/whatsnew/index.cfm?FuseAction=ShowNews&NewsID=1855>.

[117] "NASA Mission Reveals Speed of Solar Wind Stripping Martian Atmosphere." 5 Nov. 2015. NASA Mars Exploration 17 May 2016 <http://mars.nasa.gov/news/whatsnew/index.cfm?FuseAction=ShowNews&NewsID=1869>.

[118] "Mars Mission: India creates history as Mangalyaan successfully enters Mars orbit in first attempt." 24 Sept. 2014. The Economic Times 22 Mar. 2015 <http://economictimes.indiatimes.com/news/science/mars-mission-india-creates-history-as-mangalyaan-successfully-enters-mars-orbit-in-first-attempt/articleshow/43299562.cms>.

Infrared Imaging Spectrometer, the Mars Exospheric Neutral Composition Analyser, and the Lyman Alpha Photometer.[119] MOM observes the surface of Mars and its atmosphere and exosphere extending up to 80,000 km for a detailed understanding of the planet's evolution, especially its related geological and possible biogenic processes.[120] On 11 November 2015, ISRO announced the first science results of the mission, noting that science data collected by all instruments of the spacecraft was currently being studied and prepared for publication. The Methane Sensor for Mars (MSM) instrument required further calibration of its measurements, but so far has contributed to other studies including a published paper in which the dust patterns in the high-altitude regions of Valles Marineris were discussed. MOM's Mars Exospheric Neutral Composition Analyser (MENCA) conducted mass spectrometry measurements in the outer layers of the atmosphere and detected a number of species of particles in the range of 1–300 amu (atomic mass unit) and also studied the variation in exospheric composition with altitude. Science data from MOM's Thermal Infrared Spectrometer is being used to map the surface composition of Mars, and has already shown a temperature dip to 10–15° in zones of Martian dust storms. MOM's Lyman Alpha Photometer has made over 80 measurements of Mars' ratio of deuterium to hydrogen to trace atmospheric loss at Mars, and will enable the Lyman Alpha intensity to be mapped as a function of altitude. And lastly, MOM's Mars Colour Camera provided images of the backside of Mars' moon Deimos, inaccessible by the other spacecraft currently orbiting Mars, finding that the anti-Mars side of Deimos is smoother than the near side without any major craters or surface irregularities.[121]

2.3.4 *Saturn Science*

The Cassini-Huygens mission, a joint NASA, ESA and ASI mission, was launched in 1997. Reaching Saturn in 2004, Cassini went on to drop the Huygens probe onto Saturn's moon, Titan. The renamed Cassini Solstice Mission was supposed to end in June 2008, however, funding was provided to allow continued operation to provide new insights on Saturn and its moons; it is now slated to explore Saturn until 2017.[122]

---

[119]Laxman, Srinivas. "Mars Orbiter Mission activates all science instruments as NASA, ISRO form joint Mars working group." 1 Oct. 2014. Planetary.org 22 Mar. 2015 <http://www.planetary.org/blogs/guest-blogs/2014/10010914-mars-orbiter-mission.html>.

[120]"Indian Space Research Organisation | Mars Orbiter Mission." Indian Space Research Organisation 31 Mar. 2014 <http://www.isro.org/mars/home.aspx>.

[121]"India's Mars Orbiter Mission delivers first Science Results, looks at future Challenges." 11 Nov. 2015. Spaceflight 101 17 May 2016 <http://spaceflight101.com/mom/indias-mars-orbiter-mission-delivers-first-science-results-looks-at-future-challenges/>.

[122]Mason, Betsy. "Cassini Gets Life Extension to Explore Saturn Until 2017." 3 Feb. 2010. WIRED 18 Dec. 2012 <http://www.wired.com/wiredscience/2010/02/cassini-life-extension-2017/>.

In late January 2015, new research published in the journal Geophysical Research Letters, provided a glimpse of how Saturn's moon Titan would behave if it orbited around the Sun directly. Data collected by Cassini-Huygens magnetometer instrument on 1 December 2013 while Titan was observed beyond the region of Saturn's magnetosphere[123] showed that Titan interacted with the solar wind similar to Mars, if Mars was moved to the distance of Saturn. That finding surprised researchers as they could use the same tools to study both celestial bodies' response to solar wind, rather than having to develop a unique approach for the complex chemistry of the Titan's dense hazy atmosphere.[124] Another study of Titan's surface found that two large depressions holding 'lakes' of liquid methane and ethane likely formed over 50 million years by raining hydrocarbons that slowly eroded the dissolvable minerals lining the pools, in a process that's similar to the creation of caves and sinkholes on Earth.[125] And by 11 November 2015, new observations of Titan's South Pole made by Cassini-Huygens' Composite Infrared Spectrometer (CIRS) instrument showed the substantial growth of an ice cloud first seen in 2012. As Titan enters its 7.5 year winter season, the build-up of these southern clouds indicates that the direction of Titan's global circulation of gases is changing.[126]

By March 2015, Cassini-Huygens mission data had provided scientists with the first clear evidence that Saturn's icy moon Enceladus showed signs of present-day hydrothermal activity that may resemble that seen in the deep oceans on Earth, i.e. where hot water from the interior meets the relatively cold water at the ocean bottom.[127] Research using data from the Cassini-Huygens mission also revealed that a global ocean lies beneath Enceladus' icy crust, as researchers found a tiny, but measurable wobble as it orbits Saturn; that wobble would have been dampened significantly if the moon's surface and core were rigidly connected.[128] By October 2015, Cassini-Huygens had made two close flybys of Enceladus, along with a final close flyby in mid-December. The first of those flybys began on 14 October 2015, for a close-up look at the North Pole region of Enceladus to find signs of ancient geological activity similar to the geyser-spouting, tiger-stripe fractures in the moon's South Pole region.[129] On 28 October 2015, Cassini-Huygens made its deepest-ever dive through Enceladus' plume of ice spray, reaching an altitude of

[123]Titan spends about 95 % of the time within Saturn's magnetosphere.

[124]"Cassini Catches Titan Naked in the Solar Wind." 28 Jan. 2015. NASA Jet Propulsion Laboratory 17 May 2016 <https://saturn.jpl.nasa.gov/news/2528/>.

[125]"The Mysterious 'Lakes' on Saturn's Moon Titan." 19 June 2015. NASA Jet Propulsion Laboratory 17 May 2016 <https://saturn.jpl.nasa.gov/news/2796/>.

[126]"NASA's Cassini Finds Monstrous Ice Cloud in Titan's South Polar Region." 11 Nov. 2015. NASA Jet Propulsion Laboratory 17 May 2016 <https://saturn.jpl.nasa.gov/news/2800/>.

[127]"Spacecraft Data Suggest Saturn Moon's Ocean May Harbor Hydrothermal Activity." 11 Mar. 2015. NASA Jet Propulsion Laboratory 17 May 2016 <https://saturn.jpl.nasa.gov/news/2529/>.

[128]"Cassini Finds Global Ocean in Saturn's Moon Enceladus." 15 Sept. 2015. NASA Jet Propulsion Laboratory 17 May 2016 <https://saturn.jpl.nasa.gov/news/2542/>.

[129]"Cassini Begins Series of Flybys with Close-up of Saturn Moon Enceladus." 13 Oct. 2015. NASA Jet Propulsion Laboratory 17 May 2016 <https://saturn.jpl.nasa.gov/news/2543/>.

49 km from the moon's south polar region.[130] The dive allowed its gas analyser and dust detector instruments to directly sample the moon's plume of gas and dust-sized icy particles thought to come from the ocean below; scientists also hope to gain insights about how much hydrothermal activity is occurring within Enceladus.[131] The last close flyby of Enceladus occurred on 19 December 2015, to measure how much heat is coming through the ice from the moon's interior.[132] The remainder of Cassini-Huygens mission will continue to observe Enceladus, but will be from much greater distances.

The Cassini-Huygens mission was also able to get close views of Saturn's subtly-coloured moon Rhea on 9 February 2015 before moving toward a nearly equatorial orbit around Saturn for the rest of 2015.[133] It also made its final close approach to Saturn's large irregularly shaped moon Hyperion on 31 May 2015[134]; possibly the remnant of a violent collision that shattered a larger object into pieces, the sponge-resembling moon has unusually low density which comes from its porosity, weak surface gravity, and a chaotic rotation.[135] And on 16 June 2015, the Cassini-Huygens spacecraft made its penultimate close approach to the icy moon Dione at a distance of 516 km,[136] before making a final close flyby of Dione on 17 August 2015 at a distance of 474 km above Dione's North Pole, to allow the spacecraft's Composite Infrared Spectrometer instrument to map areas on the icy moon that have unusual thermal anomalies, and its Cosmic Dust Analyser to search for dust particles emitted from the moon.[137]

2.3.5 Venus Science

While ESA's Venus Express mission ended in December 2014, the data it collected is still generating new findings. The Venus Express mission was launched in 1995

[130]"Deepest-Ever Dive Through Enceladus Plume Completed." 28 Oct. 2015. NASA Jet Propulsion Laboratory 17 May 2016 <https://saturn.jpl.nasa.gov/news/2546/>.

[131]"Seven Key Facts About Cassini's Oct. 28 'Plume Dive'." 26 Oct. 2015. NASA Jet Propulsion Laboratory 17 May 2016 <https://saturn.jpl.nasa.gov/news/2799/>.

[132]"Cassini Closes in on Enceladus, One Last Time." 16 Dec. 2015. NASA Jet Propulsion Laboratory 17 May 2016 <https://saturn.jpl.nasa.gov/news/2548/>.

[133]"Saturn Spacecraft Returns to the Realm of Icy Moons." 30 Mar. 2015. NASA Jet Propulsion Laboratory 17 May 2016 <https://saturn.jpl.nasa.gov/news/2530/>.

[134]"Cassini Prepares for Last Up-close Look at Hyperion." 28 May 2015. NASA Jet Propulsion Laboratory 17 May 2016 <https://saturn.jpl.nasa.gov/news/2534/>.

[135]"Cassini Sends Final Close Views of Odd Moon Hyperion." 2 June 2015. NASA Jet Propulsion Laboratory 17 May 2016 <https://saturn.jpl.nasa.gov/news/2535/>.

[136]"Cassini Sends Back Views After Zooming Past Dione." 18 June 2015. NASA Jet Propulsion Laboratory 17 May 2016 <https://saturn.jpl.nasa.gov/news/2537/>.

[137]"Cassini to Make Last Close Flyby of Saturn Moon Dione." 13 Aug. 2015. NASA Jet Propulsion Laboratory 17 May 2016 <https://saturn.jpl.nasa.gov/news/2539/>.

and reached Venus in April 2006. It studied Venus's atmosphere, including its dynamics and chemistry, atmosphere-surface interactions, and interactions with solar wind, to address open questions such as the workings of the complex global dynamics of the planet, its cloud system, processes that govern the chemical state of the atmosphere, and the 'green-house effect' in its global climate. While observing the planet's surface is extremely difficult due to its thick atmosphere, by 18 June 2015, researchers using a near-infrared channel of the spacecraft's Venus Monitoring Camera (VMC) to map thermal emission from the surface through a transparent spectral window in the planet's atmosphere, had spotted localised changes in surface brightness between images taken only a few days apart. The four 'hotspots' observed are in locations that correspond to what is known from radar imagery to be tectonic rift zones, and are likely caused by the upwelling magma from below the crust when those rift zones fracture, releasing a lava flow. While researchers have been aware of the tectonic rift zones for some time, the hotspots help to reveal that they are hot and changing in temperature from day to day, adding to the evidence that Venus continues to be volcanically active.[138]

2.3.6 Mercury Science

Running as a partnership between ESA and JAXA, BepiColombo will be Europe's first mission to Mercury. The mission's targeted launch date is now planned to take place during a 1 month long window starting on 27 January 2017, with arrival at Mercury to remain in January 2024 for a 1-year mission, with a possible 1-year extension. The mission had been targeted for a window opening in July 2016, however delays in the procurement of critical units and the availability of some payloads persuaded mission planners to opt for a later launch opportunity in order to minimise operational risks.[139] The mission is made up of two spacecraft: the Mercury Planetary Orbiter (MPO) provided by ESA that will study the planet's geology, composition, inner structure, and exosphere, and the Mercury Magnetospheric Orbiter (MMO) provided by JAXA that will study the planet's magnetic field, atmosphere, magnetosphere and inner interplanetary space. Enduring temperatures in excess of 350 °C, BepiColombo has been designed to provide the measurements necessary to study and understand the composition, geophysics, atmosphere, magnetosphere and history of Mercury.[140] Following BepiColombo's launch, the two spacecraft will return to Earth's vicinity in July 2018, before

[138] "Hot lava flows discovered on Venus." 18 June 2015. ESA 17 May 2016 <http://sci.esa.int/venus-express/56050-hot-lava-flows-discovered-on-venus/>.

[139] "BepiColombo launch moved to 2017." 30 Mar. 2015. ESA 18 May 2016 <http://sci.esa.int/bepicolombo/55693-bepicolombo-launch-moved-to-2017/>.

[140] "BepiColombo Fact Sheet." 3 Dec. 2013. ESA 27 Mar. 2014 <http://sci.esa.int/bepicolombo/47346-fact-sheet/>.

encountering Venus in 2019 and 2020, and then receiving five gravity assists from Mercury between 2020 and 2023 which will slow the two orbiters sufficiently to be captured by Mercury's gravity in 2024.[141]

NASA's MErcury Surface, Space ENvironment, GEochemistry and Ranging (MESSENGER) mission, launched in August 2004, is the first space mission designed to orbit Mercury. The spacecraft entered into orbit about Mercury on 17 March 2011, providing reconnaissance of the geochemistry, geophysics, geological history, atmosphere, magnetosphere, and plasma environment of the planet for more than 4 years before completing its mission in 2015. MESSENGER completed its fourth manoeuvre to extend its orbital operations on 21 January 2015, followed by additional manoeuvres on 18 March, 2 April, 6 April, 8 April, 14 April, 24 April, and 28 April 2015 to delay the reduction of its minimum orbital altitude caused by progressive changes over time in response to the gravitational pull of the Sun. With its mission having been extended by a month by orbit correction manoeuvres, MESSENGER impacted the surface of Mercury on 30 April 2015. By the end of its mission, MESSENGER had completed over 4000 orbits of the planet, collecting more than 275,000 images and downlinking more than 4 Earth years of imaging and topographic data from altitudes between 5 km and 35 km. One of the latest findings from the mission has revealed evidence of an ancient magnetic field while Mercury was volcanically and tectonically active around 3.7–3.9 billion years ago.[142]

2.3.7 *Jupiter Science*

ESA's Jupiter Icy moon Explorer (JUICE) is being developed under ESA's Cosmic Vision 2015–2025 plan, with a foreseen launch date of 2022 aboard an Ariane 5 launcher and arrival in 2030. The nearly 5000 kg spacecraft will make a careful investigation of Jupiter's three biggest moons, using the gravity of Jupiter to initiate a series of close fly-bys around Callisto, Ganymede, and Europa, and then finally settle into an orbit around Ganymede for an 8-month study.[143] As all three moons are suspected of having oceans of water beneath their icy crusts, scientists are trying to understand whether there is any possibility that these moons could host microbial life.[144] On 16 July 2015, ESA selected Airbus D&S for a 350.8 million euros ($389

[141]"BepiColombo launch moved to 2017." 30 Mar. 2015. ESA 18 May 2016 <http://sci.esa.int/bepicolombo/55693-bepicolombo-launch-moved-to-2017/>.

[142]"ARTICLES." 2015 MESSENGER 18 May 2016 <http://messenger.jhuapl.edu/Resources/Articles.html>.

[143]De Selding, Peter B. "Airbus To Build ESA's Jupiter-bound Juice Orbiter." 17 July 2015. SpaceNews 11 Jan. 2016 <http://spacenews.com/airbus-to-build-esas-jupiter-bound-juice-orbiter/>.

[144]"ESA Selects 1bn-Euro JUICE Probe to Jupiter." 2 May 2012. BBC News 27 Mar. 2014 <http://www.bbc.com/news/science-environment-17917102>.

million) contract to build the Juice Orbiter. The contract was signed on 9 December 2015.[145]

2.3.8 *Solar Observation*

Continued observation of the Sun's external activity has the benefit of improving our understanding of its interior, its corona, the monitoring of solar wind and its consequences on Earth and its neighbouring planets. Coronal mass ejections (CMEs) from the Sun emit surges of charged particles in directions that may cross Earth's path and can damage satellites, impede space-based services and affect the terrestrial electrical infrastructure.

ESA's PRoject for OnBoard Autonomy (PROBA)-2 microsatellite mission will operate until at least the end of 2016.[146] PROBA-2 tracks spikes in CMEs ejecting from the Sun that have previously been seen to just skim Earth, typically bringing with them a burst of radio energy.[147] The mission has gathered over 400,000 images of the Sun, and made nearly 20 million in-situ ionospheric observations.[148] On 20 March 2015, PROBA-2 captured a near-total eclipse from orbit, with the imagery of the eclipse to be compared with views from other space missions, such as NASA's Solar Dynamics Observatory, as well as ground observations.[149] ESA is also developing the Proba-3, as a pair of satellites maintaining a fixed configuration to form a 150 m long solar chronograph to study the Sun's faint corona closer to the solar rim than previously achieved.[150]

NASA's Solar Dynamics Observatory (SDO) mission seeks to determine how the Sun's magnetic field is generated and structured, and how this stored magnetic energy is released in the form of the solar wind, energetic particles and variations in the solar irradiance. The spacecraft is comprised of three scientific experiments: the Atmospheric Imaging Assembly (AIA), the EUV Variability Experiment (EVE)

---

[145] De Selding, Peter B. "ESA, Airbus Formalize Jupiter Icy Moons Contract." 11 Dec. 2015. SpaceNews 11 Jan. 2016 <http://spacenews.com/esa-airbus-formalize-jupiter-icy-moons-contract/>.

[146] "Working life extensions for ESA's science missions." 20 Nov. 2014. ESA 1 Mar. 2015 <http://sci.esa.int/director-desk/54999-working-life-extensions-for-esas-science-missions/>.

[147] "Small Sun-Watcher Proba-2 Offers Detailed View of Massive Solar Eruption." 9 June 2011. ESA 17 Apr. 2013 <http://www.esa.int/Our_Activities/Technology/Small_Sun-watcher_Proba-2_offers_detailed_view_of_massive_solar_eruption>.

[148] "ESA's Space Weather Station Proba-2 Tracks Stormy Sun." 2 Dec. 2011. ESA 17 Apr. 2013 <http://www.esa.int/Our_Activities/Technology/ESA_s_space_weather_station_Proba-2_tracks_stormy_Sun>.

[149] "ESA minisatellites to follow Europe's solar eclipse." 13 Mar. 2015. ESA 18 May 2016 <http://m.esa.int/Our_Activities/Space_Engineering_Technology/ESA_minisatellites_to_follow_Europe_s_solar_eclipse>.

[150] "About PROBA3." 19 Nov. 2012. ESA 1 Apr. 2014 <http://www.esa.int/Our_Activities/Technology/Proba_Missions/About_Proba-3>.

and the Helioseismic and Magnetic Imager (HMI).[151] The SDO's global view of the Sun facilitates research that focuses on the previously unrecorded real fine structure of the star.[152] In research published on 23 June 2015 scientists also used images from the SDO mission and the joint Japanese Aerospace Exploration Agency and NASA's Hinode mission to study Venus' atmosphere as it transited between Earth and the Sun. Because each of the layers of Venus' atmosphere absorb light differently from one another, researchers were able to learn more about what kinds of atoms and molecules are actually in its atmosphere, by observing the planet's atmosphere in different wavelengths of light during its journey. Using images taken in X-ray and ultraviolet wavelengths, researchers were also able to penetrate the planet's seemingly opaque atmosphere and measure its apparent size to within several kilometres.[153]

The SOlar and Heliospheric Observatory (SOHO) continued to operate during this reporting period. As an international cooperation project between ESA and NASA the spacecraft was launched on 2 December 1995. SOHO orbits around the Sun in step with the Earth, at a distance of 1.5 million km from Earth, enabling an uninterrupted view of the star.[154] Its scientific objectives are to investigate the solar interior and explain the extreme heating of the solar corona and the mechanism by which the solar wind is produced and accelerated. Extended once again until 31 December 2016, the spacecraft discovered its 3000th comet on course with the Sun on 13 September 2015.[155] A paper appearing in Space Weather on 9 June 2015 described a new model for measuring the magnetic field configuration of a CME headed toward Earth significantly further ahead of time than the 30–60 min window that exists today. By using SOHO and STEREO observations of CMEs as they rapidly move away from the Sun, the model could track how the initial eruption evolves over time and describe how the CME will be configured as it approaches Earth. The Space Weather Research Center at NASA Goddard will test the model's reliability against a variety of historical events and any events that are witnessed over the next year.[156]

---

[151] "SDO | Solar Dynamics Observatory." NASA Goddard Space Flight Center 4 Mar. 2013 <http://sdo.gsfc.nasa.gov/>.

[152] SpaceNews Staff. "NASA Boasts Big Results from 5-minute Spaceflight." 28 Jan. 2013 SpaceNews 4 Mar. 2013 <http://www.spacenews.com/article/nasa-boasts-big-results-from-5-minute-spaceflight>.

[153] "Scientists Study Atmosphere of Venus through Transit Images." 9 July 2015. NASA 19 May 2016 <http://www.nasa.gov/feature/goddard/scientists-study-venus-atmosphere-through-transit>.

[154] About the SOHO Mission. "SOHO Fact Sheet." SOHO—Solar and Heliospheric Observatory 4 Mar. 2013 <http://sohowww.nascom.nasa.gov/about/docs/SOHO_Fact_Sheet.pdf>.

[155] "ESA/NASA Solar Observatory Discovers Its 3,000th Comet." 15 Sept. 2015. NASA 19 May 2016 <http://www.nasa.gov/feature/goddard/esa-nasa-solar-observatory-discovers-its-3000th-comet>.

[156] "New Tool Could Track Space Weather 24 Hours Before Reaching Earth." 9 June 2015. NASA 19 May 2016 <http://www.nasa.gov/feature/goddard/new-tool-could-track-space-weather-24-hours-before-reaching-earth>.

NASA's Solar TErrestrial RElations Observatory (STEREO) is made up of two space-based observatories, i.e. STEREO-A travelling in a smaller and faster orbit (ahead of Earth's orbit), and STEREO-B trailing behind with a larger and slower orbit.[157] They provide new insights into CMEs, including detecting and processing data that enables the tracking of CMEs headed toward Earth. While NASA mission operations lost communication with STEREO-B on 1 October 2014, as the spacecraft was about to orbit around the other side of the Sun, as of 11 December 2015 the spacecraft had entered into a region where it could once again receive radio signals without the Sun's interference. Mission operators are now trying to connect with the spacecraft, but suspect its Inertial Measurement Unit (IMU) might have caused it to spin. Nevertheless, communication with the spacecraft is expected to become easier as it is converged upon by Earth.[158] Moreover, while the STEREO-A spacecraft underwent side lobe repointing operations on 20 August and 1 December 2014, which had the spacecraft provide lower resolution data through 2015,[159] it resumed normal operations and began transmitting data at its full rate on 9 November 2015. The spacecraft resumed its normal science operations on 17 November 2015, which includes transmission of lower-resolution real-time data to monitor solar events, and intermittent high-definition images of the sun's surface and atmosphere.[160]

NASA's Interface Region Imaging Spectrograph (IRIS) satellite observes how solar material moves, gathers energy and heats up as it travels through the Sun's lower atmosphere. In addition to being where most of the Sun's ultraviolet emission is generated, this region between the Sun's photosphere and corona powers the Sun's million-degree atmosphere, and drives the solar wind. Its mission complements the SDO and the Hinode missions to explore the variable solar atmosphere and how it impacts Earth. Positioned in a sun-synchronous polar orbit, making continuous solar observations, IRIS entered the second leg of its 2-year mission on 27 June 2015.[161]

The Reuven Ramaty High Energy Solar Spectroscopic Imager (RHESSI) focuses on the basic physics of particle acceleration and energy release in solar flares. RHESSI completed its fourth annual procedure on 1 September 2014, to rejuvenate its detectors from the effects of radiation damage, and resumed collecting solar X-ray and gamma-ray data.[162] Its fifth month-long anneal will take place early in 2016, pausing the collection of science data during that time.

[157] "First Ever STEREO Images of the Entire Sun." 6 Feb. 2011. NASA 5 Mar. 2013 <http://www.nasa.gov/mission_pages/stereo/news/entire-sun.html>

[158] "Saving NASA's STEREO-B: The 189-million-mile Road to Recovery." 11

[159] "Q & A for Operations of STEREO During Superior Conjunction." 3 July 2014. NASA 1 Mar. 2015 <http://www.nasa.gov/content/goddard/q-a-for-stereo-during-superior-conjunction/>.

[160] "NASA's STEREO-A Resumes Normal Operations." 19 Nov. 2015. NASA 19 May 2016 <http://www.nasa.gov/feature/goddard/nasa-s-stereo-a-resumes-normal-operations>.

[161] "IRIS Celebrates Year Two With Ongoing Scientific Discoveries." 26 June 2015. NASA 19 May 2016 <https://www.nasa.gov/feature/goddard/iris-celebrates-year-two-with-ongoing-scientific-discoveries>.

[162] "RHESSI Resumes Observations." 1 Sept. 2014. RHESSI Science Nuggets 24 Mar. 2015 <http://sprg.ssl.berkeley.edu/~tohban/wiki/index.php/RHESSI_Resumes_Observations>.

2.3.9 Solar System Science

The dwarf planet Ceres and Near-Earth Objects (NEOs) including comets and asteroids were the central focus for exploration in 2015.

NASA's Dawn mission studies the asteroid Vesta and dwarf planet Ceres, the two largest bodies in the asteroid belt between Mars and Jupiter (with average diameters of 525 km and 950 km respectively), to characterize the early solar system and the processes that dominated its formation. Following its launch on 27 September 2007, and with the use of a gravity assist from Mars in February 2009, the spacecraft explored the 525 km in diameter 'protoplanet' Vesta for 14 months, from July 2011 to September 2012. The spacecraft began its approach phase toward dwarf planet Ceres at the beginning of the year, and inserted into orbit on 6 March 2015.[163] The Dawn spacecraft then began to spiral down to increasingly lower orbits in the following months, reaching a distance of 4400 km from Ceres on 7 June 2015, lowering to 1470 km by 25 August 2015, and dropping to its final orbit of 380 km from the surface by 26 October 2015.[164] Dawn has mapped Ceres at each orbit altitude with increasing detail; in its final orbit, the resolution of the transmitted images was 35 m per pixel. The spacecraft's visible and infrared mapping spectrometer enables scientists to identify specific minerals present on the dwarf planet by the way each mineral reflects the range of visible and infrared-light wavelengths.[165]

ESA's Rosetta mission was moved to a safe distance as it observed 67P/Churyumov-Gerasimenko's closest approach to the Sun on 13 August 2015, and remained there until mid-September to protect the spacecraft as the comet's activity settled. Rosetta had been moved away in the earlier months which made communication with Philae even harder. As the comet begins anew its 6.5 year orbit around the Sun, ESA and DLR decision makers must decide whether to study the composition of the comet's tail, requiring its orbit to be adjusted to about 1000 km from the comet's nucleus, or attempt to re-establish communications with Philae as soon as possible. At the end of 2015, Philae's operators still hoped to re-establish contact with the lander and turn its momentum wheel on, which might shift it to a better position.[166] ESA and the DLR decided to extend Rosetta's mission into September 2016, the spacecraft will then be guided to alight on the comet's surface to finish its mission.[167] However a persisting issue is in the bottleneck in access to

[163]"NASA Spacecraft Becomes First to Orbit a Dwarf Planet." 6 Mar. 2015. NASA 24 May 2016 <http://dawn.jpl.nasa.gov/news/news-detail.html?id=4503>.

[164]"Dawn Heads Toward Final Orbit." 26 Oct. 2015. NASA 24 May 2016 <http://dawn.jpl.nasa.gov/news/news-detail.html?id=4751>.

[165]"Ceres Spots Continue to Mystify in Latest Dawn Images." 22 June 2015. NASA 24 May 2016 <http://dawn.jpl.nasa.gov/news/news-detail.html?id=4633>.

[166]Crew, Bec. "Our last chance to contact the Philae comet lander probably failed." 13 Jan. 2016. Science alert 20 May 2016 <http://www.sciencealert.com/our-last-chance-to-contact-the-philae-comet-lander-probably-failed>.

[167]De Selding, Peter B. "ESA Managers Face Choice on Rosetta Priorities." 13 Aug. 2015. SpaceNews 11 Jan. 2016 <http://spacenews.com/esa-managers-face-choice-on-rosetta-priorities/>.

Rosetta and Philae data, as its images are embargoed for 6 months before being released publically.[168] Here, ESA will need to determine how to balance the interest of the mission's principal investigators to reap their return of investment on the project and have a first look at the data, with the interest of the public in accessing those images shortly after observing the landing of Philae.[169] Nevertheless, some findings from the Rosetta mission are providing insight into the origin of Earth's oceans, as Rosetta's Orbiter Spectrometer for Ion and Neutral Analysis (ROSINA) measured the comet's deuterium/hydrogen (D/H) ratio to be more than three times greater than for Earth's oceans and for other recorded Kuiper Belt comets; it also measured higher than previous Oort cloud comets as well, potentially indicating that comets in the Kuiper belt may have formed over a wider range of distances than previously thought.[170]

NASA's reboot of the Wide-field Infrared Survey Explorer (WISE) mission, this time to discover and characterize near-Earth objects (NEOs) with infrared light, was reactivated in September 2013 for 3 additional years of service.[171] The original WISE mission was completed early in 2011. Following the completion of surveying the whole sky twice in infrared light, the space telescope had been placed in 31 months of hibernation.[172] Now dubbed NEOWISE, the mission began its second year of operation at the start of 2015, after detecting and characterising 40 NEOs (8 of which being potentially hazardous asteroids (PHA)), and observing and characterising 245 previously known NEOs in the first year after the mission was restarted.[173] By November 2015, the Astrophysical Journal had published the results of the NEOWISE census of 163 comets which found that while water-ice sublimation likely drives the activity in comets when they come nearest to the Sun, carbon monoxide and carbon dioxide emissions are the main drivers of activity as the comet travels to colder reaches beyond four astronomical units (i.e. four times the Earth-Sun distance, or about 600 million km).[174] Moreover in outer solar system science, data from the WISE mission has helped astronomers to find a cluster of

[168]De Selding, Peter B. "ESA Hikes Budget, Takes Steps To Send Astronaut to Chinese Space Station." 16 Jan. 2015. SpaceNews 8 Jan. 2016 <http://spacenews.com/esa-hikes-budget-takes-steps-to-send-astronaut-to-chinese-space-station/>.

[169]De Selding, Peter B. "ESA Hikes Budget, Takes Steps To Send Astronaut to Chinese Space Station." 16 Jan. 2015. SpaceNews 8 Jan. 2016 <http://spacenews.com/esa-hikes-budget-takes-steps-to-send-astronaut-to-chinese-space-station/>.

[170]"Rosetta fuels debate on origin of Earth's oceans." 10 Dec. 2014. ESA 9 Mar. 2015 <http://www.esa.int/Our_Activities/Space_Science/Rosetta/Rosetta_fuels_debate_on_origin_of_Earth_s_oceans>.

[171]"NASA Spacecraft Reactivated to Hunt for Asteroids." 21 Aug. 2013. NASA 2 Apr. 2014 <http://www.jpl.nasa.gov/news/news.php?release=2013-257>.

[172]"NASA's Asteroid Hunter Spacecraft Returns First Images after Reactivation." 19 Dec. 2013. NASA 9 Mar. 2015 <http://www.jpl.nasa.gov/wise/newsfeatures.cfm?release=2013-373>.

[173]"NEOWISE: A Yearlong Look at the Sky." 15 Jan. 2015. NASA 22 May 2016 <http://www.jpl.nasa.gov/wise/newsfeatures.cfm?release=2015-018>.

[174]"NEOWISE Observes Carbon Gases in Comets." 23 Nov. 2015. NASA 22 May 2016 <http://www.jpl.nasa.gov/wise/newsfeatures.cfm?release=2015-355>.

stars forming in a remote location at the edge of the Milky Way galaxy, an atypical occurrence as stellar nurseries are usually found closer to the galaxy's core and at lower latitudes from the galactic plane.[175]

WISE data has also helped astronomers to view the most luminous galaxy found to date, designated as WISE J224607.57-052635.0, the galaxy belongs to a new class of objects recently discovered by WISE, i.e. extremely luminous infrared galaxies (ELIRG). Shining with the light of more than 300 trillion suns, some 12.5 billion light-years away, it is thought to have a behemoth black hole at its centre, which is heating surrounding gasses to temperatures of millions of degrees and blasting out high-energy, visible, ultraviolet and X-ray light, as it sucks in matter. In a study published in the 22 May 2015 issue of The Astrophysical Journal, three potential explanations as to why black holes in ELIRGs could reach their gargantuan size were outlined including: they were born that way; the continuous braking of the Eddington limit, wherein the pressure of light trying to blast out of the black hole regulates the amount of gas drawn in by the black hole; or the bending of the Eddington limit, wherein the black hole is not spinning fast enough to repel matter, allowing it to consume food faster than typically thought possible.[176]

Commercial interest in NEOs has increased in recent years with several private U.S. companies already investing millions of dollars to be the first to mine asteroids. As an initial step, Planetary Resources and Deep Space Industries (DSI) business models focus on developing prospecting spacecraft that will image and characterise promising asteroids. On 16 July 2015, Planetary Resources launched its Arkyd 3 Reflight (A3R) spacecraft from the ISS to validate several core technologies that will be incorporated into a future fleet of Arkyd spacecraft that will be launched into the solar system.[177] Similarly, DSI is developing its FireFly spacecraft; but rather than testing its technologies in Earth orbit, the first generation spacecraft will be launched directly toward promising asteroids. The main focus in asteroid mining is in the potential availability of rare metals that could be returned to Earth, or water that could be converted into propellant and oxygen for deeper space exploration. While initial estimates of the value of some known asteroids are upwards of $100 trillion, to reach these caches both companies will need to invest substantial resources in developing the necessary mining technology, in addition to successfully rendezvousing and securing a spacecraft onto an asteroid's surface.[178]

[175] "Living on the Edge: Stars Found Far from Galaxy Center." 27 Feb. 2015. NASA 22 May 2016 <http://www.jpl.nasa.gov/wise/newsfeatures.cfm?release=2015-079>.

[176] "NASA's WISE Spacecraft Discovers Most Luminous Galaxy in Universe." 21 May 2015 NASA 22 May 2016 <http://www.jpl.nasa.gov/wise/newsfeatures.cfm?release=2015-173>.

[177] "Planetary Resources' First Spacecraft Successfully Deployed, Testing Asteroid Prospecting Technology in Orbit." 16 July 2015. Planetary Resources 27 Nov. 2015 <http://www.planetaryresources.com/2015/07/planetary-resources-first-spacecraft-deployed/>.

[178] Calandrelli, Emily. "The Potential $100 Trillion Market For Space Mining." 9 July 2015. Techcrunch.com 26 Nov. 2015 <http://techcrunch.com/2015/07/09/the-potential-100-trillion-market-for-space-mining/>.

2.3.10 Outer Solar Science

ESA's Gaia mission completed the first year of its main scientific observations on 21 August 2015 to measure the positions and motions of a billion stars of the roughly 100 billion stars in our galaxy to create the most accurate map yet of the Milky Way. The spacecraft was launched on 19 December 2013 and underwent a 6-month long in-orbit commissioning period prior to beginning routine scientific operations on 25 July 2014 and initiating its main survey operation in the following month. While Gaia's first scheduled data release is expected in the summer of 2016, the spacecraft has already been able to measure the parallax, i.e. the apparent motion of a star against a distant background, for an initial sample of two million stars. Gaia is also measuring the proper motion of stars moving through space, but with roughly 14 measurements of each star in the sky. To disentangle the parallax and proper motions of those stars researchers correlate their recordings with the positions extracted by the Gaia's predecessor, the Hipparcos satellite, taken between 1989 and 1993.[179] On 6 November 2015, Gaia was in a unique position to view the Moon's transit across the Sun using its thermistors; while its sensitive optics must be pointed well away from the Sun and Moon, its thermistors allowed Gaia to chart how its temperature changed during the transit.[180] And from 16 to 20 November 2015, some 200 members of the Gaia Data Processing and Analysis Consortium (DPAC) met in Leiden, The Netherlands, to review the current status of preparations for its first catalogue release from ESA's billion star surveyor mission in 2016.[181]

NASA's Kepler space telescope K2 mission began its fourth campaign on 8 February 2015 to study nearly 16,000 target stars, in search of Earth-sized exoplanets in the habitable zone of other solar-like oscillating stars, where liquid water could exist on their surfaces, and to examine an array of astrophysical phenomena occurring in the M45 (Pleiades), NGC1647, and Hyades star clusters.[182] Three more campaigns were conducted in the year focusing on: M44 (Beehive), M67; North Galactic Cap; and Near Galactic Centre, NGC 6717, and Pluto.[183] After having lost two of its four reaction wheels in May 2013, the Kepler spacecraft lost its ability to precisely point at the original field of view; by May 2014, NASA approved the Kepler 2 mission which repurposed Kepler to accurately

---

179 "Gaia's first year of scientific observations." 25 Aug. 2015. ESA 22 May 2016 <http://m.esa.int/Our_Activities/Space_Science/Gaia/Gaia_s_first_year_of_scientific_observations>.

180 "Gaia's sensors scan a lunar transit." 6 Nov. 2015. ESA 22 May 2016 <http://sci.esa.int/gaia/56783-gaias-sensors-scan-a-lunar-transit/>.

181 "Gaia data processing consortium meets ahead of first catalogue release next year." 16 Nov. 2015. ESA 22 May 2016 <http://sci.esa.int/gaia/56838-gaia-data-processing-consortium-meets-ahead-of-first-catalogue-release-next-year/>.

182 "Mission Manager Update: K2 in Campaign 4." 2 Apr. 2015. NASA 22 May 2016 <http://kepler.nasa.gov/news/nasakeplernews/index.cfm?FuseAction=ShowNews&NewsID=381>.

183 "K2 Campaign fields." 16 May 2016 NASA 23 May 2016 <http://keplerscience.arc.nasa.gov/k2-fields.html>.

point at target sky fields along the ecliptic plane of Earth's orbit using the Sun to maintain its stability.[184] At the beginning of 2015, scientists had an assortment of more than 4175 candidate planets for further study, with around 1000 planets already verified.[185] In January 2015, Kepler found three nearly earth-size planets orbiting a neighbouring red M-dwarf star, EPIC 201367065, some 150 light-years away; the outermost planet orbits far enough from the star that transmitted light is similar to that received by Earth from the Sun.[186] In June 2015, the three exoplanets orbiting around Kepler-138 were among the smallest observed in both size and mass; Kepler-138b is the first Mars-size exoplanet to have both its mass and size measured, while Kepler-138c and Kepler-138d are approximately the same size as Earth.[187] And in July 2015, Kepler-452b was found to have the closest match of planet characteristics to Earth, being 60 % larger than Earth, orbiting a Sun-like star in a 385 day orbit period, placing it in the star's habitable zone.[188]

After running out of the coolant needed to chill its longer-wavelength instruments in 2009, NASA's Spitzer Space Telescope was repurposed to track exoplanets around other stars with the use of infrared light. As exoplanets cross in front of their stars, they block out a fraction of the light, allowing the size of the planet to be revealed, in addition to giving clues about the planet's atmosphere by the infrared light that they also emit.[189] Spitzer astronomers, working in collaboration with astronomers at the Polish Optical Gravitational Lensing Experiment (OGLE) in Chile, were able to find a remote gas planet about 13,000 light-years from Earth using a process known as microlensing. Microlensing occurs when the light of a distant star is magnified and brightened by the gravity of another star that passes in its foreground. Should the closer star have a planet it its orbit, the planet might cause a blip in the magnification. Observing the microlensing event of a star simultaneously from space and from Earth enables researchers to measure the star's parallax, by watching the star brighten at different times, from remote distances and

[184]"NASA STATEMENT: Two-Wheel Kepler Mission Invited to 2014 Senior Review." 4 Dec. 2013. National Aeronautics and Aerospace Administration 2 Apr. 2014 <http://www.nasa.gov/kepler/nasa-statement-two-wheel-kepler-mission-invited-to-2014-senior-review/>.

[185]"NASA's Kepler Marks 1,000th Exoplanet Discovery, Uncovers More Small Worlds in Habitable Zones." 6 Jan. 2015. NASA 22 May 2016 <http://kepler.nasa.gov/news/nasakeplernews/index.cfm?FuseAction=ShowNews&NewsID=367>.

[186]"Three Nearly Earth-Size Planets Found Orbiting Nearby Star." 16 Jan. 2015. NASA 22 May 2016 <http://kepler.nasa.gov/news/nasakeplernews/index.cfm?FuseAction=ShowNews&NewsID=370>.

[187]"Kepler-138b: a Mars-size exoplanet, now with a measured mass." 17 June 2015, NASA 22 May 2016 <http://kepler.nasa.gov/news/nasakeplernews/index.cfm?FuseAction=ShowNews&NewsID=390>.

[188]"NASA's Kepler Mission Discovers Bigger, Older Cousin to Earth." 23 July 2015. NASA 22 May 2016 <http://kepler.nasa.gov/news/nasakeplernews/index.cfm?FuseAction=ShowNews&NewsID=393>.

[189]"How Engineers Revamped Spitzer to Probe Exoplanets." 24 Sept. 2013. NASA 3 Apr. 2014 <http://www.spitzer.caltech.edu/news/1560-feature13-07-How-Engineers-Revamped-Spitzer-to-Probe-Exoplanets>.

unique vantage points. Spitzer has been used to view 22 other microlensing events in collaboration with OGLE and several other ground-based telescopes, and had viewed another 120 microlensing events by mid-2015.[190] Spitzer data also helped astronomers detect thermal emissions coming from the rocky exoplanet '55 Cancri e' which orbits a sun-like star about 40 light years away from Earth.[191] Moreover, Spitzer helped astronomers confirm the discovery of the nearest rocky exoplanet beyond the solar system transiting its star some 21 light-years away; the planet, HD 219134b, which is about 1.6 times the size of Earth, has a mass that is 4.5 times that of Earth, and orbits its star within 3 days.[192]

China launched its Dark Matter Particle Explorer (DAMPE) on 17 December 2015, to conduct a 3-year mission to shed new light on the nature of dark matter—a hypothetical kind of matter (along with dark energy) that could explain where the remaining estimated 85 % of the total mass-energy in the universe exists. Because it does not interact with ordinary matter and therefore cannot be observed directly, scientists have used three methods to investigate the elusive mass. The first method is through the use of particle accelerators, such as the Large Hadron Collider (LHC) at CERN physics research centre near Geneva, Switzerland, and accelerators in other parts of the globe, which 'create' anti-matter particles whose properties can be investigated. Another is through the use of special targets placed underground to see the traces left by colliding dark matter particles, including experiments by the China Jinping Underground Laboratory in Sichuan, China. The third investigation method measures the properties of particles that annihilate or decay in space, as with experiments conducted *inter alia* by the Alpha Magnetic Spectrometer (AMS) onboard the ISS, or the CALorimetric Electron Telescope (CALET) recently attached to Japan's Kibo module. The DAMPE experiment helps to push forward the third method by increasing the observation spectrum (i.e. nine times wider than the AMS), as well as the energy resolution (i.e. at least three times higher than international peers). Following a 500 km sun-synchronous orbit, the 1900 kg satellite will observe the direction, energy, and electric charge of high-energy particles in space. In the initial 2 years of its mission, DAMPE will scan space in all directions, followed by another year or more that will focus on areas where the potential signatures of dark matter are most likely to be observed.[193]

---

[190] "NASA's Spitzer Spots Planet Deep Within Our Galaxy." 14 Apr. 2015. NASA 22 May 2016 <http://www.spitzer.caltech.edu/news/1746-feature15-05-NASA-s-Spitzer-Spots-Planet-Deep-Within-Our-Galaxy>.

[191] "Astronomers find first evidence of changing conditions on a super Earth." 5 May 2015. University of Cambridge 24 May 2016 <http://www.cam.ac.uk/research/news/astronomers-find-first-evidence-of-changing-conditions-on-a-super-earth>.

[192] "NASA's Spitzer Confirms Closest Rocky Exoplanet." 30 July 2015. NASA 22 May 2016 <http://www.spitzer.caltech.edu/news/1792-ssc2015-02-NASA-s-Spitzer-Confirms-Closest-Rocky-Exoplanet>.

[193] Yue, Huang. "China Launches Country's First Dark Matter Satellite." 17 Dec. 2015. CRI English News 22 Jan. 2016 <http://english.cri.cn/12394/2015/12/17/4201s908753.htm>.

India successfully launched its first astronomy satellite, Astrosat, on 28 September 2015. Launched by ISRO's PSLV launcher, the 1513 kg satellite was lifted to a 650 km orbit altitude, inclined 6° relative to the equator. Over the next 5 years, the satellite will provide optical, ultraviolet, and X-ray images of black holes and other related phenomena.[194]

2.4 Satellite Applications

2.4.1 Space-Based Communications

In the lead-up to the World Radiocommunication Conference (WRC-15) held near the end of 2015, the lower part of the C-band frequency spectrum (3.4–3.8 GHz) appeared to be at risk of being allocated to terrestrial wireless broadband networks, while the upper part of the spectrum (3.8–4.2 GHz) seemed to be uncontested.[195] Due partly to its lower susceptibility to signal degradation in heavy rain than higher frequencies, the band is prized both by emerging markets and industrial markets alike.[196] While the satellite industry uses the frequency band for communications and TV broadcasting services, in addition to supporting disaster and emergency communications using mobile satellite systems, the terrestrial wireless broadband networks want to use the band for smartphones and other mobile devices to provide terrestrial internet broadband service (i.e. International Mobile Telecommunication (IMT)-Advanced service).[197] Yet coexistence in the same frequency range is unlikely, as the stronger terrestrial signals will undoubtedly interfere with or drown out the weaker signals transmitted by satellites in orbit.

Prior to the WRC held in 2007, the 3.4–4.2 GHz range of the C-band spectrum had been exclusively reserved in the ITU for satellite use. However, after that 2007 WRC meeting, while the satellite industry had achieved a modicum of success in fending off the interest of the same terrestrial proponents, it was forced to share at least part of this spectrum to terrestrial networks within the borders of several dozen nations.[198] In

---

[194] De Selding, Peter B. "PSLV Rocket Launches India's 1st Astronomy Satellite, 4 Spire Cubesats." 28 Sept. 2015. SpaceNews 5 Jan. 2016 <http://spacenews.com/pslv-rocket-launches-indias-1st-astronomy-satellite/>.

[195] De Selding, Peter B. "Satellite Sector Spoiling for a Spectrum Fight at WRC-15." 23 Mar. 2015. SpaceNews 13 Jan. 2016 <http://spacenews.com/satellite-sector-spoiling-for-a-spectrum-fight-at-wrc-15/>.

[196] De Selding, Peter B. "European Satellite Operators Told To Be Happy with Half a C-Band Loaf." 29 Jan. 2015. SpaceNews 10 Dec. 2015 <http://spacenews.com/satellite-operators-told-to-take-half-a-loaf-of-c-band/>.

[197] Zaheer, Zahid. "Op-ed | The Battle for C-band." 17 Mar. 2015. SpaceNews 12 Jan. 2016 <http://spacenews.com/op-ed-the-battle-for-c-band/>.

[198] De Selding, Peter B. "Satellite-based Aircraft Tracking Joins C-band Fight on WRC-15 Agenda." 2 Apr. 2015. SpaceNews 22 Dec. 2015 <http://spacenews.com/satellite-based-aircraft-tracking-joins-c-band-fight-on-wrc-15-agenda/>.

anticipation of the upcoming second round of terrestrial demand for the spectrum, the voice of satellite operators became more united, led by the European Satellite Operators Association (ESOA) which expanded to include other disparate operators in the Middle East, Africa, and some from within Russia.[199]

From the beginning, satellite operators were at a disadvantage in arguing their case to exclusively retain their 0.800 GHz chunk of the C-band in the ITU, since they could not provide a clear picture of how many C-band antennas are in service in affected areas, due partly to the lack of registration requirements with government authorities of satellite antennas on ground. Yet for years, the satellite sector has said that tens of thousands of C-band receive-only antennas are in operation in Latin America, Africa and Asia.[200] To give support to the unquantifiable dependence on satellite C-band signals, Vietnam and Thailand submitted government testimonials to the ITU on the importance and prevalence of C-band satellite services in their countries, and urged countries in similar situations to do the same.[201]

The WRC-15 conference was held from 2 to 27 November, at the ITU in Geneva, Switzerland. With representatives from 163 countries in attendance, the main focus of the meeting centred on the issue of C-band spectrum. With the backing of governments in all represented regions, 3.4–3.6 GHz of C-band was opened to terrestrial networks within those regions (not including Asia-Pacific). However, the remaining upper share of the spectrum stayed within the exclusive domain of satellite networks.[202]

Other topics that were not resolved in the WRC-15 included U.S. and German desire for Ku- or Ka-band satellite spectrum for use on unmanned aerial vehicles (UAVs) on transoceanic or trans-continental routes. While the U.S. had rallied around 50 governments to support their position, concerns by other governments ranged from the amount of access it would give to civil airspaces, to safety, and to the difficulties in allocating fixed satellite spectrum to UAVs.[203] The WRC-15 also took no action in regard to the power output of constellations of non-geostationary satellites whose signals might interfere with signals from geostationary satellites. Based on rules enacted by the ITU in 2000, frequency priority is given to

[199]De Selding, Peter B. "Satellite Group Expands Ahead of C-band Fight at WRC." 16 Mar. 2015. SpaceNews 22 Dec. 2015 <http://spacenews.com/satellite-group-expands-ahead-of-c-band-fight-at-wrc/>.

[200]De Selding, Peter B. "Satellite Sector Spoiling for a Spectrum Fight at WRC-15." 23 Mar. 2015. SpaceNews 13 Jan. 2016 <http://spacenews.com/satellite-sector-spoiling-for-a-spectrum-fight-at-wrc-15/>.

[201]SpaceNews Editor. "Editorial | Mobilized for WRC-2015." 5 Aug. 2015. SpaceNews 13 Jan. 2016 <http://spacenews.com/editorial-mobilized-for-wrc-2015/>.

[202]De Selding, Peter B. "Satellite Industry Held its Ground at Global Spectrum Conclave." 27 Nov. 2015. SpaceNews 7 Jan. 2016 <http://spacenews.com/satellite-industry-held-its-ground-at-global-spectrum-conclave/>.

[203]De Selding, Peter B. "U.S., Germany Face Uphill Fight for UAV Spectrum." 13 Nov. 2015. SpaceNews 7 Jan. 2016 <http://spacenews.com/u-s-germany-face-uphill-fight-for-uav-spectrum/>.

geostationary satellites.[204] But with the growth of low Earth orbit constellations by SpaceX and OneWeb, and to a lesser extent O3b, there is a chance that signals by geostationary satellites will face interference from the higher signal output of those constellations.

2.4.2 Space-Based Positioning, Navigation and Timing Systems

The development of GNSS systems continued during this period, with the relevant actors increasing their efforts to complete their full satellite constellations.

While Russia's Glonass GNSS constellation is in the midst of fleet replenishment, no Glonass satellites were launched in 2015. In 2014, Russia successfully launched its GLONASS-M satellites on 23 March and on 14 June; it also launched its next-generation model GLONASS-K satellites on 30 November 2014. The -K series will succeed the -M predecessor, and will most likely incorporate significant technical improvements, including a new more accurate timing device and a non-pressurized structure, bringing operational performance close to U.S. and European standards. The constellation requires 24 operational satellites to provide complete global navigation coverage, with three sets of eight satellites operating on three orbital planes.[205]

On 17 December 2015, Europe inserted its 11th and 12th Galileo satellites into medium Earth orbit. While in 2014, one of the four in-orbit validation (IOV) satellites suffered irreparable power degradation from an unknown anomaly, and two satellites were placed in a bad orbit by a Soyuz Fregat upper stage, potentially limiting their complement of Galileo services, the European Commission will be able to debut initial Galileo services (i.e. open service, SAR, and PRS) with the remaining nine fully operational satellites by late 2016. An Ariane 5 ES launcher will place four satellites in orbit by November 2016, followed by two satellites on a Soyuz launcher around the same period, and two more Ariane 5 ES launches in mid-2017 and 2018, bringing the constellation to a total of 26 satellites. Galileo was designed to be a 30 satellite constellation, with 24 operational satellites, 6 in-orbit spares, and 2 more spares on the ground; however, to give the system a safety margin, the European Commission has recently opened an invitation to tender for 8 more Galileo satellites.[206]

---

[204] Foust, Jeff. "Low Earth Orbit Constellations Could Pose Interference Risk to GEO Satellites." 26 Oct. 2015. SpaceNews 24 Dec. 2015 <http://spacenews.com/low-earth-orbit-constellations-could-pose-interference-risk-to-geo-satellites/>.

[205] "Glonass System." Glonass.it 12 May 2012 <http://www.glonass.it/eng/glonass-story.aspx>.

[206] De Selding, Peter B. "Successful Galileo Launch Paves Way for Service Debut." 17 Dec. 2015. SpaceNews 6 Jan. 2016 <http://spacenews.com/successful-galileo-launch-paves-way-for-service-debut/>.

China began the third phase of its Beidou (Compass) satellite navigation system when it launched its first next-generation Beidou satellite into inclined geosynchronous orbit on 31 March 2015.[207] As the satellite tested its new navigation signals and inter-satellite links, it was joined by two more Beidou 3 satellites in MEO orbit on 25 July 2015, followed by another geostationary Beidou-3 on 29 September 2015. By the end of 2015, 18 Beidou satellites were in operation providing regional navigation coverage; the Beidou GNSS system will expand to global coverage as it deploys its entire 35 satellite constellation by 2020. As Beidou signals expand into new markets and are marketed on commercial navigation devices, their non-integration with GPS-receivers may make GNSS use more chaotic and result in negative economic impacts for the industry overall.[208] While Beidou GNSS receivers can also receive signals, GPS, Glonass, and Galileo, accessing Beidou signals on Western GNSS receivers, may prove more challenging, unless integrated without regional protectionist measures.[209]

Japan's regional Quasi-Zenith Satellite System (QZSS) will likely expand to a seven satellite constellation, from its previous four satellite configuration, following Japan's Cabinet Office budget for fiscal year 2015, which earmarked 61 % of its ¥22.3 billion budget for the expansion.[210] Whereas the QZSS system was initially designed to augment the regional accuracy of the GPS signal, talks on 10 December 2015 focussed on Japan becoming a full member of the U.S. GPS system on a more equal basis in the form of a new cooperation framework—which was still being coordinated by the end of 2015.[211] The move is meant to strengthen Japan's position in light of China's expansion in the Asia-Pacific maritime region, and North Korea's missile launch activity.[212]

India launched its fourth India Regional Navigation Satellite System (IRNSS-1D) on 28 March 2015. The satellite joined the nearly identical IRNSS-1A and IRNSS-1B in inclined geosynchronous orbits; IRNSS-1C was placed in

---

[207]Ferster, Warren. "China Launches Next-gen Beidou Navigation Satellite." 4 Apr. 2015. SpaceNews 8 Jan. 2016 <http://spacenews.com/china-launches-next-gen-beidou-navigation-satellite/>.

[208]Johnson-Freese, Joan. "Found in Space: Cooperation." 9 Oct. 2015. China US Focus 22 Jan. 2016 <http://www.chinausfocus.com/foreign-policy/u-s-china-space-cooperation-a-welcome-dialogue-begins/>.

[209]De Selding, Peter B. "China Official: Beidou Gear Will Receive GPS, Glonass, Galileo Signals." 6 Feb. 2015. SpaceNews 12 Jan. 2016 <http://spacenews.com/china-official-beidou-gear-will-receive-u-s-russian-and-european-gnss-signals/>.

[210]Rajagopalan, Rajeswari Pillai. "Op-ed | Japan's Space Policy Shift Reflects New Asian Realities." 23 Feb. 2015. SpaceNews 10 Dec. 2015 <http://spacenews.com/op-ed-japans-space-policy-shift-reflects-new-asian-realities/>.

[211]Foust, Jeff. "Japan Seeks To Become Full Partner with U.S. in Space." 11 Dec. 2015. SpaceNews 7 Jan. 2016 <http://spacenews.com/japan-seeks-to-become-full-partner-with-u-s-in-space/>.

[212]GPS World staff. "QZSS May Expand to Meet Japan's Surveillance Needs." 22 Aug. 2014. GPS World 20 May 2015 <http://gpsworld.com/qzss-may-expand-to-meet-japans-surveillance-needs/>.

geostationary orbit in late 2014. The seven satellite constellation is expected to be completed by mid-2016, composed of two satellites (and two spares) in geosynchronous orbit, and three satellites parked in geostationary orbit. When fully deployed, the 14.2 billion rupee system is expected to operate for 10 years.[213]

2.5 Technology Developments

New developments in space-related technologies merit discussion, as they indicate current trends in space technology and reveal the focus of relevant policy decisions. The following chapter provides an overview of recent activities, spanning all major space faring powers and institutions, both established and emerging ones.

2.5.1 Propulsion

On 16 December 2015, ISRO successfully tested the PSLV upper stage ability to relight after deploying six satellites into LEO orbit. In the test, the launcher's fourth stage engine was briefly reignited, firing for nearly 5 s. While launch systems such as ESA's Ariane 5 and the ILS Proton M already have a dual launch capability, enabling the release of payloads into the same orbit, ISRO aims to harness the capability of lifting multiple payloads into space and placing them in different orbits. Mastering that technology is no easy feat, since it is very hard to cool down a rocket engine in a short time before reigniting it once again. After 17 min into the launch, the upper stage was shut down for a 50 min cooling interval before being ignited once again.[214]

2.5.2 Spacecraft Operations and Design

Despite the notable success of the European Space Agency's (ESA) Rosetta mission in converging with and observing Comet 67P/Churyumov-Gerasimenko and its successful landing of Philae on its surface in November 2014, the spacecraft continues to face seemingly insurmountable challenges as it captures multiple views of the comet and tries to position itself to potentially receive further signals

[213] Jayaraman, K.S. "India Launches Fourth Navigation Satellite." 30 Mar. 2015. SpaceNews 1 Jan. 2016 <http://spacenews.com/india-launches-fourth-navigation-satellite/>.

[214] "Isro Successfully Tests Multiple Burn PSLV Engine on Wednesday." 17 Dec. 2015. Gadgets360° 17 Feb. 2016 <http://gadgets.ndtv.com/science/news/isro-successfully-tests-multiple-burn-pslv-engine-on-wednesday-779156>.

from Philae. By late March 2015, as the comet continued toward its 13 August 2015 perihelion with the Sun, Rosetta underwent a series of increasingly close flyby manoeuvres to examine the increasing amount of dust and gas ejected from the comet as its surface temperature rose. On one risky manoeuvre on 28 March 2015, which brought Rosetta to within 14 km of the comet's surface, its close proximity to the growing cloud of dust particles resulted in Rosetta mistaking them for false stars, disorienting the spacecraft and breaking its high-gain antenna's communication link with ESA. Rosetta's signal link with ESA returned as it orbited further away from the comet, but as it continued to encounter the false-star readings, Rosetta was briefly placed in safemode, which shut down its scientific instruments as its navigation instruments were being reconfigured. By 30 March 2015, Rosetta was taken out of safemode and relocated to an orbit of around 400 km from the comet, before being manoeuvred to an orbit of 140 km from the comet in the following weeks.[215]

On 13 June 2015, the Philae lander awoke again from its protracted hibernation on Comet 67P/Churyumov-Gerasimenko, and transmitted fresh data to ESA and the DLR for a total of 85 s. While the comet journeyed closer to the Sun, the lander had awoken earlier and had been collecting data, but was unable to communicate with Rosetta until then.[216] Remarkably, Philae's foibled landing in a shadowed enclave on the comet had the unexpected benefit of sheltering the lander from high temperatures as the comet travelled toward the Sun. Had Philae landed in its intended spot temperatures reaching upward of 40 °C would have permanently ended Philae's operation by March 2015—long before the comet's closest encounter with the Sun in August 2015. In Philae's case, temperatures ranged from −150 °C at some points to reach −36 °C in June (comfortably above the minimal temperature of −45 °C it needs for operation). While it began to transmit several tens of seconds of communications with Rosetta, amid some extended gaps in communication, that connection had increased to 20 min by 9 July 2015.[217] However, in the following weeks, increasing amounts of gas and particles ejected from the defrosting comet likely affected Philae's position or changed the orientation of its transmitters as its transmission intervals after 9 July became irregular.[218]

[215]De Selding, Peter B. "Rosetta Survives Close Encounter with Comet Debris." 3 Apr. 2015. SpaceNews 8 Jan. 2016 <http://spacenews.com/rosetta-survives-close-encounter-with-comet-debris/>.

[216]De Selding, Peter B. "Philae Lander Reawakens on Comet." 14 June 2015. SpaceNews 11 Jan. 2016 <http://spacenews.com/philae-lander-reawakens-on-comet/>.

[217]De Selding, Peter B. "Scientists: Philae Landing Mishap Blessing in Disguise as Probe's Life Extended." 17 June 2015. SpaceNews 11 Jan. 2016 <http://spacenews.com/scientists-philae-landing-mishap-blessing-in-disguise-as-probes-life-extended/>; De Selding, Peter B. "Philae Comet Lander Reawakens, Phones Home." 10 July 2015. SpaceNews 11 Jan. 2016 <http://spacenews.com/philae-comet-lander-reawakens-phones-home/>.

[218]De Selding, Peter B. "Philae's Comm Problem has Lander's Operators Concerned." 20 July 2015. SpaceNews 11 Jan. 2016 <http://spacenews.com/philaes-comm-problems-has-the-landers-operators-concerned/>.

2.5.3 Suborbital Activities

On 23 November 2015, Blue Origin conducted the first successful test flight of its New Shepard suborbital launcher. Travelling at speeds reaching Mach 3.72, the New Shepard launcher reached its apogee at 100.5 km, followed by a powered vertical landing; the launcher's unoccupied crew capsule had also separated at its peak and parachuted to a landing. A similar test was conducted on 29 April 2015, which reached an altitude of 93 km but failed to make a controlled landing due to a hydraulic problem with the propulsion module.[219] Additional test launches are planned for 2016, followed by manned tests, and possible commercial service soon afterward.

Soon after Blue Origin's historical first, on 21 December 2015 SpaceX conducted the first successful landing of the first stage of its two-stage Falcon 9 launcher. Unlike Blue Origin's suborbital test, the Falcon 9 mission—which also delivered 11 commercial Orbcomm OG2 satellites into orbit—travelled at speeds reaching Mach 10, and involved a series of three "boostback burns," to return the first stage to its landing pad in Cape Canaveral, Florida.[220] Prior to this success, SpaceX had conducted two unsuccessful controlled vertical landing attempts on a floating barge in 2015, the attempt in January 2015 ending in an explosion as the booster had run out of the hydraulic fluid needed to control the launcher's grid fins, while the second attempt in April 2015 landed too hard onto the drone barge.[221] Despite the 'black swan' reusable launcher event, an initial assessment by European rival Arianespace in October 2015 highlighted that the cost in energy of returning the stage, the launcher's refurbishment, and the fact that reuse means a smaller production run and thus higher per-unit costs would require SpaceX to keep a launch rhythm of 30 times per year to justify the work.[222] Moreover, the Falcon 9's marketability and its refurbishment reliability are other concerns, especially for a customer that must balance the unknown risk of launching a satellite on a Falcon 9 that has been reused several times over, rather than being the launcher's first payload.[223]

---

[219] Foust, Jeff. "Blue Origin Flies — and Lands — New Shepard Suborbital Spacecraft." 24 Nov. 2015. SpaceNews 7 Dec. 2015 <http://spacenews.com/blue-origin-successfully-flies-new-shepard-suborbital-vehicle/>.

[220] Wall, Mike. "Wow! SpaceX Lands Orbital Rocket Successfully in Historic First." 21 Dec. 2015. Space.com 7 July 2016 <http://www.space.com/31420-spacex-rocket-landing-success.html>.

[221] Lynn, Amy. "Why the reusable rocket competition between SpaceX and Blue Origin isn't actually a competition." 29 Jan. 2016. Science Alert 7 July 2016 <http://www.sciencealert.com/here-s-why-the-reusable-rocket-competition-between-spacex-vs-blue-origin-isn-t-actually-a-competition>.

[222] De Selding, Peter B. "With Revenue Looking Up, Arianespace Seeks To Bring Ariane 5 Costs Down." 21 Oct. 2015. SpaceNews 13 Jan. 2016 <http://spacenews.com/with-revenue-looking-up-arianespace-seeks-to-bring-ariane-5-costs-down/>.

[223] De Selding, Peter B. "With Eye on SpaceX, CNES Begins Work on Reusable Rocket Stage." 5 Jan. 2015. SpaceNews 30 Dec. 2015 <http://spacenews.com/with-eye-on-spacex-cnes-begins-work-on-reusable-rocket-stage/>.

2.5.4 Other Technologies

Developments in general technology and science continued to advance in 2015, with implications reaching beyond the space sector.

The UK is funding 90 % of ESA's 70 million euros commitment to developing a new generation satellite payload called Eutelsat Quantum, which will bring satellites closer to becoming fully software defined. Normally the practically frozen designs of satellites make manufacturers hard pressed to keep up with the generational leaps taken by terrestrial telecommunications technology. The Eutelsat Quantum payload is intended to change that with the ability to change a satellite's coverage and on-board power with respect to how demand for the satellite develops over an average 15-year lifecycle. For instance, a flexible payload on a large communications satellite could be used to provide capacity to different regions without having to change its orbital slot; it could also be used to follow a single or multiple drones, ships, or aircraft, depending on the needs of the customer. The contract for the Eutelsat Quantum satellite was signed on 9 July 2015, valued at 180 million euros; however Eutelsat's contribution was not made clear at the time. The Eutelsat Quantum is expected to launch in 2018.[224]

By August 2015, NASA's lunar rover prototype, Resource Protector (RP15) had completed a series of tests designed to simulate an entire terrestrial mission. RP15's hardware was built by the developers of NASA's Robonaut, while its software came from NASA's Lunar Atmosphere and Dust Environment Explorer Mission. The tests involved the use of its drill on the ground, the collection of samples, and other distributed operations. Expected to cost less than $250 million, the solar-powered rover will carry a neutron spectrometer to dig for water ice deposits in the polar region of the Moon's surface; it will also make brief excursions into the permanently shadowed regions in search of ice deposits, to be analysed on return to sunlit regions. With the RP15 now entering Phase B, NASA aims to carry out the mission by 2020, and is exploring potential partnerships either with JAXA using a crushable pad landing system (eliminating the need for landing legs), or by partnering up with one of the competitors in the Lunar X Prize competition.[225]

---

[224]De Selding, Peter B. "Eutelsat, ESA Taking a 'Quantum' Leap Toward Fully Software-defined Satellite." 9 July 2015. SpaceNews 13 Jan. 2016 <http://spacenews.com/eutelsat-esa-taking-a-quantum-leap-toward-fully-software-defined-satellite/>.

[225]Foust, Jeff. "NASA Tests Lunar Rover Prototype with Eye Toward Flying Real Thing." 21 Sept. 2015. SpaceNews 11 Jan. 2016 <http://spacenews.com/nasa-tests-lunar-rover-prototype-with-eye-toward-flying-real-thing/>.

Part II
Views and Insights
Edited by Arne Lahcen

Chapter 3
Issues and Perspectives on Space Launch Vehicles' Development

Jean-Marc Astorg

3.1 Introduction

The genesis of orbital launch vehicles finds its roots in the ballistic missiles developed in the 1950s by the United States and the Soviet Union. In the wake of the Cold War, the two superpowers transformed their missiles into space carrier rockets, and then plunged themselves into a frenetic race to the Moon. With the development of commercial space applications in the 1970s—first telecommunications, then Earth observation and meteorology and, finally, navigation—launchers evolved into an area of predominantly commercial battles fought over by both public and private operators. The interests of states remain, however, closely intertwined, with launchers ultimately serving as instruments of political sovereignty to ensure independent access to space, a prerogative, so far, of only a few great powers. Space carrier rockets are extremely complex systems requiring a seamless mastery of advanced technologies, including solid or liquid propulsion, pilotage and guidance, light structures, and automatic operations. Countries that decide to provide themselves with an indigenous launch capacity are fully aware of the long-term investments required and of the need to support an industry that has a limited market.

In 1973 France, which in 1965 became the third space power with the launch of Asterix by the Diamante rocket, proposed to its European partners the development a new rocket named Ariane to provide Europe with an autonomous access to space. The need for such autonomy was evident in light of the U.S. refusal to launch the Franco-German Symphonie telecommunication satellite, if it was for commercial purposes. With the successful launch of Ariane 1 on 24 December 1979, Europe enjoyed the availability of a heavy-lift launcher to satisfy its internal needs and

J.-M. Astorg (✉)
CNES, Paris, France
e-mail: jean-marc.astorg@cnes.fr

C. Al-Ekabi et al. (eds.), *Yearbook on Space Policy 2015*, Yearbook on Space Policy, DOI 10.1007/978-3-7091-4860-0_3

compete on the emerging international market. In 1980, the Arianespace undertaking was created by CNES to commercialise Ariane's services and, thanks to the commercial fiasco of the Space Shuttle, to which the US had turned all its efforts, Ariane's commercialisation swiftly proved successful.

Since then, Europe has affirmed itself as a leader in the provision of commercial launch services, particularly owing to the steady commitment of European countries in supporting the construction a robust industry and the steady development of the subsequent generations of Ariane's rocket family. The most recent evolution is the heavy-lift launcher Ariane 5, which is capable of placing two heavy satellites of 5 tonnes each into geostationary orbit. In recent times, however, the satellite launch sector has been experiencing profound transformations that have forced Europe to react in order to defend its prominent position.

3.2 The Current Range of European Launcher: Ariane 5, Soyuz and Vega

Currently, Arianespace operates three launch vehicles from the Guiana Space Centre:

- The heavy-lift launcher Ariane 5, which entered into service in 1996, has so far achieved 72 successful launches in a row. It is designed to place two heavy telecommunications satellites into geostationary transfer orbit. It performs 7–8 launches a year with noteworthy reliability and quality of service, although at a price clearly above current market conditions.
- The medium-lift launcher Soyuz ST, for which Europe built a launch complex in French Guiana following the cooperation agreement signed with Russia in 2003. Operational since 2011, it has performed 15 launches from the Guiana Space Centre, and is fundamentally purposed to launch the Galileo satellite navigation constellation and other European institutional satellites (Earth observation, science). Following the Russian Federation's decision to build a new launch base in Vostochny and the decision to not renew the agreement for exclusive commercial exploitation accorded to Arianespace, a climate of uncertainty weighs on the future of Soyuz from French Guiana. It performs 2–3 missions a year.
- The lightweight Vega, whose development in 1997 was strongly advocated by Italy, realised its maiden flight in 2012 and has had six successful launches since then. The Vega rocket is specifically designed to launch commercial or scientific missions with a mass lower than 1.5 tonnes in Low Earth Orbit. It is launched 2–3 times a year.

With this array of launch vehicles, Arianespace currently enjoys the availability of a fleet covering almost the full spectrum of the launch market in terms of both mass and orbits. The company remains the leading service provider on the open commercial market. This is a market of roughly 25 satellites per year, against the 75 satellites launched on a captive basis by such countries as the Unites States, Russia, China, etc. It is, however, noteworthy that in Europe there is no clear preference for the launch of European institutional satellites. The lack of a predictable launch business base has proved to be Ariane's Achilles heel and has put the company on an uneven footing vis-à-vis its competitors, which on the contrary enjoy a guaranteed internal market with advantageous conditions for their own operators.

Against this constraining backdrop, in 2009 France decided to propose a renewal of the European launchers' fleet. Such a renewal was essential in the light of Ariane 5's primary focus on the commercial market, of the ensuing need to reduce launch prices, and of the unfeasibility of upgrading the rocket so as to further reduce such prices. In 2009, the idea of a new European launcher—Ariane 6—was formalised, but it was only in 2014 that its development was officially sanctioned thanks to the impetus of France and in the wake of increasing commercial competition.

3.3 The Current Global Context: Fierce Competition and a Guaranteed Captive Market for Ariane's Competitors

Owing to the development of new launch vehicles throughout the world, in recent years the launch service sector has indeed witnessed a hardening of competition.

The main threat has come from California's Silicon Valley with the creation in 2002 of a new company devoted to the development of new low-cost launchers for the future colonisation of Mars: SpaceX. Initially snubbed by its future competitors, but strongly supported by NASA, SpaceX nearly disappeared in 2008 due to three consecutive failures of its Falcon rocket, but has nowadays positioned itself as a major player in the sector, with 19 launches of its new heavy-lift launcher Falcon 9 (of which one failed in June 2015). Thanks to sound design choices, simplified architecture, centralised and streamlined production, and significant support from the U.S. government through the procurement of institutional launches paid well above current market conditions, the launch price of Falcon 9 on the commercial market is lower than that of Ariane 5 by a factor of two. This poses a major threat to Ariane 5, which is almost exclusively devoted to meeting commercial market needs. In the span of a few years, SpaceX has *de facto* replaced International Launch Services (ISL)—which commercialises the Russian heavy rocket Proton—as the main competitor of Arianespace.

International competition is, however, not limited to Space X. Shaken by the achievements of SpaceX, the United Launch Alliance (ULA), the traditional American operator for U.S. institutions (DoD, NASA), has decided to kick-start the development of a new rocket, Vulcain. It is expected to enter into service in 2020 and to simultaneously replace both the Atlas 5 launcher (hindered by the utilization of the Russian RD180 rocket) and the Delta 4, which remain, however, excellent launchers in terms of reliability.

Russia on its side, has brought into service the new launch base of Vostochny and continues to pursue the development of its Angara family of launchers, while maintaining a tailored pricing policy for its Proton heavy rocket on the commercial market, despite regular failures.

In Asia, Japan has decided to develop a new launch vehicle, H-3, with the declared objective of halving the costs of the current H-2 rocket, while China successfully continues to pursue the renewal of its Long March family. The two new versions, the Long March 7 and the Long March 5, could become formidable competitors were China to have access to the satellite launch market using American components. India, on its side, is developing the new GSLV Mark 3 launcher with its own technologies.

In this context of intense worldwide renewal of launch vehicles, increasing competition and price drop, Europe has had to react to maintain its leading position on the commercial market. This was the major item on the agenda of the ESA Conference at Ministerial level that was held in Luxemburg in December 2014.

3.4 The Luxembourg Decision of December 2014

Thanks to France's impetus, and in the wake of the June 2014 announcement of the creation of a joint venture by Airbus and Safran, the two major industrial actors in the launch sector, the ESA Ministerial Conference of December 2014 eventually took some important decisions intended to provide Arianespace with the necessary tools to better cope with the increasing global competition: an upgraded launcher—Ariane 6—that will be 50 % cheaper than Ariane 5, and a guaranteed market of five institutional launches per year at European level. At the Ministerial, ESA Member States also decided to proceed with the development of a new version of the Vega rocket: named Vega-C, it will be more efficient and will better meet the market needs of European small satellites. Ariane 6 and Vega C will make use of a common engine to reduce costs.

The Ariane 6 concept proposed by Airbus Safran Launcher (ASL) and adopted by the 12 participant states (with France financing 52 % of the 3.4 billion euros of development costs), is not particularly innovative if compared to Ariane 5: it will make use of a number of heritage elements, although it will be built by ASL and its partners under a completely new and rationalised industrial organization. Replacing the scattered production activities that were agreed at the beginning of the Ariane 5 programme in 1987, the production of Ariane 6 will be concentrated in clusters of excellence where industrial capacity will be fully utilised so as to reduce fixed production costs. A new launch complex will be built by CNES in Guiana to reduce operational costs by halving campaign duration and significantly streamlining and automating operations. Finally, with the aim of further supporting cost reduction, Ariane 6 will feature some technological innovations that are the result of research activities financed by European space agencies since the beginning of the 2000s. Ariane 6 will come in two variants: Ariane 62, which will be used for the launch of European institutional satellites and Ariane 64, which like Ariane 5, will retain double launch capacity, but will be half as costly.

To attain these economic objectives, it has also been decided to introduce profound changes in the overall governance. New risk-sharing principles between the public and private sectors were approved and then implemented in the development contracts signed in August 2015. Industry has committed to bearing the risks relating to the commercial exploitation of Ariane 6 and to take responsibility for the launcher design, thus acquiring greater autonomy during the development phase in order to better execute the exploitation stage.

The Ariane 6 programme will be subject to a final review in September 2016, but its development is currently well underway with the target of a maiden flight in 2020. Yet, the notion of a European preference clause for launch services remains to be developed by key European countries and the European Commission. This is exactly where the paradox of Europe's policy on access to space most starkly emerges: European countries invest several billion euros in the development of a launch vehicle, which is, however, not systematically used.

3.5 A Long-Term Trend: The Reduction of Costs for Accessing Space

Although with Ariane 6 and Vega C, Europe will posses a range of advanced and cost-effective launch vehicles, the current pressures to reduce costs for accessing space are likely to remain a major trend in the long run. The main tools currently pursued to reduce costs are: standardization of production, streamlining of industrial organization, widespread utilisation of 3D printing, automation of operations and processes, and partial or full reusability. At present, a number of operators are indeed betting on the reusability of some rocket stages to reduce costs. Reusability is not a new idea, as it has always appeared clear that building a system as complex as a launch vehicle for a 30 min mission is not an optimal solution. However, mastering the technologies required for returning a rocket stage is an extremely complex feat that requires considerable work. Not surprisingly, previous reusability efforts have all proven that expendable launchers remain the most economic solution. Even if some operators have nowadays successfully demonstrated the technical feasibility of reusability, it remains generally acknowledged that its economic viability can be only proven with through regular use over years and with an economic calculation that takes into account the real costs of recovery operations, the impact on launcher reliability and the technical performance of a reusable launch system.

Against this rapidly evolving scenario, France has defined its strategy vis-à-vis the future: investing in the development of engines much cheaper than those of

Ariane 6 thanks to the utilisation of new propellants and 3D printing. These low-cost engines will be eventually used to equip future European launchers, which will be either expendable or reusable. This is the underlying objective of the Prometheus project that France recently proposed to its European partners.

The race to space thus continues at a global level with the objective of significantly reducing the costs of accessing space and fully seizing the opportunities offered by this ever-changing sector. Indeed, were a substantial reduction of costs to be achieved, a number of missions would become feasible and space programmes worldwide could indeed witness a dramatic upsurge. This is exactly what is at stake for space transportation in the coming years.

Chapter 4
Space at a Crossroads: Government Programmes Can Benefit from Commercial Space and Other Changes If Long-Standing Challenges Are Addressed

Cristina T. Chaplain and Laura D. Hook

4.1 Introduction

Government-built satellites provide vital capabilities to national security, government operations, the science community, and the economy. For many years, U.S. government-built satellites have paved the way in space technology advancements. They have also been expected to operate under harsher conditions than their commercial counterparts and to be much more secure, adding to the time and money it takes to develop, produce, and launch them. At the same time, the U.S. Government Accountability Office (GAO) has found that government space programmes have also been characterised by large cost overruns and schedule delays. On top of inherent risks, agencies have faced many challenges in producing realistic estimates, adhering to high standards for quality, and employing sound programme management practices. As troubled space programmes have become less tolerable in times of growing fiscal pressures, the U.S. government is beginning to embrace the use of commercially built satellites as well as commercial-like business practices. While "going commercial" seems attractive, particularly given the rise of many new innovative suppliers, GAO's work indicates that the U.S. government must overcome significant hurdles before it can successfully adopt new ways of procuring space capabilities.

C.T. Chaplain (✉) • L.D. Hook
U.S. Government Accountability Office, Washington, DC, USA
e-mail: ChaplainC@gao.gov; HookL@gao.gov

C. Al-Ekabi et al. (eds.), *Yearbook on Space Policy 2015*, Yearbook on Space Policy, DOI 10.1007/978-3-7091-4860-0_4

4.2 The GAO and Space

The GAO is an independent, nonpartisan agency that works for Congress. Often called the "congressional watchdog", the GAO investigates how the federal government spends taxpayer dollars. Most of the GAO's work is done at the request of congressional committees or subcommittees, or is mandated by public laws or congressional committee reports. For example, we have reviewed the largest space programmes from the U.S. Department of Defence (DOD) and the National Aeronautics and Space Administration (NASA) (such as DOD's Evolved Expendable Launch Vehicles (EELV), the Global Positioning System, and NASA's James Webb Space Telescope and human exploration programmes), as well as cross-cutting issues such as cost estimating, quality management, space protection, programme terminations, and acquisition best practices. These reviews have examined ways the government can better leverage the commercial sector as well as the effect trends in the commercial sector are having on government space acquisitions. In recent years, we have focused particularly on DOD's efforts to introduce competition and commercial-like contracting practices into the EELV programme as well as DOD's decisions on whether to break up large satellites into smaller ones and rely on commercial satellites to host government payloads, and NASA's efforts to introduce more commercial approaches for acquiring crew and cargo services. We have also been examining the challenges DOD faces in adopting commercial practices and procurement methods for communications bandwidth, ground stations, and other services.

By "going commercial", we generally mean the government partnering with commercial companies, working under commercial-like contracting arrangements, and/or choosing to buy a capability as a service from the commercial sector rather than developing and owning it. We tend to define commercial suppliers fairly broadly, to include new companies breaking into the space arena and putting their own capital at risk as well as companies that are more established and serve a mix of private and government customers. We generally do not consider traditional government space contractors in this category but sometimes these contractors have a commercial side to them. "Commercial-like arrangements" are those in which the government pays a fixed price for a capability and does not impose the standard technical and/or financial reporting requirements specified under the Federal Acquisition Regulation. Technology, design, safety, and security requirements, and other risks, as well as the availability of a product in the commercial marketplace, help determine whether such an arrangement should be applied. In space, "going commercial" can also mean that instead of acquiring a whole satellite, the government could develop a sensor, which would be flown on a commercial satellite, or procuring the same capability (bandwidth or ground control, for instance) in the form of a service from commercial providers (see Fig. 4.1). All of these concepts are being explored by the government today. Some have been tried before with mixed success.

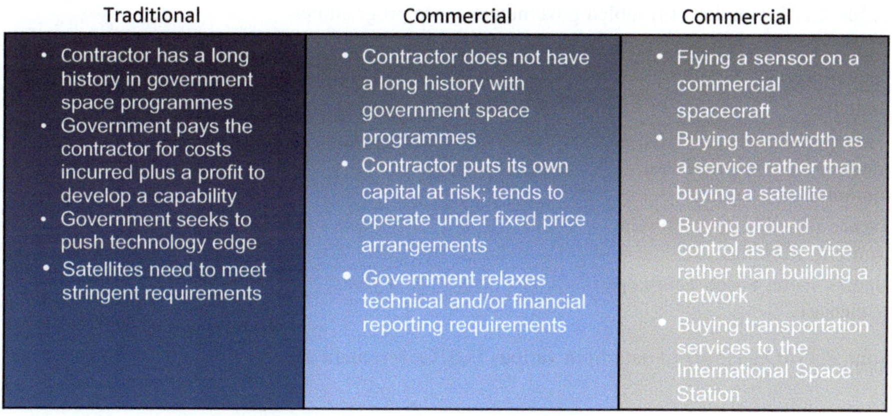

Fig. 4.1 Spectrum of traditional and commercial arrangements. Source: GAO analysis

4.3 "Going Commercial" Is Becoming an Attractive Option for the U.S. Government

Right now, the U.S. government is at a crossroads and the idea of adopting commercial approaches or relying on commercial satellites and launch providers has become more attractive. All agencies involved in space—NASA, DOD, etc.—have found it difficult to execute space programmes within budget and schedule, and not just because space can be more demanding than other large-scale development efforts. In some cases, schedule delays have added up to years, and cost overruns have been in the billions of dollars. Budgetary pressures make it more likely that troubled programmes will either be terminated or curtailed. They also make it difficult for agencies to initiate new programmes. When combined with the realisation that space is becoming more contested and that the commercial sector may offer viable alternatives to acquiring capabilities, all agencies involved in space are increasingly thinking about "going commercial".

While government space programmes have long been characterised by large cost overruns and schedule delays, the programmes that began in the early 2000s were particularly troublesome—first, because agencies took on more programmes than they could afford; second, because programmes were undertaken at a time when the government dramatically reduced its oversight in favour of empowering contractors; and third, because space programmes tended to be overly ambitious (Table 4.1).

The types of management and oversight problems the GAO found included: poor planning, optimistic cost estimating, funding gaps, lax oversight, poor contractor performance, parts quality problems, frequent programme manager turnover, and incentives that did not effectively incentivise contractor performance. Many programmes in the early 2000s were also hampered by the practice of starting programmes with immature technologies, which invariably created an environment

Table 4.1 Examples of troubled government space programmes

Satellite	Percent programme cost growth	Schedule delay	Examples of large-scale space programmes cancelled since 2000
Advanced Extremely High Frequency System (protected communications)	116	5 years	• Space Radar • Transformational Satellite Communications System • National Polar-orbiting Operational Environmental Satellite System • Constellation programme (Human spaceflight)
Space Based Infrared System (missile warning)	288	9 years	
James Webb Space Telescope (science)	77	4 years	

Source: GAO analysis of data from various U.S. Government agencies

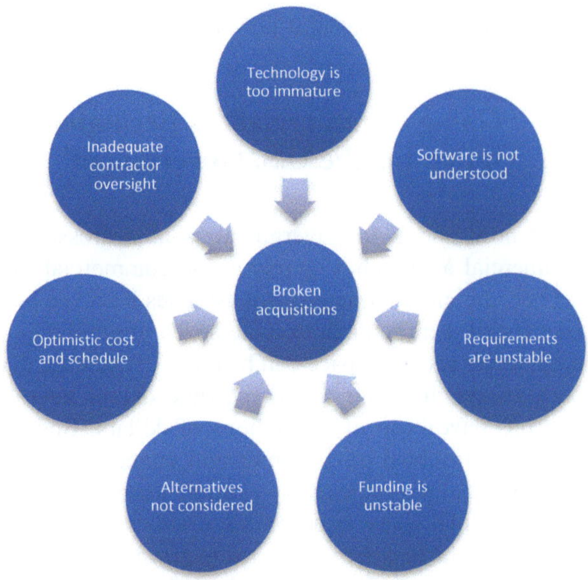

Fig. 4.2 Common problems identified by GAO. Source: GAO analysis

where technical and design problems were being addressed while the programme was moving through integration, testing, and even production. DOD satellites also tended to be monolithic—attempting to satisfy the needs of many and getting the greatest possible capability out of satellites in light of the high cost of launching them. While this approach helped prevent duplication, it further complicated satellite design and, without clear lines of authority, conflicts between satellite users created funding problems and hampered decisions on requirements. In commenting on our reports, space agencies tended to argue that inherent technical, design, and engineering risks made space projects costly and difficult. But more often than not, our reports found that management and oversight problems were the real drivers behind cost and schedule growth (see Fig. 4.2).

In recent years, space agencies have taken a step back, re-examined their practices, rededicated themselves to high quality standards, reformed cost estimating practices, embraced realism, scaled down programmes to make them more executable and embraced practices that bring more discipline to programme management. And in some cases, we have seen real progress. For instance, we reported in 2015 that NASA's portfolio of major projects—excluding the James Webb Space Telescope programme—experienced just a 2.4 % cost increase compared to a 19 % increase in 2010.[1] However, we still find programmes are experiencing the same types of problems seen in the past. In September 2015, for example, we reported that the Air Force has experienced significant difficulties in developing the Global Positioning System (GPS) Next Generation Operational Control System (OCX) and has consistently overstated progress to the Office of the Secretary of Defence compared to advisory independent assessments it received. At the time of our review, it needed $1.1 billion and 4 years more than planned to deliver OCX due to poor acquisition decisions and slow recognition of development problems.[2] In early 2015, we reported that NASA's Space Network Ground Segment Sustainment project's contractor provided unrealistic estimates that, despite project officials being aware of this issue prior to committing to the project, led to the need for a re-baselining shortly after the project was confirmed.[3] Specifically, the contractor underestimated costs for information technology infrastructure and overestimated its ability to meet its staffing requirements. As a result, the project's costs are expected to exceed the agency's committed baseline cost by 30 % and the project's committed final acceptance review date is expected to occur 23 months after the originally scheduled date, slipping from June 2017 to May 2019.

While acquisition problems have played into decision-making, they alone have not caused government agencies to consider relying on commercial space suppliers instead of traditional suppliers. For instance, growing threats to satellites from adversary attacks, such as anti-satellite weapons and communications jamming, and environmental hazards, such as orbital debris, have begun to change DOD's thinking about satellite acquisitions. With the realization that a single launch failure, on-orbit problem, or adversary attack on one of these large satellites could result in the loss of billions of dollars of investment and a significant loss of capability, DOD is now considering alternative approaches for acquiring space-based capabilities and plans to make decisions on acquisition approaches over the next 2–3 years. One such approach involves disaggregating, or breaking up, large satellites into multiple, smaller satellites or payloads.

---

[1]U.S. Government Accountability Office (2015) NASA: Assessments of Selected Large-Scale Projects. GAO, Washington, D.C.

[2]U.S. Government Accountability Office (2015) GPS: Actions Needed to Address Ground System Development Problems and User Equipment Production Readiness. GAO, Washington, D.C.

[3]U.S. Government Accountability Office (2015) NASA: Assessments of Selected Large-Scale Projects. GAO, Washington, D.C.

Table 4.2 Highlights of DOD's business case for disaggregating satellites

• Smaller, less complex satellites may shorten development and production schedules and avoid delays that contribute to cost overruns.
• A more continuous satellite production rate could reduce per-unit costs and provide a more predictable programme baseline.
• Smaller satellites may allow for lower-cost launch options, such as greater use of commercial launch vehicles, as available, and combining lighter, smaller payloads onto a single launch vehicle or attached to a host satellite.
• Less complex satellites may reduce risk in research and development, integration, and launch, thereby reducing overall costs.
• Building more, less-complex satellites may provide the Department of Defence DOD the opportunity to use commercial products and systems that have already been tested in the market. DOD may be able to use innovative business practices—such as fixed price contracting and service leasing—more frequently and potentially improve overall acquisition performance.
• DOD may be able to solicit competition separately for payloads and satellite buses to enable more tailored acquisition.
• Smaller programmes with stable production rates may allow for stable funding profiles, avoiding large funding spikes typical of large programmes that sometimes result in budget cuts to pay for other priorities in constrained budget environments.
• Demand for more satellites may stimulate new entrants and competition to lower acquisition costs.

Source: GAO analysis of data from various U.S. Government agencies

Another involves relying on commercial satellites to host government payloads. DOD proponents of these changes have also recognised that the complexity, size, and mass of today's satellites increases the inherent risk of developing them and that disaggregating satellites could reduce this risk. Disaggregation could also enable DOD to solicit competition separately for payloads and satellite buses, thus expanding the field of suppliers who can compete for programmes.[4] Other potential benefits cited by disaggregation proponents are highlighted below (Table 4.2).

Acquisition cost was a factor in establishing fixed price arrangements for the commercial crew and commercial cargo programmes, as such an arrangement enabled NASA to afford a more traditional programme to build larger rockets and crew capsules for long-distance space travel while spurring a new industrial base at the same time (Table 4.3).

[4]U.S. Government Accountability Office (2014) DOD Space Systems: Additional Knowledge Would Better Support Decisions about Disaggregating Large Satellites. GAO, Washington, D.C.

Table 4.3 NASA's commercial crew's approach for obtaining crew transportation systems

• NASA's engineers and aerospace specialists work closely with companies to develop crew transportation systems that can safely, reliably and cost-effectively carry humans to low-Earth orbit, including the International Space Station, and return safely to Earth.
• Interested companies are free to design the transportation system they think is best. For the contracts phase of development and certification, each company must meet NASA's pre-determined set of requirements.
• The companies are encouraged to apply their most efficient and effective manufacturing and business operating techniques throughout the process.
• The companies own and operate their own spacecraft and infrastructure.
• The partnership approach allows NASA engineers insight into a company's development process while the agency's technical expertise and resources are accessible to a company.

Source: NASA

4.4 Industry Changes Also Make "Going Commercial" More Attractive

The GAO has not reported on the growth of the commercial space industry as a whole, but we have reported recently that the U.S. commercial space launch market is expanding. According to Federal Aviation Administration (FAA) data, for example, over the last 3 years, orbital launch revenue generated by U.S. companies increased from none in 2011 to $1.1 billion in 2014.[5] More broadly, according to the Space Foundation, 2014 was a good one for the global space economy. Consisting of launch and ground services, satellite manufacturing, satellite television and communications, government exploration, military spending, and other interests, the global space economy grew by 9 % in 2014, reaching a total of $330 billion worldwide. The remainder was composed of government investments in space, which experienced a combined growth of 7.3 % in 2014.[6]

The commercial side is increasingly associated with "disruptors"—new groups or individuals that bring innovative, nimble, and responsive and sometimes radical approaches to developing space capabilities. These innovators may have funding sources that do not rely on the government. They also tend to favour contracting approaches that shift the risk from the government to the contractor but require a more hands-off role from the government. They are developing technologies that could produce cost savings, such as those that focus on reusability and new fuel types. These innovators tend to take approaches that can address DOD's concerns about resilience. For instance, some design their satellites to have shorter lives because they are easily replaced and frequently upgraded with technology refreshers (Fig. 4.3).

[5]U.S. Government Accountability Office (2015) Federal Aviation Administration: Commercial Space Launch Industry Developments Present Multiple Challenges. GAO, Washington, D.C.

[6]Space Foundation (2015) The Space Report. Space Foundation, Colorado Springs, CO.

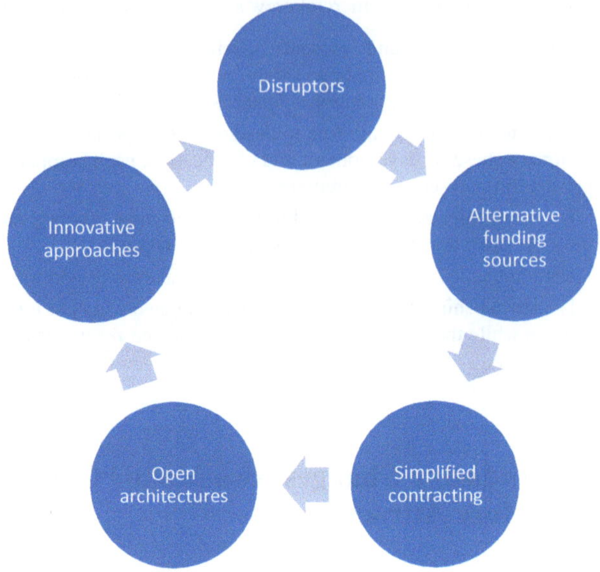

Fig. 4.3 What commercial space can offer government space. Source: GAO analysis

The GAO and others have pointed out numerous ways in which government space programmes can benefit from the products, practices, and approaches being developed or adopted by commercial space companies. In the area of satellite control operations, for example, we reported in 2013 that making future Air Force satellite control operations interoperable would allow one operator to use a single terminal to control numerous satellites, similar to commercial practices.[7] This could reduce costs associated with purchasing multiple types of software and training the operators on each system, as well as potentially reducing the number of staff required, since one person could operate multiple types of satellites more easily. One example put forward by an industry group study estimated that increasing interoperability and automation could allow one Air Force satellite control operations group to reduce its operations personnel by 45 %. We also suggested software products used by the commercial sector could benefit government satellite operations.

Both the GAO and the U.S. Department of Commerce have emphasised that placing a government payload on a commercial satellite could cost a fraction of the amount of building, launching, and operating an entire satellite.[8] In assessing decisions on disaggregation, we reported that the potential demand for more satellites may stimulate new entrants and competition to lower acquisition costs.

[7]U.S. Government Accountability Office (2013) Satellite Control: Long-Term Planning and Adoption of Commercial Practices Could Improve DOD's Operations. GAO, Washington, D.C.

[8]U.S. Government Accountability Office (2013) 2013 Annual Report: Actions Needed to Reduce Fragmentation, Overlap, and Duplication and Achieve Other Financial Benefits. GAO, Washington, D.C.

In assessing purchases of satellite bandwidth, we have repeatedly recommended that DOD adopt strategic sourcing techniques that are being adopted by leading commercial companies.[9]

Because the government is only beginning to adopt some of these practices and techniques, we have not assessed and substantiated savings and efficiencies in a comprehensive way. But we have seen a few examples of savings and potential savings. For instance, the Federal Aviation Administration's Wide Area Augmentation System involves two satellite-based sensors carried on commercial satellites. This hosted payload arrangement was designed to improve the accuracy of GPS signals for aircraft navigation and landing. The FAA conducted a lease versus buy analysis at the beginning of the programme and found that a lease would be more cost-effective than the purchase of a satellite, saving $260 million over the 21-year life cycle. Further, NASA was unable to quantify the exact cost savings from using hybrid networks for satellite control. One commercial company that provides services to NASA estimated the use of commercial networks reduced NASA's operations and maintenance cost by about 30 % with very low mission risk.

4.5 Without Addressing Long-Standing Challenges, Going Commercial Could Have Negative Consequences

Despite the potential for significant savings, there are still unresolved questions about whether some changes being considered offer a better path forward. For instance, government agencies have used commercial-like contracts in the past to procure launch services and satellites based on the assumption that the government could leverage predicted growth in the commercial market. When this growth did not materialise, contracting strategies had to be revamped to more traditional arrangements. But because the programmes began without government insight into costs, it was difficult to execute these arrangements later on. In some cases, the government attempted to apply commercial approaches to programmes that sought to make technology leaps only to find that contractors could not deliver results for the fixed amount they were being paid and that progress was difficult to track without the same level of insight that the government would have under traditional arrangements. Too often in space, "commercial" translated into the government taking a too hands-off approach, losing control over such things as requirements setting and quality assurance.[10] This history has made some agency and congressional officials hesitant to embrace newer efforts.

[9]U.S. Government Accountability Office (2013) Defense Satellite Communications: DOD Needs Additional Information to Improve Procurements. GAO, Washington, D.C.

[10]U.S. Government Accountability Office (2010) Briefing on Commercial and Department of Defense Space System Requirements and Acquisition Practices. GAO, Washington, D.C.

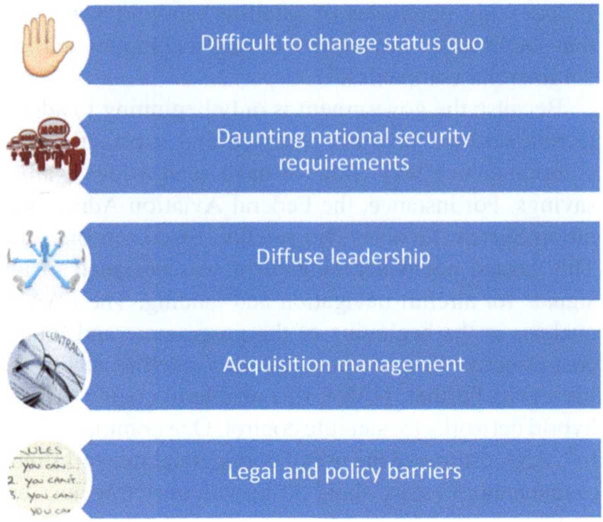

Fig. 4.4 Barriers to going commercial or making any significant changes to space acquisition management. Source: GAO analysis

There are also substantial barriers to adopting commercial-like approaches in government or engaging non-traditional suppliers. These barriers are more prominent on the military side and tie into the government's unique requirements, disconnects among the agencies involved with space, and long-standing acquisition problems (see Fig. 4.4). Addressing these challenges is essential for the government to successfully "go commercial".

Some of the new approaches being considered, such as disaggregating satellites, require the government to make significant cultural and process changes in how it acquires space systems—for instance by relying on new contractors, relinquishing control to providers who host government payloads on commercial satellites, using different contracting methods, and executing smaller but more numerous and faster-paced acquisition programmes. It will likely require the government to be more flexible and agile when it comes to satellite acquisitions, especially with regard to coordinating satellite delivery with interdependent systems, such as user equipment. Yet, we have reported that DOD's culture has generally been resistant to changes in space acquisition approaches and that fragmented responsibilities have made it very difficult to coordinate and deliver interdependent systems. At present, although some improvements in leadership have been made, there is still no single person or organization that is held accountable for balancing needs against wants, ensuring coordination among the many organizations involved with space systems acquisitions, and ensuring that resources are directed where they are most needed.[11] These challenges pose barriers for even the best-run efforts and they likely will be exacerbated if DOD decides to disaggregate or make other significant changes to its space systems.

[11]U.S. Government Accountability Office (2014) Space Acquisitions: Acquisition Management Continues to Improve but Challenges Persist for Current and Future Programmes. GAO, Washington, D.C.

Our past work has also found that businesses find it difficult to break into the DOD space system acquisitions environment because, among other reasons, the DOD space community is risk averse—DOD officials see unproven companies as risky for expensive space programmes because delays, problems with technologies, or other issues can have significant consequences. There is a perception that programme managers prefer to award space-related contracts to large contractors, in whose quality, practices, and longevity they have confidence.[12]

The government typically also has more demanding requirements than the commercial sector. For example, DOD uses radio frequencies resistant to atmospheric scintillation in polar regions and under challenging conditions such as in dense foliage and adverse weather. DOD typically has more demanding standards for radiation-hardened parts, such as microelectronics, which are designed and fabricated with the specific goal of enduring the harshest space radiation environments, including nuclear events. For some programmes, companies may need to create separate production lines and in some cases special facilities.[13]

Both DOD and NASA also remain on GAO's high-risk list because of problematic acquisition management. Unless the change being adopted takes the government out of development altogether (for instance, the government chooses to stop developing a certain type of satellite and instead buys the capability as a service from the private sector), poor management practices will likely hamper future programmes no matter which suppliers are involved.

In addition, we consistently find shortcomings in data needed to implement and validate new approaches. For example, at the time of our 2013 review, DOD was unable to identify all funding for satellite control operations across all DOD satellite programmes. Programmes have not needed to keep track of budgets by separating satellite control operations funding out from other satellite mission funding since the focus has been on dedicated ground control networks. However, without knowing how much it spends on basic satellite control operations for all of its satellites, DOD cannot calculate the potential savings or perform a cost/benefit analysis of any future changes to satellite control operations. Our reports have also pointed to numerous legal and policy challenges. For instance, DOD does not have authority to adopt techniques such as long-term leasing for procuring bandwidth. Federal law and policy also have limited the government's access to some hosted payload and ride sharing options.

[12]U.S. Government Accountability Office (2010) Space Acquisitions: Challenges in Commercializing Technologies Developed under the Small Business Innovation Research Programme. GAO, Washington, D.C.

[13]U.S. Government Accountability Office (2010) Briefing on Commercial and Department of Defense Space System Requirements and Acquisition Practices. GAO, Washington, D.C.

4.6 GAO's Recommendations Can Help Remove Some Barriers

Adopting GAO's recommendations would help remove the barriers relating to acquisition management that stand in the way of engaging new suppliers and adopting new approaches. Our recommendations span every phase of the acquisition process and are focused on such things as ensuring that decisions are based on the right knowledge, requirements are achievable within available resources, and risks are appropriately managed and mitigated.

We have recommended that requirements and resources (for example, time, money, and people) be matched at programme start. In the past, we have found that programmes, such as NASA's Constellation programme and Commercial Crew Programme, have not had sufficient funding to match demanding requirements. Funding gaps can cause programmes to delay or delete important activities and thereby increase risks, and can limit the extent to which competition can be sustained. Realistic cost estimates and assessments of technical risk are particularly important at programme start. Space programmes have historically been optimistic in estimating costs (although recently DOD and NASA have been making strides in producing more realistic estimates). The commitment to more realistic, higher confidence cost estimates would be a great benefit to any new development programme and enable Congress to ensure its commitment is based on sound knowledge. We have also found that imposing overly ambitious deadlines can cause an array of problems. For instance, they may force programmes to overlap design activities with testing and production. The many setbacks experienced by the Missile Defence Agency's ground-based midcourse defence system, for example, are rooted in schedule pressures that drove concurrent development. While our past recommendations focused on establishing a sound business case for each new programme, we have also suggested that more government-wide planning is needed to ensure new programmes have a healthy start. For instance, more robust science and technology planning and funding for future launch technologies might have helped avoid the U.S.'s prolonged reliance on the Russian-made RD-180 engines for the Atlas V launch vehicle.

We have also made recommendations aimed at adopting knowledge-based practices during execution. Programmes should use quantifiable data and demonstrable knowledge to make go/no-go decisions, covering critical facets of the programme such as cost, schedule, technology readiness, design readiness, production readiness, and relationships with suppliers. Our work has tied the use of such metrics to improved outcomes. In addition, programmes should place a high priority on quality, for example, holding suppliers accountable to deliver high-quality parts for their products through such activities as regular supplier audits and performance evaluations of quality and delivery, among other things. Prior to EELV, DOD experienced a string of launch failures in the 1990s due in large part to quality problems.

The GAO has also identified best practices for acquiring a space-based capability as a service. These would be applicable to acquiring launch services as well as satellite bandwidth. We have particularly emphasised the importance of comprehensive spend analysis for efficient procurement. Spend analysis provides knowledge about how much is being spent for goods and services, who the buyers are, who the suppliers are, and where the opportunities are to save money and improve performance. Private sector companies are using spend analysis as a foundation for employing a strategic approach to procurement. Companies do not take a one-size-fits-all approach to individual service purchase decisions.

4.7 Agencies Will Still Need to Fix Problems That Go Beyond Acquisition Management

Space agencies have already been adopting many of our acquisition management recommendations and while we have measured some improvements, it is clear more needs to be done to tackle the more elusive problems affecting government space programmes. For example, in DOD clearer lines of leadership and authority are needed to improve requirements setting, ensure ground systems get funding priority so they can be better aligned with satellite systems, and resolve conflicts between the many stakeholders involved with satellite acquisitions. Moreover, leaders need to be more actively involved in challenging resistance to new approaches and new suppliers. Lastly, as long as the government has unique requirements for its space systems, there will be barriers to commercial companies. In this case, if successful, NASA's commercial crew programme could serve as a model for other government programmes, as NASA is focused on breaking down requirements barriers by defining high-level system integration requirements for its performance-based contract. Overall, the barriers should not discourage the government from trying new approaches, but it is important for the government to make such changes incrementally and with the right knowledge about alternatives, costs, risks and feasibility. Our work in the next few years should help assess the extent to which this is happening and what more the government must to do to facilitate change.

Chapter 5
Virgin Galactic: Creating Safe, Reliable, and Frequent Access to Space

Richard DalBello

5.1 Introduction

Virgin Galactic, a privately funded company based in the United States of America, is on track to become the world's first commercial spaceline. Virgin Galactic is currently developing SpaceShipTwo, a reusable suborbital vehicle that will reach space altitudes on frequent, affordable, and safe suborbital voyages. SpaceShipTwo is being designed both for human astronauts and for automated payloads, with customers ranging from tourists to researchers and educators.

Virgin Galactic is also developing a space transportation service to provide an affordable, dedicated ride to orbit for smaller payloads. No longer will small satellite users be forced to make a choice between the limitations of flight as a secondary payload or paying dramatically more for a dedicated launch vehicle. Virgin Galactic's satellite launch vehicle, LauncherOne, is a two stage, air-launched liquid propulsion (LOX/RP) rocket. This system is designed to conduct operations from a variety of locations, allowing customers to select a wide range of launch windows and azimuths.

5.2 SpaceShipTwo

Virgin Galactic's missions to space are designed for affordability and it is our goal to transform the human exploration of space. Since the space programme began, only 552 people have flown in space (see Fig. 5.1). Currently, Virgin Galactic has nearly 700 confirmed customers raising the possibility that the roster of humans

R. DalBello (✉)
Virgin Galactic, Las Cruces, NM, USA
e-mail: Richard.DalBello@virgingalactic.com

C. Al-Ekabi et al. (eds.), *Yearbook on Space Policy 2015*, Yearbook on Space Policy, DOI 10.1007/978-3-7091-4860-0_5

Fig. 5.1 Global astronauts as of 2016

who have flown to space will very quickly be dominated by those who have flown through Virgin Galactic.

Based on the historic SpaceShipOne—which safely carried human beings into space in 2004 and won the $10M Ansari X PRIZE—Virgin Galactic's vehicles have been designed with the intention of setting new standards for human spaceflight safety, frequency, flexibility, and cost. Our suborbital spaceflight system consists of two vehicles: WhiteKnightTwo ("WK2", Fig. 5.2) and SpaceShipTwo ("SS2", Fig. 5.3). WhiteKnightTwo is a four-engine, dual-fuselage jet aircraft capable of a range of high-altitude, heavy lift missions. However, the WhiteKnight's primary mission is to serve as the mothership for the SpaceShip. WhiteKnightTwo's signature design features are its twin fuselage configuration; single-piece, carbon composite high main wing; and four jet engines. The twin boom configuration allows for a spacious central attachment area between the two fuselages for SpaceShipTwo. SpaceShipTwo is designed to safely and routinely cross the border between the Earth's atmosphere and space with a crew of two pilots and as many as six astronauts. SpaceShipTwo can also be configured as a research vehicle for a wide variety of subjects such as microgravity and upper atmospheric research.

SpaceShipTwo, VSS "Unity", was recently completed by The Spaceship Company in Mojave, California. Unity, will now begin a thorough and rigorous test flight programme before beginning operational flights from Spaceport America in New Mexico. SpaceShipTwo is powered by a hybrid rocket motor—one that combines elements of solid rockets and liquid rocket engines—that uses a solid fuel source and a liquid oxidiser. Hybrids are particularly well suited for SpaceShipTwo because they can be shut down safely and quickly at any point during the flight, unlike solid motors, but do not require the cryogenic propellant tanks or complex plumbing needed by liquid rocket engines.

Fig. 5.2 The WhiteKnightTwo aircraft is shown here during a test flight

Fig. 5.3 SpaceShipTwo "Unity"

SpaceShipTwo uses a unique 'feather' system to safely re-enter the Earth's atmosphere. The twin tail booms of the spacecraft can be rotated up, allowing the spacecraft to mimic the performance of a capsule or of a winged vehicle at the appropriate parts of its trajectory. Each of those vehicle types on its own offers important advantages: winged vehicles provide gentle deceleration and precision landing accuracy but require extremely high precision and advanced thermal protection systems. Capsules, on the other hand, require less precision and less advanced thermal protection, but subject astronauts to much harsher decelerations

Fig. 5.4 Mated vehicle pair (SS2 and WK2)

and less accurate landings. By changing its configuration in flight, SpaceShipTwo can benefit from the advantages of both types. In addition to its unique re-entry system, SpaceShipTwo's cabin has been designed to maximise safety and comfort and is designed explicitly to optimise its passengers' experience. A dozen windows line the sides and ceiling of the spacecraft, offering each spaceflight participant the ability to view the black skies of space as well as stunning views of the Earth below. Exposure to G-forces during SpaceShipTwo's ascent and descent is safely and comfortably managed thanks to systems such as our custom-designed, articulated seats, which are upright during rocket boost and reclined during re-entry (Fig. 5.4).

5.3 Human Spaceflight

Since the beginning of the space age, ordinary citizens have dreamed of experiencing outer space. As the world's first commercial spaceline, Virgin Galactic desires to lead this market. The provision of suborbital space experiences, for 'space tourism' and suborbital research, will be the company's first product to market.

Throughout five decades of human spaceflight, national space agencies have captured the attention of the world, inspiring untold numbers of entrepreneurs, inventors, ordinary citizens, and entire new industries. However, the government space agencies of the world are not empowered to provide the means for ordinary citizens to become astronauts. This means that until now, the world's more than seven billion inhabitants have had no means whatsoever to fly into outer space, regardless of their wealth or talents.

Virgin Galactic is dedicated to changing that paradigm by dramatically reducing the cost of safely taking humans into outer space, thereby opening the market to those people ineligible or uninterested in becoming professional government astronauts. Although the exact ticket prices for space tourism flights have not been publicly listed, the price paid for the most recent ticket sold has widely been estimated at US$52 million,[1] a number made credible by the publicly listed price per Soyuz seat paid by NASA, US$70 million.[2] Virgin Galactic's voyages to space are less than one half of 1 % of the government cost; and although the two voyages offer very different experiences in terms of flight duration, both allow customers to become astronauts, to experience true microgravity, and to see the Earth from space.

5.4 Microgravity Research

Microgravity environments are often associated with space-based applications, however, microgravity platforms were used for science and industrial applications long before the creation of national space programmes. From Galileo's use of the Leaning Tower of Pisa to observe objects in freefall to the construction of shot towers in the 1700s to create perfectly round lead shots for use in firearms, microgravity environments have been used for research and commercial applications. Even though free fall is still used to achieve weightlessness, additional platforms now exist that allow further access to these microgravity environments. Reusable Suborbital Launch Vehicles (rSLV) such as Virgin Galactic's SpaceShipTwo, will allow up to 4 min of high-quality microgravity without the costly price tag of on-orbit flights.

NASA's Flight Opportunities programme has already chartered flights of SpaceShipTwo in order to fly research payloads into space and researchers are proposing a wide variety of experiments for these vehicles. With SpaceShipTwo offering a new way to access the microgravity utilization environment, Virgin Galactic looks forward to partnering with NASA to open new frontiers in science, research, and technology.

[1]Klotz, Irene. "Singer Sarah Brightman to Train for Orbital Spaceflight." *Discovery News*. 5 September 2014. Web. 8 September 2014.

[2]Wall, Mike. "NASA to Pay $70 Million a Seat to Fly Astronauts on Russian Spacecraft." *Space.com*. 30 April 2013. Web. 8 September 2014.

5.5 LauncherOne

The private sector has played a dominant role in the development of new applications of the space environment. Today, tens of billions of dollars of private capital have been invested in the fields of satellite communications and earth observation. New commercial applications focused on areas such as weather data collection, satellite servicing, and in-space manufacturing are on the horizon.

Increasingly, commercial satellite service providers are looking to smaller satellites to solve traditional challenges. By harnessing the power of new generations of microprocessors and significant advances in power collection and propulsion, satellite engineers are devising ways to package robust instrument packages into smaller and smaller form factors. Many of these concepts have been successfully demonstrated in space using rideshare opportunities, hosted payloads, cubesats deployed from the International Space Station, or other similar, low cost means of accessing space. However, there is not yet a flexible, routine, and affordable way for small satellite operators to place their spacecraft in their desired orbits at prices commensurate with the budgets and business plans of such missions.

Virgin Galactic is leveraging the technical and business successes of its suborbital SpaceShipTwo programme to initiate a small satellite launch service. Virgin Galactic has made significant progress towards designing a new vehicle that will give satellite developers and operators a better option for carrying their small satellites into orbit. This new service is called LauncherOne ("L1", Fig. 5.5). Virgin Galactic's new LauncherOne vehicle is dedicated to the task of lowering the cost and increasing the frequency of space access for payloads in the 150–500 kg weight range.

The LauncherOne system consists of three major elements: the launch vehicle, carrier aircraft, and ground segment. The carrier aircraft is a modified 747-400 that

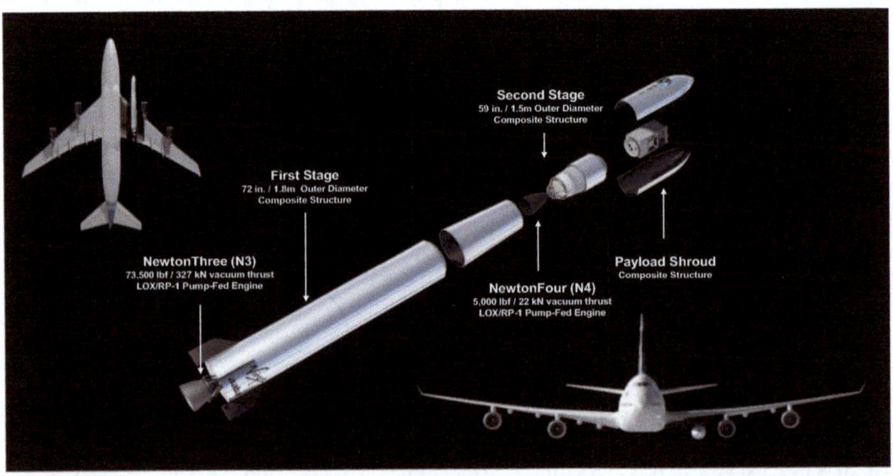

Fig. 5.5 Artist's concept showing LauncherOne

will carry the launch vehicle under the port (left) side wing between the fuselage and inboard engine. The carrier aircraft will provide electrical power, purge gasses, health monitoring, and control of the launch vehicle. The ground segment consists of equipment to load propellants on the launch vehicle, ground stations to gather and distribute telemetry, and a mission control centre to monitor the launch operations.

LauncherOne is an air-launched two-stage rocket that is carried to an altitude of approximately 35,000 ft by the carrier aircraft. Launching from a carrier aircraft instead of being tied to a fixed ground launch pad allows LauncherOne to operate outside the normal weather and fixed location constraints associated with ground launch systems. Freed from these constraints, LauncherOne missions can be tailored to meet the requirements of the customer in a quick and affordable manner. LauncherOne will use an autonomous flight termination system, so operations of the vehicle can be completely independent of existing launch range safety services.

Once released from the carrier aircraft, LauncherOne will fire its single main stage engine, a 73,500 lbf, LOX/RP-1 rocket engine. After stage separation, the single upper stage engine, a 5000 lbf LOX/RP-1 rocket engine will carry the satellite(s) into orbit. Both the main and upper stage engines are being designed, tested, and built by Virgin Galactic. At the end of this sequence, LauncherOne will deploy our customers' satellite (or satellites) into their desired orbit. Both stages of LauncherOne will be safely deorbited, while the carrier aircraft will return to a predetermined airport, where it can be quickly prepared for its next flight.

At present, Virgin Galactic is working towards initial test flights of the LauncherOne system in 2017. Virgin Galactic will operate LauncherOne under a Federal Aviation Administration (FAA) license through the Office of Commercial Space Transportation (AST). Initial launch sites include West Coast and East Coast sites within the United States. Additional launch site options are planned.

Chapter 6
The Future of Satellite Applications: The End-User Perspective

Stefaan De Mey

6.1 Benefits from Space Revisited

Serving primarily geostrategic and military objectives, space has traditionally been the playground of political giants. But it was also a field for major scientific successes and technological breakthroughs. In the early days of the space age the benefits from space for citizens and the economy received less attention, although it is not that society did not reap massive benefits. Meteorology without space is now unthinkable. Military GPS gradually finding its way into the civilian world brought about remarkable change in how we move around and how we do business. And any textbook on space applications will rightfully praise the commercial success and pervasiveness of Satcoms.

Today, bringing the benefits of the investments in space back to the citizens is becoming more and more an objective in itself. A new Space Policy for Europe is one of the 23 key European Commission initiatives for 2016. In early 2016, European Commission (EC) Commissioner Elżbieta Bieńkowska announced an ambitious strategy for the uptake of space data and services in the private and public sectors.[1] It will be an emerging priority for the EC, to be incorporated in EC's new Space Policy. Whether or not this is driven by a necessity to better justify the public investments in space in the wake of the financial and economic crises, is of lesser importance. The fact remains that satellite based applications have reached a

[1]Bieńkowska Elżbieta (2016) Welcome message by the European Commissioner for Internal Market, Industry, Entrepreneurship and SMEs, European Commission. Presented at the 8th annual Conference on European space policy. Europe as a global space player, European Commission – Charlemagne Building, Brussels, 12–13 Jan 2016. https://ec.europa.eu/commission/2014-2019/bienkowska/announcements/european-space-strategy-creates-jobs-and-growth_en. Accessed 21 Mar 2016.

S. De Mey (✉)
Eurisy, Paris, France
e-mail: stefaan.de.mey@esa.int

C. Al-Ekabi et al. (eds.), *Yearbook on Space Policy 2015*, Yearbook on Space Policy, DOI 10.1007/978-3-7091-4860-0_6

critical turning point. The space field has leveled out. Tangible benefits have come within reach for small public and private players. Although only the tip of the iceberg has emerged, satellite applications are already touching everyone's everyday life.

Geopolitical objectives continue to have weight. Every self-respecting superpower still has the ambition to remain or become part of the select group of space faring nations. Space shields and intercontinental ballistic missiles still affect the geopolitical agenda. The difference is that down-to-earth applications are no longer surfing those geopolitical waves. Instead, there is ever-increasing access to affordable space platforms. Growing capabilities open unprecedented opportunities. This paves the way for embedding satellite data, services, and applications at the heart of the IT society. It also induces a tension between private and public initiatives and investments.

In this context, this contribution highlights some points of view on the future of satellite applications. The insights and observations presented stem mainly from Eurisy's direct work with end-user communities. Eurisy, an association of space agencies and governmental bodies in space affairs in Europe, raises awareness of emerging satellite applications that can help professional communities in many sectors of application: from transport to risk management, from habitat protection to energy, from climate change to the Internet of Things, to name a few. On the basis of its fieldwork with end-users, Eurisy provides feedback to decision-makers on possible measures to overcome obstacles to the diffusion of space-derived innovation in society.

6.2 Space Serving Society: Understanding Demand Is Key to Becoming Part of the Solution

6.2.1 Looking Beyond the Space Segment Downstream Value-Added Chain: The User Perspective

A 2014 Booz & Co report details the different segments of the global 2012 space economy. The study values the worth of the upstream (satellite manufacturing and launchers) at 16 billion euros, that of the midstream (satellite operations and ground equipment) at 19 billion euros, plus 107–140 billion euros for the downstream (consumer and value added satellite services and related consumer equipment).[2] The financial figures illustrate the multiplier effect of investments in space. What makes it even better is that the impact of satellite applications on society and the economy does not end with the reported downstream space segment revenues. To better understand the impact of space beyond the downstream, it is helpful to

[2]Booz & Company (2014) Evaluation of socio-economic impacts from space activities in the EU. Final Report.

understand how satellite applications are seen from the end-user perspective as opposed to the classical top-down view.

A traditional top-down approach classifies satellite applications into four different functional groups: communication, EO, navigation, and meteorology.[3] This traditional classification has its merits. It conveniently splits functional capabilities while at the same time relating each to homogeneous user groups: satellite communication users, EO users, satnav users and the meteorology user communities.[4] The idea of matching satellite applications to user communities is excellent, but in practice this top-down approach usually only reflects the value-added chain to the level of intermediary users. The value-adders in this segment often share a professional culture and language. They are familiar and concerned with (satellite) technical specifications, even though they may have complementary skills relative to each other. They feel at ease talking in technical terms about Satcoms, EO and navigation. By contrast, the vast majority of end users of services usually belong to a specific professional community: that of the sector in which they operate. These beneficiaries often do not have a technology background and are usually not concerned about how a solution is achieved. For them, the solution is a black box (often just a simple map). An individual farmer does not see him/herself as belonging to the EO, navigation or meteorology user communities, even if he/she uses EO-derived information to make decisions on fertilising a plot of land, or an EGNOS-equipped tractor for precision farming, or weather information for planning field work.

This example also reveals another point. Satcom infrastructure ensures connectivity, EO satellites provide (geo-spatial) data, and satnav provides absolute location, relative movement and time. The assumption that (end) user communities are using only one of these capabilities at a time does not match reality. Often a solution will combine several technological capabilities. Even if the end-user can break down the solution into its functionalities, he/she might not be aware of how these are achieved. Even more so, the end-user does not necessarily know, or in any case could not care less, about whether these functionalities are satellite enabled or not. In fact, many of the building blocks of such solutions are terrestrial technologies.

Therefore, the traditional classification linking each satellite segment to a user group appears to be inadequate when dealing with end-users. Rather than the satellite segment dictating the user community, it should be the latter that dictates the technologies to be used as part of solutions for them. When approaching user

---

[3]Pelton JN, Madry S, Camacho-Lara S. Satellite Applications Handbook: The Complete Guide to Satellite Communications, EO, Navigation, and Meteorology. In: Handbook of Satellite Applications. P 3–17. Springer New York. Editors: Pelton JN, Madry S, Camacho-Lara S. ISBN: 978-1-4419-7670-3 (Print) 978-1-4419-7671-0 (Online).

[4]Note that the meteorology satellites category is actually not referring to a specific functional capability but directly to a specific user community. In fact the meteorology satellites are largely overlapping with the EO satellites. The separate category is justified because the satellites are serving a very specific and mature user community, also closely involved in the design, development and operations of the satellites.

communities, their needs should be the reference point—not the (satellite enabled) technical capability. Or as Dr Ian L. Thomas from Earth Observation Consultants International proposed: "Routine user involvement in the use of satellite data is a key element for bringing the benefits of space to society. To me the key to advancing user involvement is to really seek to understand what the users actually and routinely need and to then work in partnership with them to explore its delivery."[5]

Space agencies are gradually taking this approach. On a European level, the ESA ARTES Integrated Applications Promotion (IAP) programme is a good example. The user is central and user communities are approached with tailored efforts that combine at least two different satellite segments. A thematic approach is also central, with various prescribed themes such as eHealth, security, train control and monitoring, unmanned aircraft system operations, digital cinema, augmented reality, renewable energy, coordinating rescue services, offshore oil and gas field monitoring, and media.[6]

National strategies are also developed in that direction. The Dutch roadmap 2015–2020 is one such example.[7] It makes vertical integration, beyond the space community, propriety for sectors such as water, agro & food, logistics, and energy among others. In an effort to improve the country's market position in these sectors internationally, space decision makers foresee an active involvement of the global Dutch champions that operate in these sectors, along with Dutch knowledge institutes.

Looking beyond the intermediate users, it is obvious that satellite enabled solutions are generating benefits far beyond the downstream. A quantitative estimation of these benefits across sectors is not easy. EARSC, the European Association of Remote Sensing Companies, has recently published a series of case studies quantifying the impact of EO along the full value-added chain in given sectors.[8] The studies show the economic impact of satellite-enabled services from the upstream all the way down and far beyond the space downstream segment. One of the examples studies the impact of satellite data on winter navigation in the Baltic region. In this case, the thousands of tons of fuel saved by icebreakers in the logistic chain are not the bottom line. The report illustrates how the biggest gains lie further downstream, with the ports and the factories being served, followed by the

[5]Eurisy (2015) EO uptake: the Holy Grail? – Interview with Dr Ian L Thomas. http://www.eurisy.org/article-eo-uptake-the-holy-grail_14. Accessed 27 Apr 2016.

[6]European Space Agency. Integrated Applications Promotion programme. https://artes-apps.esa.int/opportunities. Accessed 27 Apr 2016.

[7]Lucht- en Ruimtevaart Nederland. Topsector HTSM – Roadmap Space 2015–2020. http://www.hollandhightech.nl/nationaal/innovatie/roadmaps/space/roadmap-space-issue2-2-2015-2020. Accessed 27 Apr 2016.

[8]European Association of Remote Sensing Companies (2016) Assessing the detailed economic benefits derived from Copernicus Earth Observation (EO) data within selected value chains": Copernicus Sentinels' Products Economic Value. Case reports – Forest management in Sweden & Winter navigation in the Baltic. http://earsc.org/library/. Accessed 27 Apr 2016.

dependent businesses and the general public. Aggregating the average benefits from EO since 2004, when it came into use, the authors calculate a total benefit between 223 million euros and 1.2 billion euros for Finland and Sweden.

6.2.2 *Serving User Communities Better: Horizontal Integration at the Level of the Offer*

Passionate about science, exploration and technological advancements, the space community has traditionally operated in a rather isolated "space bubble", using its own standards, methodologies and practices. The nature of a space programme—a multi-annual development cycle, high cost, and low production quantities—justifies this culture. But down-to-earth consumer oriented businesses require short product cycles, high volumes, and low costs and prices.

Bringing space closer to society requires aligning these practices. Various initiatives have already been taken to give the downstream room to work on innovative solutions in a business oriented setting and mindset. Some of the pioneering examples in Europe are the ESA business incubators all across Europe, the "Communauté de Compétences Techniques Applications"[9] and "Boosters"[10] initiatives fostered by CNES in France, and the Satellite Catapult initiative in the UK.[11] In addition, some regions have embedded space in their smart specialisation strategy supported by EU regional policies. During its 2014 Italian EU Presidency, the Italian government launched a European Innovation Partnership on Space (EIP "Space") initiative. The initiative aims at offering opportunities to space companies to develop, innovate and access new markets, by tackling the key EU societal challenges.[12]

The pro-active set-up of satellite data warehouses is another initiative going in this direction. With Copernicus, Europe aims to foster a new industry that uses the free and open data and services to build and commercialise added value services. By procuring data centrally and making it available for free even before the operational phase of Copernicus, governments have sought to kick start their industry. The Netherlands Space Office has been leading the way with a national satellite data warehouse to mutualise the data and speed up application

---

[9]CNES. Les Communautés de Compétences Techniques. http://cct.cnes.fr/. Accessed 15 Apr 2016.

[10]Frank Niedercorn (2016) La France veut booster les start-up autour des applications du spatial. LesEchos.fr. http://www.lesechos.fr/journal20160415/lec2_pme_et_regions/021824393384-la-france-veut-booster-les-start-up-autour-des-applications-du-spatial-1214102.php#xtor=CS1-33. Accessed 30 Apr 2016.

[11]Satellite Applications Catapult . https://sa.catapult.org.uk/programmes/our-programmes/. Accessed 15 Apr 2016.

[12]NEREUS. EIP "Space" brief communication note. http://www.nereus-regions.eu/sites/all//events/EIP/EN_final.pdf. Accessed 15 Apr 2016.

development.[13] The warehouse can be considered as a Public-Private Partnership (PPP): the government pays for the data infrastructure whereas industry invests in service development and market introduction.

All these efforts have already led to various new applications with a significant impact on society and the environment e.g. for disaster management, environment monitoring and transport. Nevertheless, the uptake in Europe is still below expectations, especially compared to the levels attained in the U.S.[14] More effort will be needed for satellite-based services to reach their full potential. An important condition in achieving this is for the space community to look not only downward, beyond the limit of the downstream, to the end-user communities, but also to enlarge the base of intermediate users, horizontally.

To make the point, let us focus on the satellite communication sector. Over the years, the Satcom business has grown into a real success story and has become the largest commercial space domain. Vertically, the share of Satcoms in the end-to-end space value chain is impressive. Thanks to a sound downstream, in 2012 the global revenues of the Satcom industry reached 104 billion euros in a space industry worth 176 billion euros.[15] The Satcom services that generate this revenue cover a multitude of applications: from maritime communication, mobile satellite phones, emergency communications, to broadband. This is an impressive position when we take the vertical view, i.e. looking at the share of the Satcom downstream within the space industry; but when we take a horizontal view, i.e., when we consider the share of Satcoms in the global telecom sector, the story is completely different. The same revenue of billions of euros is completely overshadowed by the total 3.7 trillion euros worth of the telecom business.[16] It goes to show that although the Satcom value chain is long, established and mature, and although new applications such as satellite radio and broadband only arrived in the last decade, Satcoms have not been able to significantly raise their share in the huge telecom market, and to become a truly mainstream solution.

How to explain the dominance of terrestrial solutions? Do they offer inherently better performance, at a better price? Not necessarily. In many cases satellite solutions may provide similar or better performance and value for money. But the space sector is living on a proverbial different planet with slow product cycles and high development costs. It gives established technologies a head start to maintain competitive advantages through continuous and fast innovation. An eminent example is the sector of freight rail wagons or inter-modal tank containers. The assets

[13]Nederland's Space Office. Satellietdataportaal. http://www.satellietdataportaal.nl/. Accessed 15 Apr 2016.

[14]European Commission Directorate General for Internal Policies, Policy Department A: Economic and Scientific Policy (2016) Space Market Uptake in Europe. www.europarl.europa.eu/RegData/etudes/STUD/2016/569984/IPOL_STU(2016)569984_EN.pdf. Accessed 15 Apr 2016.

[15]Booz & Company (2014) Evaluation of socio-economic impacts from space activities in the EU. Final Report.

[16]Booz & Company (2014) Evaluation of socio-economic impacts from space activities in the EU. Final Report.

concerned typically travel for very long periods all over the globe without their own energy source and in harsh conditions. Traditional monitoring technologies based on cellular networks for example have never been able to guarantee an easy and seamless follow up for these kinds of transport. Despite that, they remain a prominent competitor to Satcom solutions, whose performance is inherently much better. The key to counteract this is to get satellite enabled technologies out of the exclusive niche of the space sector and increasingly make them part of the standard toolset of a broader community of innovators and service providers. For this to become reality, the space sector will need to adapt its processes and focus on the user communities. In 2014, ESPI Director Peter Hulsroj made a similar observation for the EO sector.[17] Downstream space sector providers often develop user-oriented applications in complete isolation from each other, and without taking into account what already exists on the market, or even pre-operational solutions. This probably leads to duplication, limited cross-fertilisation and underuse of existing functionalities. It is advisable instead to create a lively ecosystem of users and developers around the wealth of data coming from satellites or, even better, around data coming from all possible sources.

Satcom against cellular technology, fibre and other terrestrial solutions. GPS against Wi-Fi mapping, SIM card tracking, cellular triangulation. Earth Observation data against aerial, UAV, and ground measurements. Only when put in direct competition with and in a similar business oriented environment, has the satellite solution a chance to become a mainstream solution for the different functionalities wherever it has the potential to be better. In this spirit, government space policy makers point to a need to give access to the downstream space segment to new entrants (new start ups and existing enterprises) from all possible other sectors. Although a good first step, this is not enough. Instead, a real Copernican shift in mindset is necessary: space has to become a new entrant in the broader societal sectors.

All this is probably only a matter of time anyway. Private industry seems to have become aware of the opportunities. After integrating part of the vertical value chain (from launchers to applications) in a commercial setting, private industry is now also horizontally integrating a private satellite ecosystem into the world economy. An example of this is the rebranding by Google of Skybox into Terra Bella, literally moving the interest from the sky to the Earth.[18] Terra Bella still plans to launch a dozen satellites in the sky over the next few years, however the centre of gravity becomes the Earth. A wide array of geospatial data sources (not only satellite imagery), machine learning capabilities, and experts will enable the company to

[17]ESPI (2014) Executive Brief. An App Store for Earth Observation Tools and Applications? http://www.espi.or.at/images/stories/dokumente/executive_briefs/ESPI_Executive_Brief_5.pdf. Accessed 15 Apr 2016.

[18]Emil Protalinski (2016) Google rebrands Skybox as Terra Bella, will launch 'more than a dozen satellites' over the next few years. VentureBeat.com. http://venturebeat.com/2016/03/08/google-rebrands-skybox-as-terra-bella-will-launch-more-than-a-dozen-satellites-over-the-next-few-years/. Accessed 15 Apr 2016.

meet the economic, societal, and humanitarian challenges and further their business this is how space gets fully integrated into the Google machine!

6.3 The Big Data and Interconnectivity Gordian Node

Referring to Google in the previous paragraph brings us almost seamlessly to the topic of big data. Some call satellite imagery the ultimate big data.[19] We can confirm. Big data is characterised by high volumes, high velocity, and high variety.[20] All the big data elements can be found in satellite imagery. But space big data goes beyond the pixels in optical images. Big Data from space refers to Earth and Space observation data collected by an almost endless variety of space-borne and ground-based sensors.[21]

6.3.1 Volume, Velocity and Variety: Satellite Data Is Big Data

6.3.1.1 High Volume

Satellite programmes are designed to observe and track whatever happens, often on a global scale, producing high volumes of data. Archived data is reaching the exabyte (10^{18}) scale. In 2015, the Sentinels—the satellites launched by the EC as part of its Copernicus programme, generated an average of 3 Terabytes of core products daily. It is projected that this figure will increase to more than 6 TB a day by the end of 2016. The full Sentinel-1A production available online (March 2016) encompasses over 450,000 products.[22] According to Copernicus Climate Change &

[19]Natasha Léger (2014). Satellite Imagery is the Ultimate Big Data. LBXjournal.com. http://www.lbxjournal.com/articles/satellite-imagery-ultimate-big-data/260516. Accessed 15 Apr 2016.

[20]De Mauro A, Greco M, Grimaldi M (2015) What is big data? A consensual definition and a review of key research topics. In: AIP Conference Proceedings 1644, 97 (2015); doi:10.1063/1.4907823.

[21]Soille P and Marchetti PG (2016). Proceedings of 2016 conference on Big Data from Space (BiDS'16), Auditorio de Tenerife, Santa Cruz de Tenerife, Spain, 15–17 March 2016. Publications Office of the European Union. ISBN: 978-92-79-56980-7. http://publications.jrc.ec.europa.eu/repository/handle/JRC100655. Accessed 30 Apr 2016.

[22]Bargellini P, Laur H (2016) Big Data from Space: the Copernicus contribution. Presentation at the 2016 conference on Big Data from Space (BiDS'16), Auditorio de Tenerife, Santa Cruz de Tenerife, Spain, 15–17 March 2016. http://congrexprojects.com/custom/16M05/bids/ALL/01_0945_Bargellini.pdf. Accessed 30 Apr 2016.

Atmosphere Monitoring Service (ECMWF) the data reanalysis amounts to 6 Petabytes. If the data were stored on CDs, they would tower 11 km high.[23]

6.3.1.2 High Velocity

Big Data is produced continuously at an ever-increasing pace. Often it is a matter of seconds. Indeed, for critical decisions knowledge must be extracted in (near) real-time. An excellent example of such a service based on satellite data is CleanSeaNet. With this service the European Maritime Safety Agency (EMSA) delivers daily satellite imagery to coastal authorities that have a legal obligation to monitor oil spills in their territories. One of the "customers" is the Central Command for Maritime Emergencies (CCME), a joint institution of the German Federal Government and the five German Coastal States.[24] The CCME receives approximately 450 satellite images per year, each covering between 900 and 1600 km^2, including some areas in neighbouring countries. The information also contains a classification of the dark spots detected according to the likelihood of them being oil spills, and their potential impact. SAR images, results of oil spill and vessel detection analysis, and relevant auxiliary data (wind and swell detection, vessel traffic information, nautical charts, meteorological, oceanographic information and oil drift models) are made available to coastal states through the web-based and user-tailored CleanSeaNet portal within 30 min of the capture of the images. Moreover, by combining information on potential oil spills and their position relative to vessels, the CCME is also able to identify the potential polluters, further discouraging illegal behaviour.

6.3.1.3 High Variety with Variable Accuracy and Quality

The variety of data types and sources and their complementarities is what makes big data both valuable and complex. Satellite data is delivered by sensors acting over various frequencies of the electromagnetic spectrum. The variety lies not only in the different types of data. Accuracy and quality also differ, which adds to the challenge of extracting information and meaning from big data.

---

[23]Copernicus ECMWF on twitter. 10 Nov 2015. https://twitter.com/copernicusecmwf/status/664038602783920129. Accessed 15 Apr 2016.

[24]Eurisy (2015) Central Command for Maritime Emergencies: remote sensing to intervene on oil spills. In: Eurisy (2015) Operational uses of satellite-based applications in the public sector: a case-study review. ISBN 978-2-9551847-0-7.

6.3.2 Data to the User or User to the Data?

Virtually all satellite derived signals and data can be geo-tagged. In addition, satellite navigation services enable other data sources to be geo-tagged. This brings satellite data and services to the heart of the broader (geospatial) big data ecosystem. Whereas for the space community big data is a relatively new challenge, for the broader IT sector big data is rather old news. Embedding satellite data in the overarching big data ecosystem is therefore crucial. By doing so, space data too could benefit from the constant progress in hardware and software developments, high performance computing, cloud computing, data visualisation, machine learning algorithms, and all other (r)evolutions that are at the core of the many innovative developments related to big data.

On the European scene, various channels have been set up to distribute Copernicus data. In complement to that, from Finland to Australia and from the UK over Poland to Greece, ESA is rolling out so called collaborative ground stations in an effort to facilitate the distribution of satellite data as close as possible to user communities.[25] These investments are certainly useful and needed to support the proliferation of free and open data access initiatives. They help boost the number of users. A sharp increase of users has indeed been observed on the ESA open access data hub: 26,680 self-registered users, of which 18,000 are for Sentinel products.[26] Some 3.7 million Sentinel-1 products have been downloaded to date, i.e. 4.5 Petabytes of data (status on 10 March 2016). In 2015, 30 TB of products were disseminated on average a day. By the end of 2016, an estimated 100 TB of products will be disseminated daily.

With this level of use, the introduction of complementary data access solutions becomes a must. A further unlimited linear expansion of ground infrastructure bringing the data to the users will however create further fragmentation and become counterproductive. Future solutions need to reduce fragmentation and mutualise costs related to data exploitation. A big data problem requires a big data solution. A future solution needs therefore to interconnect platforms that enable moving users to the data. These platforms ideally should be able to bring together data from various sources (space and non-space data) with collaborative tools (processing

---

[25]European Space Agency, Sentinel Collaborative Ground Segment. http://www.esa.int/Our_Activities/Observing_the_Earth/Copernicus/Sentinel_Collaborative_Ground_Segment. Accessed 30 Apr 2016.

[26]Bargellini P, Laur H (2016) Big Data from Space: the Copernicus contribution. Presentation at the 2016 conference on Big Data from Space (BiDS'16), Auditorio de Tenerife, Santa Cruz de Tenerife, Spain, 15–17 March 2016. http://congrexprojects.com/custom/16M05/bids/ALL/01_0945_Bargellini.pdf. Accessed 30 Apr 2016.

tools, data mining tools, user tools, ...) and the corresponding computing and hosting resources required.[27]

6.3.3 From Big Data to Connected Data and IoT

Today, with smartphones equipped with various sensors and carried around by almost all individuals all over the globe, crowdsourcing offers promising perspectives. For the collection of data from a large group of people to be of value, timeliness is of the essence. In regions with connectivity issues (low or no coverage by fibre and mobile networks) that prevent data transmission, Satcom can become an indispensable solution.

Exploiting this big data however also means collecting and combining the data sources and turning them into actionable information before (re) distributing the results. Turning the data into value requires not only IT and computing skills, but also expert knowledge in the thematic sector concerned. Improving the link between thematic research and practical applications is key. An excellent example of a scientific initiative with links to a practical application can be found at Berkeley University. Recently, scientists from the Berkeley Seismological Lab released an application exploiting the smartphone's accelerator (more commonly used for gaming!) and GPS sensors to record the ground shaking, with the goal of creating a worldwide seismic detection network that could eventually warn users of impending jolts from nearby earthquakes.[28]

Combining crowdsourcing and EO adds another satellite-derived dimension. An interesting example of an emerging experimental application is the urban air pollution monitoring service set up by PlumeLabs, Twitter and DigitasLBi.[29]

And with Twitter we are coming close to yet another Silicon Valley tycoon starting to find its place in space. Facebook's internet.org initiative aims to make the Internet available to every person on Earth. To achieve this, the team is exploring a variety of technologies, including high-altitude long-endurance planes, lasers, but also its own constellation of satellites. But there is more. For this project, to reveal the locations of more than 2 billion disconnected people, Facebook technologists have used artificial intelligence algorithms similar to what allows Facebook to recognise faces in photos uploaded to its platform. Thanks to these algorithms, Facebook analyses satellite imagery over 21.6 million km^2. One can only guess that

[27]Bargellini P, Laur H (2016) Big Data from Space: the Copernicus contribution. Presentation at the 2016 conference on Big Data from Space (BiDS'16), Auditorio de Tenerife, Santa Cruz de Tenerife, Spain, 15–17 March 2016. http://congrexprojects.com/custom/16M05/bids/ALL/01_0945_Bargellini.pdf. Accessed 30 Apr 2016.

[28]UC Berkeley (2016) New app turns smartphones into worldwide seismic network. http://news.berkeley.edu/2016/02/12/new-app-turns-smartphones-into-worldwide-seismic-network/. Accessed 14 Apr 2016.

[29]Plumelabs. https://www.plumelabs.com/. Accessed 15 Apr 2016.

Facebook could use the same artificial intelligence skills to further extract value and business intelligence from satellite imagery, with no limits in sight.[30]

In addition, after connecting all people, Facebook also shows an ambition to become a platform for the Internet of Things by connecting all domestic and other appliances connected to the Internet.[31] Or how one day you might become friends with your coffee machine!

6.4 Public Authorities: Main Users of Satellite Applications?

6.4.1 Satellite Applications in Support of a Public Mission

Some public authorities are using satellite enabled services and data. On a European level the EC uses satellite applications to acquire independent and timely information relevant for implementing EU policies and regulations. For instance, the EC relies on satellite data and services to identify and measure the main crop areas in Europe, estimate production early in the year and check the validity of farmers' applications for EU subsidies.[32] Indeed, satellites not only support the EU Common Agricultural Policy but, in addition, contribute to the Digital Agenda for Europe and the 'Integrated Industrial Policy for the Globalisation Era' with regard to environmental observations sharing and standard setting.[33]

Also, as illustrated by Eurisy's "Satellites Going Local" publications, public authorities at national and sub-national levels use operational satellite services based on EO, satnav and Satcom in various fields.[34]

---

[30]Dillow C (2016) What Happens When You Combine Artificial Intelligence and Satellite Imagery. Fortune.com. http://fortune.com/2016/03/30/facebook-ai-satellite-imagery/. Accessed 14 Apr 2016.

[31]Janakiram MSV (2015) Why Facebook Might Emerge As The Biggest Internet of Things Company. Forbes.com. http://www.forbes.com/sites/janakirammsv/2015/04/14/why-facebook-might-emerge-as-the-biggest-internet-of-things-company/#736f9bc17438. Accessed 14 Apr 2016.

[32]European Commission Joint Research Center. Agricultural Monitoring. https://ec.europa.eu/jrc/en/research-topic/agricultural-monitoring. Accessed 14 Apr 2016.

[33]European Commission Joint Research Center, Agricultural Monitoring. https://ec.europa.eu/jrc/en/research-topic/agricultural-monitoring. Accessed 14 Apr 2016.

[34]Eurisy (2011–12014) Satellites Going Local. http://www.eurisy.org/publications.php. Accessed 04 Apr 2016.

6.4.1.1 Satellite Applications in Support of (EU) Policies and Regulations on Regional and Local Levels[35,36]

Many EU policies and regulations have a direct impact on economic development, environment, climate change, transport, energy, agriculture, health and urban planning—all of which concern regional and local authorities. Satellite services help public authorities to respond and adapt to such policies and regulations: e.g. EO data is used to map habitats in Wales at a quarter of the cost of previous maps; solar energy production in the city of Lyon (France) is evaluated and monitored at the modest cost of 1 % of the energy annually produced; the Alsace region relied on EO to avoid a 17 million euros penalty threat and EC court action for inadequately protecting an endangered species of hamster.[37,38]

6.4.1.2 Satellite Applications Improving Public Services

Satellite data and services not only provide valuable tools for public authorities to comply with and implement policies and regulations. Eurisy has also documented several cases where satellite services are either more cost effective, or provide better quality, or both, in delivering public services to citizens.[39] For example the region of Flanders (Belgium) saves 140,000 euros annually in maintenance costs thanks to a satellite navigation system that regulates traffic lights; the city of Diemen (the Netherlands) monitors soil resilience on its territory at a quarter of the cost of its annual maintenance budget.

6.4.1.3 Satellite Applications for Regional Economies

Last but not least, satellite applications also support efforts by public authorities to grow local economies. On the Channel island of Alderney, an EGNOS solution allowing more frequent flights to the island made the local airport an engine for the

[35]Eurisy (2015) Operational uses of satellite-based applications in the public sector: a case-study review. ISBN 978-2-9551847-0-7.

[36]Fiore GM, De Mey S, Jolly C, Greenaway B (2015) Satellites for Society: reporting on the operational uses of satellite-based services among public administrations in Europe, In: Proceedings of the 66th International Astronautical Congress, Jerusalem, Israel, 2015.

[37]Europe1.fr (2009) La France doit protéger son Grand hamster sous peine d'amende. www.europe1.fr/politique/la-france-doit-proteger-son-grand-hamster-sous-peine-d-amende-34907. Accessed 15 Apr 2016.

[38]Court of Justice of the European Union, Luxembourg (9 June 2011) Judgment in Case C-383/09, Commission v France 1, Press Release No 58/11. http://curia.europa.eu/jcms/upload/docs/application/pdf/2011-06/cp110058en.pdf. Accessed 15 Apr 2016.

[39]Eurisy (2015) Operational uses of satellite-based applications in the public sector: a case-study review. ISBN 978-2-9551847-0-7.

local economy and welfare.[40] Another example is the regional programme "A wireless territory", operational since 2004. This initiative has made use of satellite telecommunications and wireless technologies to offer broadband to local authorities and their associations, schools, mountain and hill communities.[41] Users were reticent to use the technology at first, but satellite broadband proved a good solution in the mix, and a first-choice for remote areas, such as mountain huts. Broadband has often made the difference between local SMEs in inaccessible areas staying in business or not. The economic and social impacts are considerable.

6.4.2 *Public Authorities a Junior Partner for the Satapps Industry?*

Public customers are a vital buyers' group in the satellite applications and data market. For EO, a recent survey conducted by the European Association of Remote Sensing Companies (EARSC) indicates that about 65 % of the market relies on public customers.[42] The high relative market share does not necessarily mean a general and wide uptake of EO data by public customers. The use of satellite-based services in Europe has indeed not yet reached its full potential, especially at regional and local levels.

To understand better how public authorities adopt and use satellite applications and the impact of these tools on their work, in 2015, Eurisy launched an online survey. Eurisy has recently published an analysis of replies from 106 European public authorities so far.[43]

A first observation from the survey is that satellite-based services are available at relatively small cost, but that it is still difficult to quantify their benefits.[44] Only 25 % of the respondent public authorities carried out a formal cost-benefit assessment of the satellite-based services used. This probably means that the evaluation of these services is mainly empirical, experience-based.

---

[40]Eurisy (2015) Use of EGNOS to support approaching and landing at Alderney Airport. In: Eurisy (2015) Operational uses of satellite-based applications in the public sector: a case-study review. ISBN 978-2-9551847-0-7.

[41]Eurisy (2014) Piedmont: Stimulating the regional economy through broadband access. http://www.eurisy.org/good-practice-piedmont-stimulating-the-regional-economy-through-broadband-access_79. Accessed 15 Apr 2016.

[42]European Association of Remote Sensing Companies (EARSC) (2015) A Survey into the State and Health of the European EO Services Industry.

[43]Eurisy (2016) Satellites for Society: Reporting on operational uses of satellite-based services in the public sector. ISBN 978-2-9551847-1-4.

[44]61 % of respondents declared spending less than 5 % of their annual budget, whereas 41 % of users even declared spending as less as 1 % of their budgets to adopt/implement a satellite-based services. For operating these services, 73 % of respondents spend less than 5 % of their annual budget, while 62 % of them report expenses of less than 1 % of their budget.

Respondents used satellite services mostly for environmental protection, in agriculture, risk management, and transport. Though public administrations confirm being familiar with satellite services, and despite the fact that various satellite applications are mature, the need to further develop tailor-made solutions remains.[45] These results show that public authorities are not only mere users of satellite services, but also participate in their development, improvement and provision.

In the past, the uptake of satellite services by public authorities was driven by top-down initiatives. Case study observations, however, suggest that the merits of the technology alone do not necessarily convince the users to adopt satellite services.[46] Similarly, top-down measures to stimulate uptake are not sufficient unless they are accompanied by grassroots support to the potential end-users in their appropriation of these tools. Therefore, governments should continue to encourage innovation in these fields, not only by stimulating the creation of new satellite services, but also by fostering the use of existing products that have already been proven to work.

Moreover, national governments, ministries, EU institutions, and other cluster organisations should support user communities in sharing knowledge among peers. Such exchange platforms could prove fundamental in stimulating uptake on a wide scale. To support the uptake of those operational satellite solutions whose benefits are proven through end-users' experience and which help fulfil European policy obligations, such services should be referenced in European directives and regulations, as well as in European transversal funding schemes.[47]

To further increase user pull, supporting end-users more directly may further help new service development and align offer and demand. Already in 2010 Eurisy advocated dedicating part of the space budget direct to potential end-user communities, via non-space funding mechanisms familiar to them (Life+, regional funds and so on).[48] The so-called pre-commercial procurement (PCP) and public procurement of innovative solutions (PPI) funding schemes go in that direction. These funding schemes allow public authorities to test new solutions, participate in their

[45]In the survey, 40 % of respondents found the needed satellite-based services available on the market whereas 37 % of them needed these services to be developed or adapted to respond to their specific needs.

[46]Secara T, Reetz S, De Mey S (2012) Satellite applications take-up in Europe: An analysis of three case studies with regional authorities. In: Proceedings of the 63rd International Astronautical Congress, Naples, Italy.

[47]Eurisy (2012) Achieving Europe 2020: Policy coordination in support for Cities, Regions and SMEs – Facilitating access to the benefits of satellite services. http://www.eurisy.org/data_files/publications-documents/2/publications_document-2.pdf. Accessed May 2016.

[48]Eurisy (2010) Creating sufficient user pull to secure the benefits of satellite services for society. http://www.eurisy.org/data_files/publications-documents/8/publications_document-8.pdf. Accessed 15 Apr 2016.

development and eventually procure them when they reach maturity.[49] In 2015 PCP and PPI schemes covered thematic areas ranging from health and aging, to transport and e-government, to sustainability, education and ICT. PCP and PPI are technology neutral, so satellite enabled solutions can compete alongside terrestrial technologies in all fields. In addition, in 2016, the EC also started experimenting with a specific PCP call for Earth Observation solutions for the public sector.

6.5 The Stars Are the Limit

Space programmes have always been dominated and directed by governmental initiatives. Nevertheless, also in the past, private capital from successful entrepreneurs and philanthropists such as James Lick, Andrew Carnegie and Daniel Guggenheim among others has played a critical role in advancing and investing in space exploration capabilities.[50]

To date the benefits from space for citizens and the economy are omnipresent in our society. Bringing tangible benefits from investments in space remains high on the political agenda. In parallel, with increasing opportunities for commercial applications and exploitation, private actors are increasingly entering the field. Private investors also bring a more commercial logic and approach into the traditional space sector. This accelerates the horizontal integration of space into overall commercial activities.

While embedded in the ecosystem of bigger big data and the Internet of Things, the challenge remains for the geospatial value-added chain to produce actionable geospatial products, business intelligence and services. These would be part of on an entirely new set of disruptive business applications for a wide range of industries and sectors. Public authorities, including regional and local, remain a senior partner for both co-developing and procuring geospatial products and satellite enabled services.

Interestingly enough, while private investors are certainly looking for profitable ventures in the downstream application segment, this new dynamism seems also to again attract private capital to upstream exploration activities. Private investment in space tourism and planetary mining has an understandable commercial logic, even if investments are huge and risks unpredictable. But how about a recent initial

---

[49]Lieve Bos (2015) Horizon 2020 Support to Innovation Procurement. Presented at the EARSC Workshop PCP/PPI in Horizon2020 Projects on Earth Observation, Brussels, 3 Dec 2015. http://earsc.org/file_download/292/PCP+PPI+under+Horizon+2020-background+%26+experiences+in+other+sector+%28L.+Bos%29.pdf. Accessed 15 Apr 2016.

[50]NASA (2014) Emerging Space – The Evolving Landscape of 21st Century American Spaceflight. http://www.nasa.gov/sites/default/files/files/Emerging_Space_Report.pdf. Accessed 9 May 2016.

100 million dollar investment by Yuri Millner, supported by Stephen Hawking and Mark Zuckerberg, in a project that aims within a generation to develop a nanocraft, on a sail pushed by a light beam, capable of reaching the Alpha Centauri star system in a record time of 30 years?[51] The stars are the limit.

[51]The Breakthrough Initiatives (2016) http://breakthroughinitiatives.org/News/4. Accessed 10 May 2016.

Chapter 7
Efficient Access to Space: Implications for Space Astronomy

Shuang-Nan Zhang

7.1 Introduction

Space exploration has lasted for about half a century, and astronomical observations with space telescopes and other instruments have allowed us to understand the laws of the nature and the universe itself much deeper, wider and in ever greater detail. For example, many important and ground-breaking new phenomena and astronomical objects have been discovered by space astronomical observations, and two Nobel prizes in physics have been awarded to the results of space astronomical observations. However, access to space still remains very expensive, such that only a very limited number of space astronomical facilities with much contrived capabilities have been sent to space, in comparison to the numerous powerful experiments and observational facilities built on the ground. This certainly hinders our ability to further study the laws of nature and explore the universe.

For quite some time, Cheap Access To Space (CATS) has been advocated by industries, mostly driven by commercial space applications. Indeed many technological advances have made access to space more affordable than ever before; however, scientific space applications have still remained very expensive. Given the technologies available, the high costs of access to space, especially for military and scientific applications, may not be reduced significantly in the near future. For example, space science missions costing from hundreds of millions of euros to billions of euros are still quite common, and thus the number of space science missions is still very small. Many distinguished scientists have spent their whole careers studying and proposing space missions, but unfortunately without success!

S.-N. Zhang (✉)
Key Laboratory of Particle Astrophysics, Institute of High Energy Physics, Chinese Academy of Sciences, Beijing 100049, China
e-mail: zhangsn@ihep.ac.cn

C. Al-Ekabi et al. (eds.), *Yearbook on Space Policy 2015*, Yearbook on Space Policy, DOI 10.1007/978-3-7091-4860-0_7

Here we propose a new concept: Efficient Access to Space (EATS), in contrast to CATS. In EATS, the demands for requirements for access to space can be categorised into three groups: **luxury, good and fair**; the costs for each group can be significantly different from the others. Even for scientific applications, the concept of EATS can also be applied, to enable significant overall cost reduction in scientific exploration or significant increase of opportunities for scientific exploration and more important scientific discoveries to be made with the same overall costs.

7.2 Telescopes on the Ground and in Space

As shown in Fig. 7.1, the need to make astronomical observations in space is mostly due to the fact that the earth's atmosphere only opens windows to several very narrow bands (i.e., high frequency radio waves, near infrared and visible light) out of the very wide range of wavelengths/frequencies of electromagnetic radiations from many kinds of astronomical objects and the universe itself, from very low frequency radio waves, to microwaves, to high frequency radio waves, to infrared, to visible light, to ultraviolet, to X-ray and to gamma-rays.

Even in these several narrow bands, the atmosphere is still partially opaque and also produces significant disturbances to incoming astronomical signals. Therefore multi-wavelength exploration of the universe must rely on access to space. In the following, only optical telescopes are used as an example to compare those on ground and in space, because optical telescopes can be built and can work in both environments.

7.2.1 Optical (Visible Light) Telescopes on Ground and in Space

In Fig. 7.2, several pictures of the world's largest ground based optical telescopes are displayed. Without going into detail on these telescopes, we simply summarise the situation as: 10 telescopes with aperture diameters larger than 8 m and 42 telescopes with aperture diameters larger than 3 m.

In comparison, currently there is no single optical (visible) telescope in space with an aperture diameter larger than 3 m. In Fig. 7.3, the world's major optical telescopes in space are listed; so far the most well known and perhaps scientifically also the most productive optical telescope is the Hubble Space Telescope (HST),[1] which is the oldest and largest optical telescope in space with an aperture diameter of merely 2.4 m.

[1]Hubble Space Telescope. https://en.wikipedia.org/wiki/Hubble_Space_Telescope

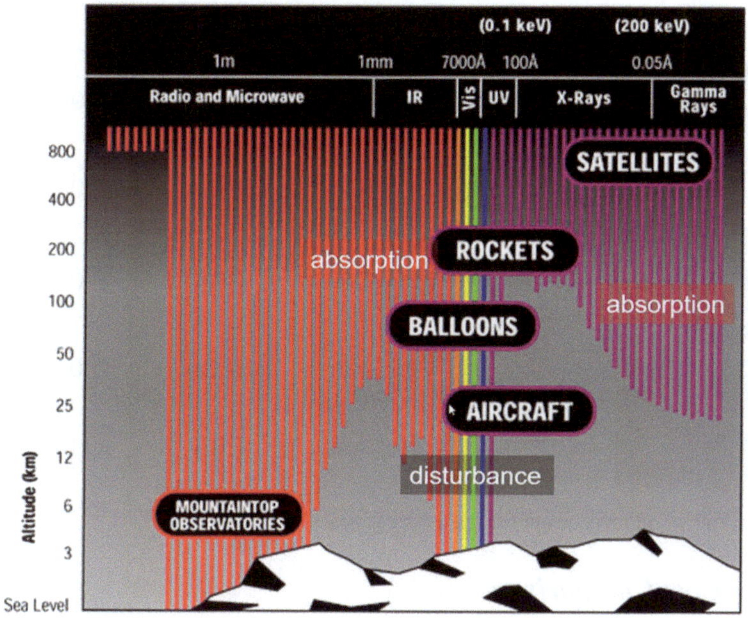

Fig. 7.1 Attenuation of electromagnetic radiations by the earth's atmosphere

Fig. 7.2 Pictures of the world's largest ground based optical telescopes

Photo	Name	Space Agency	Launch	Aperture
	Hubble	ESA&NASA	1990	2.4 m
	Kepler	NASA	2009	1 m
	MOST	CSA	2003	15 cm
	Swift	NASA	2004	30 cm
	Gaia	ESA	2013	~1 m

Fig. 7.3 The world's major optical telescopes in space

7.2.2 Cost Comparison

To make a fair cost comparison between ground-based and space-based telescopes and experiments, we take the above-mentioned HST and the well-known AMS02 experiment[2] on board of the International Space Station (ISS)[3] as two archetypical examples. Our basic conclusion is that it is about 2–3 orders of magnitude more expensive to do astronomical observations and scientific research in space than to do that on the ground with instruments of similar sizes or weight scales.

The original HST cost about 1 billion euros, and consequent repairs cost around an additional 1 billion euros. In comparison, an optical telescope on the ground with the same aperture and similar focal plane instruments typically costs only 10–20 million euros, roughly 100 times cheaper than HST.

AMS02/ISS also cost about 1 billion euros, even without including the launching service. AMS02 is the most expensive, largest and heaviest single experiment ever sent to space, weighing about 8.5 tons. In comparison, the CMS experiment[4] at CERN is one of the world's largest experiments on ground, which cost about 0.5 billion euros and weighs about 12,500 tons. This means the cost of AMS02 is about 3000 times that of CMS per unit weight!

[2]The Alpha Magnetic Spectrometer (AMS-02). http://www.ams02.org/

[3]International Space Station. https://en.wikipedia.org/wiki/International_Space_Station

[4]The Compact Muon Solenoid experiment at the LHC – CMS. http://cms.web.cern.ch/org/cms-public

7.3 Why Are Space Telescopes So Expensive?

Without any detailed statistics, the total costs of space science missions may be roughly divided into four parts: about 20 % for launcher and launching service, about 40 % for satellite platform, about 30 % for payloads (mostly scientific instruments), and about 10 % for mission operation and data analysis (including scientific research). The total cost of a mission with a space telescope would be only about three times as expensive as a similar telescope on the ground, if a space telescope were built in the same way as on the ground. Apparently this is far from true. Fig. 7.4 illustrates why space telescopes are expensive:

(1) A launcher is needed for sending a telescope into space, which is an additional cost, compared to a ground-based telescope of a similar kind.

(2) For a given orbit, the cost of a launcher is roughly proportional to the weight of the whole spacecraft, including payload and satellite platform. Therefore one would try to make the platform as light as possible, which adds costs. Then the light platform has to pass space qualification tests, including tough mechanical qualification tests, which adds structural supports in the platform and thus additional costs.

(3) Since the whole platform is already quite expensive, no organisation can afford a failure of the mission, both financially and politically. Therefore very stringent quality control is imposed, which adds costs to the platform.

(4) In turn, the "cannot fail" requirement and the additional costs from such requirements escalate all the way from the platform to payloads, to launching and back to the rocket, and so on.

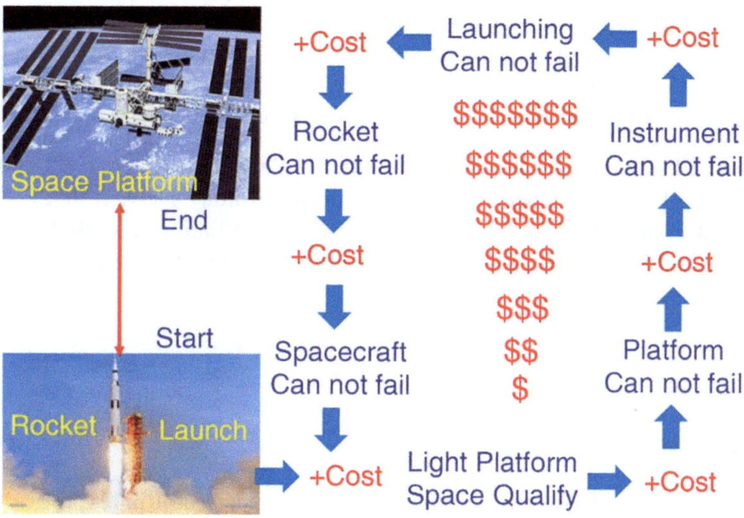

Fig. 7.4 Process of the cost escalation of space missions, due to the "cannot fail" requirement

Eventually, starting from a not-so-expensive launcher, we end up with a mission costing 2–3 orders of magnitude more than building a similar telescope on ground.

7.4 Serendipity Is the Nature of Exploration

7.4.1 Major Astronomy Discoveries

Most pioneering astronomy discoveries come as surprises, which drive our understanding of the universe and its laws. For example, of the about ten Nobel prizes awarded to the results of astronomy research, only one was a planned goal, i.e., the COBE discovery of the perfect blackbody spectrum of cosmic microwave background radiation, predicted by the Big Bang model of the evolution of the universe. All other Nobel Prize winning astronomical discoveries are either completely serendipitous, unexpected, or opposite to the predictions of the existing models or theories at the time. Here several remarkable examples are listed:

- **Cosmic rays** (discovery by Hess in 1912, Nobel Prize in Physics awarded in 1936[5]): the original goal was to use hot ballooning to study attenuation of radioactivity with atmospheric height, because at the time the Earth was known to produce natural radioactivity. Surprisingly, Hess found that the radioactivity recorded by his instrument increased rapidly after the ascended above about 1 km.
- **Cosmic microwave background** (discovered by Penzias and Wilson in 1965 and Nobel Prize in Physics awarded in 1978[6]): they built a very sensitive radio antenna, but kept receiving unknown signals from all directions in the sky, not just from the plane of the Milky Way, even after removing all possible sources of antenna noise. The unknown signals are just the relics of the Big Bang that created our Universe about 14 billion years ago.
- **Pulsar** (discovered by Jocelyn Bell—then a PhD student—in 1967 and Nobel Prize in Physics awarded to her PhD advisor Antony Hewish in 1974[7]): The original goal was to study interstellar radio scintillation, but Bell discovered unexpected periodic radio signals from a pulsar, a rapidly spinning neutron star.

[5]The Nobel Prize in Physics 1936. http://www.nobelprize.org/nobel_prizes/physics/laureates/1936/

[6]The Nobel Prize in Physics 1978. http://www.nobelprize.org/nobel_prizes/physics/laureates/1978/

[7]The Nobel Prize in Physics 1974. http://www.nobelprize.org/nobel_prizes/physics/laureates/1974/

– **Cosmic X-ray sources** (discovered in 1960s by Riccardo Giacconi and Nobel Prize in Physics awarded in 2002[8]): The original goal was to study X-ray emissions from stars like the Sun, but after two failures his team in the third rocket flight experiment unexpectedly discovered new populations of X-ray emitting objects, including black holes and neutron stars.
– **Cosmic neutrinos** (discovered by Raymond Davis Jr. and Masatoshi Koshiba in 1987 and Nobel Prize in Physics awarded in 2002[8]): The original goals of Davis Jr.'s and Koshiba's experiments were to observe the solar neutrinos and study the predicted decay of protons in the grand unified theory of physics, but instead they discovered neutrinos coming from a supernova explosion, SN1987A.

7.4.2 Expensive Access to Space Limits Our Discoveries

All the above listed pioneering, but unexpected, astronomy discoveries were made because of the additional capabilities of those experiments and observational facilities beyond the originally designed goals. This ensured that even if the original goals could not be achieved, because of our limited understanding of the laws of nature and the complexity of the universe, unexpected but even more important discoveries could be made. In other words, those unexpected discoveries might not be made, or might be at least delayed significantly, if these experiments and observational facilities were designed to meet exactly and solely the goals of these programs.

However, because of the high costs of space science missions, primarily due to the non-negligible cost of a launcher and then the escalated costs from the platform to the instruments and eventually back to the launcher itself, scientists and engineers must develop extremely contrived instruments, including with respect to their weight and complexity. Therefore in many cases, they have to justify each and every specification of the instruments, which must match their pre-determined scientific goals. Additional capabilities, though necessary for serendipitous discoveries that are often far more important than achieving the pre-determined goals, are often considered luxury and reduced or even eliminated in the end. Because of the "cannot fail" policy, we are often forced to use mature technologies and cannot utilise more risky but more powerful and innovative technologies.

Therefore expensive access to space limits the potential of the discoveries of our space science missions. This problem cannot be solved simply by increasing the total budget, since scientists always want to do more experiments and build more facilities, such that for each experiment and facility, we still always face the same problem of not including adequate additional capabilities for serendipitous discoveries.

---

[8]The Nobel Prize in Physics 2002. http://www.nobelprize.org/nobel_prizes/physics/laureates/2002/

7.5 Efficient Access to Space

The long advocated CATS might be misleading, because "cheap" may be perceived as low or reduced quality and reliability, given the available technologies. We therefore propose the concept of EATS: Efficient Access To Space, in which space applications can be categorised into three different kinds:

- **Luxury**: expensive but with very high reliability or very low failure rate.
- **Good**: moderate cost but with high reliability or low failure rate.
- **Fair**: low cost but with fair reliability or tolerable failure rate.

Examples of the **"luxury"** category could include critical military applications, manned space flights, and missions with high political significance, such as China's lunar exploration program. For these space missions, the "no failure" requirement can be applied, although some unexpected technical imperfections or even mission failures might still happen, due to the nature of space flights and space exploration. Space science missions, in principle, should not belong to this category.

The **"good"** category could include most major space science missions, major communication satellites, weather satellites, earth observation satellites, navigation satellites, etc. These satellites are usually quite expensive and perform some important functions, but also normally have redundancies or can be replaced later without causing major losses or disasters. Specifically for space astronomy, it is suggested that the "L" or "M" class missions in ESA and "Observatory" or "Midex" class missions in NASA should belong to this category. This way the total cost for each mission can be reduced by a large fraction; therefore, for a fixed total budget, total missions launched can be increased by the same fraction. However, it is quite possible that the failure rate will increase by the same fraction, thus resulting in an overall gain. Then specifically for each mission, the slightly reduced safety margin will permit the use of new (and thus slightly risky) technologies, resulting in a large increase in the impacts and results of each successful mission. In turn, space technologies and space science techniques will be advanced faster than before.

The last category, i.e., the **"fair"** category, could include all the rest of space missions, such as the "M" and "S" class missions in ESA and the "Explorer" class and smaller/cheaper missions in NASA. By allowing the safety margin at each stage of the mission to be reduced fairly, the total cost for each mission could be reduced significantly and even by a large factor, thus permitting many more missions to be launched for the same budget. In the meantime, new technologies and techniques could be applied widely and frequently; this is particularly important for space science research, because new technologies and techniques always bring in new capabilities and thus significantly increase the discovery space enabling the achievement by each mission of richer and more unexpected breakthroughs. Combined with more frequent access to space, advances in space technologies and space science techniques will be speeded up much more quickly than previously, which in turn will allow much quicker scientific advances from space exploration, as already evidenced by the fact that major astronomical discoveries are mostly not predicted and serendipitous.

The main conclusion is that by implementing EATS, space utilisation, including for military and commercial applications and more importantly for scientific exploration, can be globally more effective even with the same total budgets/investments.

Acknowledgements The author thanks the invitation of the ESPI Director, Peter Hulsroj, to attend the 2015 Autumn Conference, as well as the warm hospitality and kind support of ESPI. This research work is partially supported by the National Basic Research Program ("973" Program) of China (Grants 2014CB845802), the National Natural Science Foundation of China (Grants No. 11373036 and 11133002), the Qianren start-up grant 292012312D1117210, and the Strategic Priority Research Program "The Emergence of Cosmological Structures" (Grant No. XDB09000000) of the Chinese Academy of Sciences.

Chapter 8
Potentially Transformative Technologies for Twenty-First Century Space

Leopold Summerer

8.1 Introduction

Space as the discipline to demonstrate economic and political might and as military high ground has spurred human ingenuity to invent new concepts and develop new technologies necessary to take full advantage of it. This 'fourth space', following land-, sea- and air-spaces needed to be made reachable, navigable and useable. Driven by governmental ambitions and means, space agencies have thus developed the technologies to reach space (chemical rockets), to navigate in it (electric, chemical, nuclear and direct solar propulsion, position and trajectory determination), to use its resources (e.g. to generate electrical power via photovoltaic cells) and to communicate. In doing so space agencies created and shaped new markets and new industries, typically extending the scope of those already present in the aeronautical sector.

Created by governments for governmental needs, some of these have already transitioned from governmental to private sector driven markets, starting with telecom applications in the 1990s. Stimulated by US policy decisions in the early 2000s, the private sector has entered virtually all space domains except deep space exploration. Against this background the present paper attempts to explore potentially game changing technologies for space applications.

In one of his excellent 2010 BBC Reith lectures focussed on surviving the twenty-first century and entitled "What we will never know", UK Astronomer Royal Sir Martin Rees, traced the origins of all major transformative innovations that shape our lives and economies to three key advances in science and technology that were innovations by themselves.

L. Summerer (✉)
European Space Agency, Noordwijk, The Netherlands
e-mail: Leopold.Summerer@esa.int

C. Al-Ekabi et al. (eds.), *Yearbook on Space Policy 2015*, Yearbook on Space Policy, DOI 10.1007/978-3-7091-4860-0_8

As with most transformative innovations before, the pervasive impact of their initial discoveries was not evident, except to a few visionary scientists. All of these three innovations gestated in the 1950s.

In 1953, Watson and Crick published their seminar nature paper on the "Molecular Structure of Nucleic Acids: A Structure for Deoxyribose Nucleic Acid", in which they described for the first time the double helix, the biological structure of heredity, creating the foundation of the new scientific disciplines of genetics, molecular biology and biochemistry. Spurred by new manipulation methods such as the CRISPR interference technique[1] the full impact of this discovery is likely still to come, contributing to predictions calling the twenty-first century the biological century.

In 1958 Kilby and Noyce constructed the first integrated circuit based on the invention of the transistor in 1947, building the precursor of the silicon chip, arguably one of the most transformative inventions of the twentieth century. Similar to the space sector, the IT industry has been driven by a stimulating and fruitful interplay between governmental needs creating early markets, targeted governmental research funding, and private sector dynamics.[2]

In 1957 the first artificial satellite, Sputnik, added space as a fourth activity area of humanity after the land, sea and airspaces, and triggered the space race which channelled an unprecedented amount of defence-type public funding into the new discipline and leaped mankind onto the moon and into a future which would otherwise have come much later. It was not only a triumph of human engineering and organisational ingenuity, motivating generations to embark on Science, Technology, Engineering and Mathematics (STEM) studies and research, but the images of Earth's fragile biosphere and the inhospitable lunar environment also transformed humanity's view of our home planet and led to new social and political movements such as environmentalism.

Contrary to the impact of the integrated circuits and molecular biology, the impact of space is largely invisible to everyday life and work, though similarly pervasive and fully integrated into modern economies. All three, especially when they converge and the borders between the disciplines that they have created overlap, still have substantial transformative power, giving rise to new waves of innovation, creating new markets, new disciplines such as in the fields of info-tech, nano-tech, bio-tech and eco-tech, which are central to addressing global challenges

[1]Horvath, P., Barrangou, R., 2010. CRISPR/Cas, the Immune System of Bacteria and Archaea. Science 327, 167–170. doi:10.1126/science.1179555.

[2]An excellent account of this fruitful interplay is described in a 1999 report by the US National Research Council: Hughes, T. et al., 1999. Funding a revolution: government support for computing research (Report from the National Research Council (US). Committee on Innovations in Computing No. 0-309-52501-2), Lessons from History. National Academies Press, Washington D.C., US. More recently Mazzucato has published convincing evidence on the important role of governments in the innovation process: Mazzucato, M., 2011. The entrepreneurial state. Soundings 49, 131–142. doi:10.3898/136266211798411183.

such as ageing populations, climate change and adapting our economies and societies to the reduced availability of abundant resources.

This has been evident at the intersections between the information technologies based on the power of integrated circuits, and biology. Similarly, first the introduction of miniaturised integrated circuits into space systems and more recently the adoption of working methodologies and approaches from the IT sector into space have started to transform the space sector, initiating what might be called "Space 4.0" or "New Space".

Against this background, the following section describes three themes that all have the potential to transform the space sector: autonomy in Sect. 8.2.1, addressing ionising radiation in Sect. 8.2.2, and in space resource utilisation and very large structures in Sect. 8.2.3. This list is not intended in any way as a forecast, but is provided as the result of a combination of the analysis of research trends by the ESA Advanced Concepts Team[3] and an analysis of space sector trends. Section 8.2.4 describes an inherently non-exhaustive list of technology advances with the potential to change the way space systems are designed and used.

In each of these themes, the paper purposefully does not include information on the current technology roadmap and harmonisation activities coordinated with ESA Member States and with industry and described in detail in the ESA Space Technology Master Plan,[4] which constitute the aggregated and consolidated space technology developments by ESA, its Member States and Cooperating States, the EU and industry "with the aim of optimizing public funding and guiding developments to ensure the right technology is at the right maturity at the right time".[5]

The detailed role of the Advanced Concepts Team in the preparation of the future at ESA is described in dedicated publications.[6] For the context relevant to this paper, it is reminded that the ACT only provides input to the ESA internal strategy process.

---

[3]For further information on ACT research topics and trends: Summerer, L., 2013. Thinking tomorrows' space – research trends of the ESA advanced concepts team 2002–2012. Acta Astronautica. doi:10.1016/j.actaastro.2013.11.002; Summerer, L., 2012. Evaluating research for disruptive innovation in the space sector. Acta Astronautica 81, 484–498. doi:10.1016/j.actaastro.2012.08.009 and www.esa.int/act (last accessed 3 June 2016).

[4]For updated information on the ESTMP: http://www.esa.int/Our_Activities/Space_Engineering_Technology/About_strategy_and_harmonisation (accessed 3 June 2016), ESA "European Space Technology Master Plan" 12th edition, European Space Agency June 2015.

[5]Dordain, J.J., 2015 "Foreword to the 12th edition of the European Space Technology Master Plan", June 2015.

[6]Summerer, L., 2012. Evaluating research for disruptive innovation in the space sector. Acta Astronautica 81, 484–498. doi:10.1016/j.actaastro.2012.08.009; Summerer, L., Izzo, D., Naja-Corbin, G., I., D.-B., 2010. The seeds of disruptive innovation. ESA Bulletin 144, 34–45.

8.2 Potentially Transformative Themes for the Space Sector

8.2.1 Autonomy

The quest for autonomy of space systems has a long history. Autonomy is understood as a substantial step further than automation by giving space systems the ability to take decisions without human interference based on high-level objectives and sensory input. It includes both control system theory and Artificial Intelligence (AI), which are both necessary but individually not sufficient for autonomy.

Traditionally, space systems tend to be expensive, one-off devices that require taking relatively high levels of risk to achieve the main mission goals. Therefore, any non-essential additional risks are avoided. In terms of control systems, this typically results in full remote control for all critical elements except where it cannot be achieved, such as some phases of planetary entry descent and landing in which case one uses automation, understood as predetermined decisions that make the system act exactly as programmed. Autonomy is thus necessarily associated with a loss of control, of predictability and also of insight into the decision process, which limits post-accident failure investigation.

At the same time, spacecraft are predestined for applications of autonomous, 'intelligent' systems: the remoteness and limitations imposed by the absolute limit of the speed of light for communication and the inherently unknown environments exploration missions face provide strong incentives for conceiving spacecraft able to adapt, react and learn from their interactions with these unknown environments. Space agencies therefore typically embraced the prospects of AI and employed AI departments and research teams up to the 1980s and 1990s. Given the poor track record of AI research and the multiple false promises, why should AI be expected to deliver more on its promises now?

The answer attempted here is threefold.

First, the basic reasons for the need for autonomy have not changed: spacecraft are far away and the speed of light remains a solid barrier making effective communication with probes in the outer solar system painstakingly slow. Efforts to introduce high speed and data-rate deep space communication including optical communication links are considered as a confirmation of this underlying need.

Second, the incentives for autonomy have increased. The cost of spacecraft and launch costs have been decreasing, enabling more complex and larger constellations of spacecraft and at the same time increasing for operations the available percentage of total cost. This trend is further aggravated with long life times. Operation costs are to a large extent human resources costs but the operations of very large constellations such as those proposed for Low Earth Orbit telecom applications ('megaconstellations') require a radically different, less human intensive operations approach. In addition, the miniaturisation of sensors has enabled a substantial increase in the quality and volume of data on-board.

Third, the combination of availability of large structured data, new algorithms and commercial incentives in high-value added industries such as smart phones, digital assistants, gaming industry and transport systems has triggered a renaissance in the field of artificial intelligence. Major industries have identified autonomy as one of the driving differentiating elements and have thus been starting to invest substantial R&D funding into the field, which was so far dominated by academic and defence driven funding. The investments of the car industry into autonomous driving have channelled substantial additional money and capabilities into the research field. The takeover by the car-sharing company Uber of the majority of researchers from Carnegie Mellon's National Robotics Engineering Centre, one of the leading US autonomy and robotic labs, in 2015 can be seen as symptomatic of a broader trend.[7] The advances achieved by combining the power of machine learning with the data processing capacities of Google has enabled computers to win against the best humans in playing Go, the last high profile popular board game that was considered to be still dominated by human intelligence for several years longer.[8] Driven by their economic potential, smart assistants are expected to be gradually introduced into high value added domains such as the legal, medical and engineering professions.

While some levels of AI are already used for some tasks such as trajectory and schedule optimisation, (failure) pattern detection and payload autonomy,[9] the combination of the three elements listed above is expected to lead to substantially different spacecraft, requiring space agencies and space industry to change their system engineering approaches from the initial design, to quality control and operations. Exploratory research by ESA's Advanced Concepts Team in this field has demonstrated the potential of the introduction of AI to a range of space

[7]Ramsey, M., MacMillan, D., 2015. Carnegie Mellon Reels After Uber Lures Away Researchers. Wall Street Journal. 31 May 2015.

[8]Silver, D., Huang, A., Maddison, C.J., Guez, A., Sifre, L., van den Driessche, G., Schrittwieser, J., Antonoglou, I., Panneershelvam, V., Lanctot, M., Dieleman, S., Grewe, D., Nham, J., Kalchbrenner, N., Sutskever, I., Lillicrap, T., Leach, M., Kavukcuoglu, K., Graepel, T., Hassabis, D., 2016. Mastering the game of Go with deep neural networks and tree search. Nature 529, 484–489. doi:10.1038/nature16961.

[9]This has been successfully demonstrated in the ESA Proba-1 technology demonstration spacecraft, cf. Teston, F., Vuilleumier, P., Hardy, D., Bernaerts, D., 2004. The PROBA-1 Microsatellite, in: Proc. of SPIE Vol. pp. 132–140. And Bernaerts, D., Bermyn, J., Teston, F., 2002. PROBA (Project for On-board Autonomy), in: Rycroft, M., Crosby, N. (Eds.), Smaller Satellites: Bigger Business? Springer Netherlands, Dordrecht, pp. 53–68.

applications such as surface exploration,[10] antenna design,[11] multi-disciplinary optimisation of spacecraft and mission design, very large formations and swarms,[12] human-robotic interactions,[13] navigation in complex gravity fields, autonomous landing,[14] robotics[15] and robotic vision[16] and active space debris removal.[17]

8.2.2 *Ionising Radiation*

On leaving the protection of Earth' magnetic field, humans and spacecraft are exposed to two sources of radiation: galactic cosmic rays (GCR) and solar particle events (SPE). GCR consist of high-energy protons, helium nuclei, and heavy nuclei with energies extending to more than 10 GeV/amu, while SPEs are largely protons with energies up to a few hundred MeV with a small helium and heavy ion component. GCR interaction with the Earth's atmosphere and magnetic field creates trapped proton and electron belts important for missions in low-Earth orbit (LEO). The sun's magnetic field modulates the GCR over an approximately 11-year cycle with change of more than a factor of two in flux rates at the highest

[10]Nicol, C., Ellery, A., Cloutis, E., Lynch, B., O'Connor, L., and de Croon, G.C.H.E., Scent of Science: Autonomous Source Localization for Exploration, *European Space Agency, the Advanced Concepts Team*, Ariadna Final Report (11-6301), 2013. And Papadopoulos, E. G., Kontolatis, I., and Paraskevas, I.S., Space Gaits, *European Space Agency, the Advanced Concepts Team*, Ariadna Final Report (12-5201), 2013.

[11]Versloot, T.W., Barker, D.J., and Otero One, X., Optimisation of Near-Field Wireless Power Transfer Using Evolutionary Strategies, *The 8th European Conference on Antennas and Propagation – EUCAP 14, The Hague, NL*, The Hague, Netherlands 2014.

[12]Broecker, B., Caliskanelli, I., Tuyls, K., Sklar, E., and Hennes, D., Social Insect-Inspired Multi-Robot Coverage, *AAMAS – 14th International Conference on Autonomous Agents and Multiagent Systems*, pp. 1775–1776, 2015 and Izzo, D., Simoes, L.F., and de Croon, G.C.H.E., An evolutionary robotics approach for the distributed control of satellite formations, *Evolutionary Intelligence*, **7**(2), pp. 107–118, 2014.

[13]Claes, D., Hennes, D., and Tuyls, K., Towards Human-Safe Navigation with Pro-Active Collision Avoidance in a Shared Workspace, *DEMUR – IROS Workshop on On-line Decision-Making in Multi-Robot Coordination*, 2015.

[14]Sanchez-Sanchez, C., Izzo, D., and Hennes, D., Optimal Real-Time Landing Using Deep Networks, *6th International Conference on Astrodynamics Tools and Techniques (ICATT)*, 2016. And Izzo, D. and de Croon, G.C.H.E., Landing with Time-to-Contact and Ventral Optic Flow Estimates, *Journal of Guidance Control and Dynamics*, **35**(4), pp. 1362, 2012.

[15]Methenitis, G., Hennes, D., Izzo, D., and Visser, A., Novelty Search for Soft Robotic Space Exploration, *GECCO – Genetic and Evolutionary Computation Conference*, 2015. And Broecker, B., Caliskanelli, I., Tuyls, K., Sklar, E., and Hennes, D., Social Insect-Inspired Multi-Robot Coverage, *ARMS – Autonomous Robots and Multirobot Systems*, 2015.

[16]Qadi, A. and de Croon, G.C.H.E., Probabilistic Computing for Efficient Robotic Vision in Space, *European Space Agency, the Advanced Concepts Team*, Ariadna Final Report (12-5101), 2014.

[17]Izzo, D., Getzner, I., Hennes, D., and Simoes, L.F., Evolving solutions to TSP variants for active space debris removal, *GECCO – Genetic and Evolutionary Computation Conference*, 2015.

conditions called solar minimum from the lowest point in the cycle called solar maximum. The effective ionising dose rates and effective doses thus vary with the solar intensity cycles and in any case are a major critical and mission constraining factor not only for human exploration missions (e.g. lunar and Martian missions) but also for the re-use in space of materials and components developed for terrestrial applications.

During the solar cycle 23–24 minimum, the time to 3 % Risk of Exposure Induced Death (REID) in interplanetary space is typically less than 400 days for a 30-year-old male and less than 300 days for a 30-year-old female in the last cycle 23–24 minimum.[18]

The current approach to reducing the risks related to ionising radiation for human Mars missions are summarised by Durante et al.[19]: "*Biological countermeasures are not yet mature [...]. Among physical countermeasures, passive shielding is the only one presently available, but it is unlikely to be able to reduce the dose to an acceptable level, within the weight constraints of the launchers. Novel shielding materials can however give a significant reduction [...]. Active shielding, especially toroidal magnetic configurations are very promising, but still not mature enough for spaceflight. The best solution to the space radiation problem, as well as of the other health risks related to microgravity and isolation, is reduction of the transit time. This can only be achieved using nuclear propulsion, possibly a mixed system combining thermal and electric nuclear power (NTER), but this field also needs major developments for practical applications.*

Planners of the first mission to Mars will probably look for combinations of different approaches: passive shielding (including a storm shelter against SPE) and transit time, with the latter incorporating decisions about trajectory [...] and timing (e.g. at solar minimum to reduce the chance of SPE, or solar maximum to reduce chronic GCR exposure). Active shielding and NTER are probably the future options for later exploration missions, such as to the asteroid belt and the outer planets and their moons."

For electronic components, new technologies such as memristors[20] might change the picture due to their inherently higher radiation resistance. One of the most surprising suggestions related to countermeasures to lower the radiation risk

[18]Effective dose (expressed in Sv, the weighted equivalent of a joule of radiation energy absorbed in a kg of tissue) is used to quantify the effect of radiation on the human body.

The career effective dose before the 3 % REID limit is reached is 0.47 Sv for a 30 year-old female and 0.62 Sv for a 30 year-old male [NRC, 2008].

[19]Durante, M., 2014. Space radiation protection: Destination Mars. Life Sciences in Space Research 1, 2–9. doi:10.1016/j.lssr.2014.01.002.

[20]Fantini, A. 2015, presentation on "Radiation hardness of memristive systems" at the Workshop on Memristive systems for Space applications, ESA ESTEC, 30 April 2015, http://www.esa.int/gsp/ACT/events/workshops/ws_2015_Apr_30.html; and e.g. An overview of an overview of the effects of ionizing radiation and heavy ions on tantalum oxide (TaOJ), titanium dioxide (TiO2), and hafnium oxide (HfOx) memristive devices can be found in McLain, M.L., Marinella, M.J., 2015. Overview of the radiation response of anion-based memristive devices. IEEE, pp. 1–11. doi:10.1109/AERO.2015.7119304.

for human space-travel comes from research on hibernation. Studies have shown that during hibernation phases, animals are substantially more radiation tolerant.

As part of the research on potentially disruptive, game changing concepts, ESA has been considering torpor/hibernation since 2002.[21] Since 2014, a dedicated interdisciplinary topical team chaired by Prof. Bereiter-Hahn and Prof. Heldmaier has advised ESA on the topic. Studies have shown that hibernation would not only allow for much lower mass and potentially volume, but interestingly have also led to a better understanding of the mechanisms responsible for the absence of bone (and muscle) loss in hibernating animals. The fact that hibernation occurs in practically all branches of the mammalian phylogeny, indicates that the mechanisms are more fundamental than their separations 100 Myears ago, and it would be surprising if they were not also still present in some form in humans.[22]

8.2.3 In Space Resource Utilisation: Very Large Structures

All spacecraft so far have been constrained by the limitations related to the need to transport them from Earth into space. The fact that spacecraft need to be put into launcher fairings naturally restricts their dimensions, their mass, and their structural composition (vibration and shock environment during launch phases), which in combination result in relatively similar spacecraft morphologies (central bodies with deployable features such as solar panels, antennas, instrument booms, thermal radiators). Large structures are needed for antennas, large baseline optics, and volumes for habitats, for EO, telecom, science and exploration applications.

The combination of relatively large launchers with large fairings, the miniaturisation of space components and the use of reliable deployable elements, has enabled the gradual increase in spacecraft performances and has achieved relatively large spacecraft. There is still room for further increases before requiring robotic in-space assembly.

Today essentially all three approaches are used: 1. Miniaturisation, 2. Deployable structures (unfolding, extending, inflating) and 3. In-space robotics assembly.

Many of the mass driving requirements for space systems come from the need to survive the harsh conditions at launch undamaged. Constructing these in space

[21]Biggiogera, M., Malatesta, M., Zancanaro, C., 2004. Mammalian Hibernation Mechanisms: Relevance to a Possible Human Hypometabolic Induced State (Ariadna 03-6501). European Space Agency, Advanced Concepts Team; Rossini, L., Seidl, T., Izzo, D., Summerer, L., 2007. Beyond astronaut's capabilities: a critical review, in: Proceedings of the 58th International Astronautical Conference, Hyderabad; and Gemignani, J., Gheysens, T., Summerer, L., 2015. Beyond astronaut's capabilities: The current state of the art. IEEE, pp. 3615–3618. doi:10.1109/EMBC.2015.7319175.

[22]Gemignani, J., Gheysens, t., and Summerer, L., Beyond Astronaut's Capabilities: The Current State Of The Art, *37th Annual International Conference of the IEEE Engineering in Medicine and Biology Society, Milan, August,* 2015.

would not only free engineers to use more appropriate shapes and forms than those fitting into launcher fairings but also has the potential to reduce overall mass—at the expense of the complexity of in-space manufacturing.

So far this has been impractical, but advances in 3D/4D printing and additive layer manufacturing technology have the potential to substantially change the trade-off space. ESA has launched a new advanced manufacturing initiative, focussing on bringing the full benefits of advanced manufacturing and material science to space, including 3D printing. This will allow bringing the design and manufacturing processes of spacecraft closer to that used in the car and aeronautical industries. Under the 'design to produce' initiative, the full potential of digital manufacturing will enable reducing time and cost in the design and production of space systems. These will constitute an important first step into the area of in-space manufacturing, though further substantial advances in a number of areas are necessary.

Substantial advances have been made in collaborative robotics, high precision formation flying, and multi-agent massive formation flying. Technology developments made in the frame of the ESA Clean Space initiative on active space debris removal concepts have been advancing key technologies for in-space manufacturing. The first production of simple elements via 3D printing in microgravity has been demonstrated in an operational environment on the ISS.

It can be expected that the first in-space production attempts will rely on the use of material transported from Earth and concern simple elements. In-space resource utilisation would continue to be limited to the use of solar radiation for heat and electricity generation, potentially with substantially increased power levels first for local and then for remote though close-by use via wireless power transfer based on microwave or lasers. The re-use of elements of old spacecraft in space seems to be a natural next step, developing and relying on simple modifications to elements such as joining and later electric interfaces.

A step-change in the use of in-space resources can be expected with the start of extended surface operations. These would shift the trade-off balance from the cost and limitations of earth-to-space transportation to the complexity of the extraction and processing of in situ resources. First elements might be water on asteroids/comets for propulsion, energy storage or power. Investigations into the use of lunar regolith as base material for the 3D printing of structural elements, shelters and eventually entire habitat parts for lunar surface operations have demonstrated principal feasibility under relevant conditions.[23] These could also be applied to provide protection from radiation, micrometeoroid impacts, and thermal environment.

[23]Cesaretti, G., Dini, E., De Kestelier, X., Colla, V., Pambaguian, L., 2014. Building components for an outpost on the Lunar soil by means of a novel 3D printing technology. Acta Astronautica 93, 430–450. doi:10.1016/j.actaastro.2013.07.034; and Kading, B., Straub, J., 2015. Utilizing in-situ resources and 3D printing structures for a manned Mars mission. Acta Astronautica 107, 317–326. doi:10.1016/j.actaastro.2014.11.036.

The ongoing extension of private sector, commercially driven space activities to beyond low Earth orbit is expected to inject additional resources, dynamism and ingenuity into the development of the use of in-space resources.

8.2.4 *Further, Potential Game Changing Technologies*

While all three transformative themes described in Sects. 8.2.1–8.2.3 are the results of the combination of technological advanced with substantiated application needs, this section lists a number of potentially game changing technologies that follow a technology push logic. They are considered as exemplary from the research performed by the Advanced Concepts Team in the context of seeding disruptive innovation in the European space sector.

The efficient conversion of solar energy into electricity is one of the fundamental enabling elements of practically all space activities. Except for some deep space missions and planetary surface missions that rely on nuclear power sources, the large majority is based on photovoltaic panels. Modern multi-junction PV cells achieve efficiencies close to 40 %. A different approach has recently been demonstrated based on conversion via nano-size optical rectennas, which promises to offer new and different applications, including highly efficient wireless power transmission systems via laser.

Similarly, spacecraft operating close to the sun must deal with temperatures beyond the operating temperatures of photovoltaic cells. Recently, researchers have conceived a photon-enhanced thermionic emission concept, combining a thermionic with a photovoltaic system, which could theoretically offer similar efficiencies at such temperatures ranges.[24]

Reliable soft landing on celestial bodies remains a mission critical challenge. Recent studies on autonomous spacecraft landing use computer vision methods to improve the accuracy of state estimates used for landing.[25] Typically, these studies integrate the vision module with other exteroceptive sensors such as laser or radar altimeters. A backup emergency system or a system for micro spacecraft would benefit from a robust lightweight solution entirely based on vision and

[24]Schwede, J.W., Bargatin, I., Riley, D.C., Hardin, B.E., Rosenthal, S.J., Sun, Y., Schmitt, F., Pianetta, P., Howe, R.T., Shen, Z.-X., Melosh, N.A., 2010. Photon-enhanced thermionic emission for solar concentrator systems. Nature Materials 9, 762–767. doi:10.1038/nmat2814; Buencuerpo, J., Llorens, J.M., Zilio, P., Raja, W., Cunha, J., Alabastri, A., Zaccaria, R.P., Martí, A., Versloot, T., 2015. Light-trapping in photon enhanced thermionic emitters. Optics Express 23, A1220. doi:10.1364/OE.23.0A1220; Buencuerpo, J., Llorens, J.M., Marti, A., Cunha, J., Summerer, L., and Versloot, T.W., Photon Enhanced Thermionic Emission, *European Space Agency, the Advanced Concepts Team*, Ariadna Final Report (14-2101a), 2015.

[25]Mourikis, A., Trawny, N., Roumeliotis, S., Johnson, A., Ansar, A., and Matthies, L., Vision-Aided Inertial Navigation for Spacecraft Entry, Descent, and Landing, IEEE Transactions on Robotics, Vol. 25, No. 2, 2009, pp. 264–280.

proprioceptive sensors (e.g. gyros). Small flying animals are capable of safe and accurate landings while relying only on proprioceptive and visual information. Using the *ventral optic flow* as a measure of the translational velocity divided by the height, bees are known to use optic flow for controlling their speed and height, and when landing, they use a strategy of keeping the ventral optic flow constant.[26] Several studies have demonstrated that under some conditions, spacecraft landing control systems could be designed based on ventral optic flow and time to contact.[27]

8.3 Conclusions

Human endeavours in space have been used as the symbol of human progress during the twentieth century. Space has silently entered modern economies as an indispensable element. The twenty-first century promises to reap the fruits of these technological advances, fully bringing the benefits of using near Earth space to not only a select group of "space-faring nations" and opening a new economic sphere around Earth, allowing space agencies to concentrate more governmental resources towards the exploration of those areas that are not yet commercially viable.

Based on an analysis combining technology push and needs, this paper has described three potentially transformative domains: ways to deal with ionising radiation, the transformative power of the introduction of real autonomy based on advances in artificial intelligence and advanced robotics, and the use of advanced manufacturing techniques up to the use of resources in space, overcoming traditional restrictions linked to launchers.

The development of these will continue to involve a dynamic interplay between technology push for incremental innovation, scientific and technical discoveries that open new possibilities, the evolution of markets serving societal needs and the creation of new markets, triggered by governmental needs and governments as anchor customers, e.g. public sector infrastructure needs creating new markets for typical public sector services such as transportation, energy, waste removal, and communication.[28]

[26] Srinivasan, M., Zhang, S., Lehrer, M., and Collett, T., Honeybee Navigation en Route to the Goal: Visual Flight Control and Odometry, The Journal of Experimental Biology, Vol. 199, 1996, pp. 237–244.

[27] Valette, F., Ruffier, F., Viollet, S., and Seidl, T., Biomimetic optic flow sensing applied to a lunar landing scenario, Proceedings of the IEEE International Conference on Robotics and Automation (ICRA), IEEE, 2010, pp. 2253–2260; Izzo, D., Weiss, N., and Seidl, T., Constant-Optic-Flow Lunar Landing: Optimality and Guidance, Journal of Guidance, Control, and Dynamics, Vol. 34, No. 5, 2011, pp. 1383–1395; Izzo, D. and de Croon, G.C.H.E., Landing with Time-to-Contact and Ventral Optic Flow Estimates, *Journal of Guidance Control and Dynamics*, **35**(4), pp. 1362, 2012.

[28] A solid account on the role of governments in the creation of such markets, it is referred to Mazzucato, M., 2015. From Market Fixing to Market-Creating: A New Framework for Economic Policy (SPRU Working Paper Series No. SWPS 2015-25). University of Sussex, Sussex.

In this context, space needs to be considered as being first and foremost a "place", which follows the scheme of new places: discovery, pioneering, exploration and exploitation. The transformative, potentially game changing concepts described in this paper are expected to help transition cis-lunar space from exploration into exploitation, turning it into an extension of the terrestrial economy.

Chapter 9
Big Data from Outer Space: Opportunities and Challenges for Crisis Response

Nina Witjes, Philipp Olbrich, and Isabella Rebasso

9.1 Introduction

Shortly after the earthquake that devastated Haiti in January 2010, Google and the satellite imagery provider GeoEye released images demonstrating the level of destruction inflicted by the disaster. These images helped those on the outside grasp the severe consequences of the earthquake, but more importantly the availability of high-resolution imagery yielded opportunities for humanitarian organisations in the aftermath of this disaster. The combination of geospatial information technologies such as remote sensing, satellite navigation, internet mapping technologies with crowdsourcing initiatives or machine-learning algorithms was used to draft crisis maps and to facilitate distribution of scarce resources in support of response teams on the ground.

The Haiti earthquake is said to mark the beginning of the era of *digital humanitarianism*, which Burns defines as "the enacting of social and institutional networks, technologies, and practices that enable large, unrestricted numbers of remote and on-the-ground individuals to collaborate on humanitarian management through digital technologies."[1] The defining features of digital humanitarianism are therefore embodied in technologies of data collection, analysis and communication to improve situational awareness and effectiveness through better coordination

[1]Burns R (2014) Moments of Closure in the Knowledge Politics of Digital Humanitarianism. Geoforum, 53: 52.

N. Witjes (✉)
Munich Center for Technology in Society, Technical University Munich, Munich, Germany
e-mail: nina.witjes@tum.de

P. Olbrich
University of Groningen, Groningen, The Netherlands

I. Rebasso
Austrian Institute for International Affairs, Vienna, Austria

C. Al-Ekabi et al. (eds.), *Yearbook on Space Policy 2015*, Yearbook on Space Policy, DOI 10.1007/978-3-7091-4860-0_9

between traditional and computer-assisted approaches. Against this background, this chapter aims to scrutinise emerging opportunities and challenges for crisis response resulting from the introduction of commercial satellite imagery and Big Data analytics to humanitarian practices.

Although Big Data has been discussed extensively regarding its ontological and epistemological implications[2] as well as civil liberties and privacy issues,[3] the term *Big Data* remains rather ambiguous. Zwitter defines Big Data as "the enormous amounts of data that, using sophisticated analytics techniques, can be mined for information in order to reveal patterns and spot trends and correlations [. . .] that would not be available by looking at smaller samples."[4] Big Data can also be described by its three main properties, the three V's of Big Data: volume, velocity and variety. *Volume* refers to the amount of data; *velocity* describes the speed of data creation and collection; and *variety* points to the different types of data such as textual, audio, photographic, and video data. Furthermore, variety in the context of Big Data also refers to different sources and kinds of data such as self-generated or automatically collected data (data mining)[5] and raw data (e.g. individual social media updates) or metadata, which denotes circumstantial information such as the time, date and place of text messages or phone calls.[6]

Mayer-Schönberger & Cukier have identified three interlinked shifts related to Big Data that are set to change the way information is analysed: First, sampling becomes redundant if new data collection methods allow for n=all analyses of particular phenomena. Second, they argue that Big Data will change how we deal with exactitude as large samples are more resilient to errors and support predictive abilities especially on the macro-level. Third, in a Big Data world the search for causality will be replaced by recurrent patterns and correlations. Rather than testing previously defined hypotheses on the basis of small samples we are now able to make unforeseen connections by simply letting data speak for itself.[7] These developments are believed to make it easier to make predictions about the future such as forecasting flu outbreaks or emerging conflict.

Although the global impact of Big Data is hard to refute, grasping it in its full complexity and diverse concrete manifestations remains a challenge to policy

---

[2]Chandler D (2015) A World without Causation: Big Data and the Coming of Age of Posthumanism. Millennium, 43 (3), 833–851; Kitchin R (2014) Big Data, new epistemologies and paradigm shifts. Big Data & Society, April–June, 1–12; Frické, M (2014) Big Data and its Epistemology. Journal of the Association for Information Science & Technology, 66(4), 651–661.

[3]E.g. see Boyd D & Crawford K (2012) Critical Questions for Big Data: Provocations for a cultural, technological, and scholarly phenomenon. Information, Communication & Society, 15 (5), 662–679.

[4]Zwitter A (2015) Big Data and International Relations. Ethics & International Affairs, 29(4), 377–289.

[5]Zwitter (2015): 379.

[6]Meier P (2015) Digital Humanitarians. How Big Data is Changing the Face of Humanitarian Response. CRC Press, Boca Raton: 28–29.

[7]Mayer-Schönberger V, Cukier K (2013) Big Data: A Revolution That Will Transform How We Live, Work and Think. John Murray, London.

makers and scholars alike. In this chapter, we will zoom in one aspect, digital humanitarianism, to examine the interactions of large technological shifts in geospatial Big Data with humanitarian practices in crisis response. From a conceptual perspective, the chapter is infused by a Large Technical Systems (LTS) approach. LTSs have been defined by Joerges as "those complex and heterogeneous systems of physical structures and complex machineries which (1) are materially integrated, or coupled over large spans of space and time [...] and (2) support or sustain the functioning of a very large number of other technical systems, whose organisation they thereby link".[8] Yet, in addition to the material infrastructure the concept also accounts for human actors that not only shape and make use of it but are also affected by the technical conditions.

In this chapter we will look into the role of non-state actors, such as NGOs and commercial providers, within the system that has emerged around the rise of Big Data and the commercialisation of satellite technology to make sense of the changing face of disaster response and prevention. We will proceed as following. The first section outlines major changes in satellite technology brought about by the commercialisation of satellite technology that has taken place since the end of the 1990s. Then, the chapter illustrates the resulting opportunities for crisis response with the prominent example of the 2010 earthquake in Haiti as well as the crowdsourcing platform Tomnod. Against this background, the last section discusses the risks and challenges for crisis response when it is integrated with crowdsourcing and other Big Data analytics to cope with large amounts of satellite imagery data.

9.2 Crisis Response in a Commercial Space Age

9.2.1 Satellite Technology: New Actors, New Opportunities

Satellite imagery plays an essential role in many disaster relief efforts. However, as the body of imagery is growing, its timely analysis is becoming increasingly difficult. To illustrate, DigitalGlobe (DG) alone is currently producing about 1 billion km^2 of satellite imagery every year, i.e. 230 times the territory of the European Union (or 1000 times that of Egypt). While DG has been focusing on the very high-resolution spectrum of commercial Earth observation, it has already announced plans to complement its constellation with a number of small satellites to improve re-visit times. Such technological developments give rise to swarms of

[8]Joerges B (1988) Large Technical Systems: Concepts and Issues In: Mayntz R, Hughes P (eds) The Development of Large Technical Systems. Campus Verlag, Frankfurt am Main. p 9–37:24; see also Hughes, T P (1983) Networks of Power: Electrification in Western Society, 1880–1930. Baltimore: Johns Hopkins University Press; Mayer, M & Acuto, M (2015) The Global Governance of Large Technical Systems. Millennium, 43(2), 660–683.

much smaller satellites that come with a lower spatial resolution but have more frequent flyovers in order to image Earth on a daily basis. Innovative low-cost production guarantees that these mini satellites cost about $150,000 to $1 million, a fraction compared to their conventional forerunners, which cost between $200 million and $1 billion.[9] Moreover, they operate on a conceptual shift that stresses the resilience of the whole satellite constellation over an individual satellite so that singular malfunctions simply do not matter that much anymore. Consequently, new players, such as Skybox or Planet Labs, have been entering the market of generating and analysing satellite imagery. Skybox alone is planning on having 24 EO satellites in orbit by 2017[10] and integrating its imagery with other sources to create real-time information for a variety of fields such as traffic logistics, agriculture, and environmental monitoring. Planet Labs is on a similar mission but envisions operating a satellite constellation of more than 100 individual "doves" as the company calls their self-made nano-satellites. In essence, technological development and commercialisation of satellite imagery has increased the volume, variety and velocity of EO data. Moreover the industry is set to continue on this path, especially as the prospect of falling launch costs is expected to materialise ever more concretely. With this in mind, the following section provides insights into how these changes have already influenced the field of humanitarian action.

9.2.2 The Rise of Digital Humanitarianism

In 2011, Paul Conneally of the International Federation of the Red Cross and Red Crescent Societies contended in a talk at a TED conference that "[t]he humanitarian model has barely changed since the early twentieth century. Its origins are firmly rooted in the analogue age. And there is a major shift coming on the horizon."[11] In doing so, he was referring to an evolving "digital humanitarianism", which was largely unknown before an earthquake struck Haiti in 2010, devastating the city of Port-au-Prince and, according to official accounts, leaving over 100,000 Haitians and humanitarians feared dead.[12] The response by aid organisations to the earthquake was among the first efforts that employed high-resolution imagery for rapid

[9]The Economist (7 June 2014) Nanosats are go! Available via http://www.economist.com/news/technology-quarterly/21603240-small-satellites-taking-advantage-smartphones-and-other-consumer-technologies Accessed 22 Feb 2016.

[10]Scanlon M (2014) There's more to the small Skybox satellite than meets the eye. Much, much more Available via http://ideas.ted.com/theres-more-to-the-small-skybox-satellite-than-meets-the-eye-much-much-more/ Accessed 20 Jan 2016.

[11]Conneally P (2011) How mobile phones power disaster relief. Available via TED https://www.ted.com/talks/paul_conneally_digital_humanitarianism/transcript Accessed 14 Jan 2016.

[12]Meier P (2015): Digital Humanitarians. How Big Data is Changing the Face of Humanitarian Response. CRC Press, Boca Raton: 1; for and overview of the dispute over the death toll in the aftermath of the earthquake see O'Connor M R (2012) Two Years Later, Haitian Earthquake Death

damage assessment to such a great extent.[13] During the night after the earthquake a crisis map was created by turning to social media platforms to crowdsource information on the areas that were most affected by the disaster.[14] Experts and volunteers uncovered dozens of Tweets by Haitian witnesses describing where help was urgently needed, which they managed to chart by using the digital mapping platforms Ushahidi and Google Earth. From each geocoded Tweet, represented by a dot, a crisis map emerged which formed a real-time assessment of the devastation on the ground and the needs of the affected population. Additionally, a telephone number was put in place, to which Haitians could text pressing needs as well as further information on building damage or food shortages. The digital volunteers were soon joined by OpenStreetMap (OSM), a crowdsourced mapping platform which provides open-source street maps. OSM was supplied with high-resolution satellite and aerial imagery of the area by a number of partners, including the National Oceanic and Atmosphere Administration (NOAA) and the Rochester Institute of Technology. The remarkable speed of damage assessment was facilitated by both private companies, such as Digital Globe or Google, and public organisations, such as NOAA and the U.S. Geological Survey, who made pre- and post-earthquake satellite imagery freely available to the international community. The initial satellite imagery was complemented by high-resolution aerial photographs and field survey data.[15] The crisis map of Haiti was especially relevant in terms of raising situational awareness and critical early information, before large organisations were on site and started their operations.[16] It was used by numerous organisations such as the United Nations Disaster Assessment and Coordination, the United States Marines, and Urban Search and Rescue.[17]

The Haiti earthquake constitutes a prime example of digital humanitarianism in practice. In its aftermath, several global networks of digital humanitarian volunteers were launched to analyse the lessons-learned from Haiti and put them to use elsewhere; this includes initiatives such as the Standby Volunteer Task Force (SBTF) as well as the Digital Humanitarian Network (DHN), a United Nations (UN) co-founded interface between volunteers and humanitarian organisations. These emerging digital humanitarian networks are co-evolving as a part of an LTS that characterises their work and is becoming increasingly sophisticated.

Toll in Dispute. Available via Columbia Journalism Review http://www.cjr.org/behind_the_news/one_year_later_haitian_earthqu.php Accessed Feb 2 2016.

[13]Corbane et al. (2011) A Comprehensive Analysis of Building Damage in the 12 January 2010 Haiti Earthquake using High-Resolution Satellite and Aerial Imagery. Available via JRC http://publications.jrc.ec.europa.eu/repository/handle/JRC71936 Accessed Jan 18 2016.

[14]Meier P (2015) Digital Humanitarians. How Big Data is Changing the Face of Humanitarian Response. CRC Press, Boca Raton.

[15]Corbane et al. (2011).

[16]Morrow N et al (2011) Independent Evaluation of the Ushahidi Haiti Project. Available via https://sites.google.com/site/haitiushahidieval/news/finalreportindependentevaluationoftheushahidihaitiproject Accessed 28 Jan 2016.

[17]Morrow N et al.

The use of high-resolution satellite and aerial imagery has turned into a more common source for civil society, commercial actors and international organisations to produce real-time information in times of crisis. On the analysis side, the methods are increasingly embracing human-machine interactions such as crowd computing or microtasking. With this in mind, there seems to be a common understanding in the satellite imagery community that advanced computing and artificial intelligence in the form of data-mining and machine-learning algorithms will become a major part of digital disaster response and satellite imagery analysis in general.[18]

Today, the crowdsourcing platform Tomnod, which makes use of machine learning, automated object identification and human analysis to analyse primarily satellite imagery,[19] has become paragon focal point for digital humanitarians. Tomnod is supplied with satellite images by DigitalGlobe, which acquired the platform in 2013. Volunteers scan satellite imagery and tag certain features such as damaged buildings in the case of earthquakes. An algorithm then filters out frequently tagged regions and automatically determines the reliability before further processing, e.g. redirection to a satellite imagery expert for further analysis. In 2011 the United Nations High Commissioner for Refugees (UNHCR) partnered with Tomnod, DigitalGlobe, the SBTF and Ushahidi to locate camps of internally displaced people in Somalia, which has been disrupted by a civil war.[20] Tomnod's most prominent mission to date is the search for flight MH370, the Malaysian aircraft that mysteriously disappeared in March 2014. Over eight million volunteers joined the search and analysed over 1 million km^2 of satellite imagery.[21] Further projects have included mapping features to support rescue, recovery and analysis efforts after earthquakes in Nepal and Chile in 2015 and after tropical cyclone Winston that passed over Fiji in early 2016.

[18]Olbrich P, Witjes N (2016) Sociotechnical Imaginaries of Big Data: Commercial Satellite Imagery and its Promise to Speed and Transparency (forthcoming); Meier P (2015) Virtual Aid to Nepal. Using Artificial Intelligence in Disaster Relief. Available via Foreign Policy. https://www.foreignaffairs.com/articles/nepal/2015-06-01/virtual-aid-nepal Accessed 13 Jan 2016.

[19]Tomnod.com Available via http://www.old.tomnod.com/?q=technology Accessed 22 Feb 2016.

[20]Meier P (2 Nov 2011) Crowdsourcing Satellite Imagery Tagging to Support UNHCR in Somalia Available via http://blog.standbytaskforce.com/2011/11/02/crowdsourcing-satellite-imagery-tagging-to-support-unhcr-in-somalia/ Accessed 25 Feb 2016.

[21]DigitalGlobe (5 May 2014) Crowdsourcing Malaysia Flight #MH370 – Campaign Comes To a Close. Available via http://www.digitalglobeblog.com/2014/05/05/crowdsourcingmalaysianflightthankyou/ Accessed 25 Feb 2016.

9.3 Risks of Big Data in Humanitarian Action

The integration of commercial satellite imagery with different open-source data bears much promise for crisis response and humanitarian action. The application of machine-learning algorithms and crowdsourcing allows humanitarian actors to react to crises at an unprecedented speed. Moreover, this large technological system of satellite imagery establishes a new category of potential *global humanitarian citizens* as easy access to open-source data and satellite imagery allows people to participate in crisis response efforts all over the planet from their living rooms. However, the technological system of geospatial Big Data also poses a number of serious risks, which are potentially amplified in precarious crisis situations. In this regard, the early stage of digital humanitarian practices paired with the urgency inherent to emergencies calls for critical analysis and reflective consideration within an overall context. While not extensive, we suggest organising potential pitfalls related to geospatial Big Data in humanitarian action in three categories: data control, the digital divide, and misuse of open-source and personal data.

9.3.1 Data Control

The commercialisation of EO satellite technology in synergy with Big Data has raised considerable expectations for solving practical problems in humanitarian action such as rapid damage assessments or real-time information about road conditions. However, the basic requirement for bringing geospatial Big Data to use is access to the raw material. Commercial providers of satellite imagery continuously chip away at the former monopoly position of governments in space and have contributed to the growing circulation of satellite imagery in the public sphere. At the same time, the commercial character of satellite imagery also means that the companies aim for profit and only sell to those who can afford it, which re-produces certain imbalances. In the case of disaster response, experience has shown that companies and governments are often willing to provide free satellite imagery and new actors, such as Planet Labs, explicitly frame their work as providing access to their data for humanitarian purposes. Additionally, these efforts are facilitated by the International Charter "Space and Major Disasters"[22] which aims at supplying satellite imagery data and analysis to regions affected by natural or man-made disasters and also includes various commercial imagery providers. Yet despite the diversification of sources of satellite imagery through commercialisation, humanitarian practitioners raise the issue of "data donor

[22]International Charter "Space and Major Disasters" Available via https://www.disasterscharter.org/web/guest/home;jsessionid=7E70C331AFEC95A3FB5D9CC9E068B700.intlcharter-prod4040 Accessed 25 Feb 2016.

dependency"[23] Although a great share of Big Data sets consist of freely available user-generated content, up-to-date satellite imagery still comes with access restrictions managed by governments and private companies. As a consequence, digital humanitarian actors are at times reliant on data philanthropy[24] in order to get to work, which potentially affects the planning and execution phase in peculiar ways. At the same time, companies that often cater to the defence and security sectors see a chance of improving their image through such contributions, and "humanitarian agencies need to understand what exactly they are buying into."[25]

9.3.2 Digital Divide

When bringing issues of data control to a global scale, it also qualifies the general premise of increased accessibility to geospatial data. Only particular social groups of people possess the resources to make use of new technologies in a crisis situation, which leads to a digital divide, referring to "unequal access to and use of ICTs [. . .] due to socio-economic status, race, ethnicity, gender, age, and geographic location".[26] While educated, young males are usually overrepresented among the participants of crowdsourcing initiatives,[27] those who are most vulnerable to crisis situations, women, children, the elderly, and people of lower socio-economic status, often have less access to the technological means and capacities to use them. This digital divide exists on a variety of levels including access to technologies and skills to take advantage of them.[28] Globally, a digital divide persists between developed and developing countries, with most technologies designed and intended to be used

[23]Raymond N, Card B (2015) Applying Humanitarian Principles to Current Uses of Information Communication Technologies: Gaps in Doctrine and Challenges to Practice. Harvard Humanitarian Initiative. Available via http://hhi.harvard.edu/publications/applying-humanitarian-principles-current-uses-information-communication-technologies Accessed 22 Jan 2016.

[24]Duffield M (2013) Disaster-Resilience in the Network Age Access-Denial and the Rise of Cyber-Humanitarianism. DIIS Working Paper, No. 23. Available via www.diis.dk/files/media/publications/import/extra/wp2013-33_disaster-resilience-cyber-age_duffield_web.pdf Accessed Feb 03 2013; see also Sandvik K B, Gabrielsen Jumbert M, Karlsrud J & Kaufmann M (2014) Humanitarian technology: a critical research agenda. International Review of the Red Cross, 96 (893), 219–242.

[25]Duffield M (2013).

[26]Gilbert M (2010) Theorizing digital and urban inequalities. Information, communication & society 13(7):1002.

[27]Bott M, Young G (2012) The Role of Crowdsourcing for Better Governance in International Development. The Fletcher Journal of Human Security 27: 54.

[28]DiMaggio P, Hargittai E, Neuman W R, & Shafer S (2004) Digital inequality: From unequal access to differentiated use. In Neckerman K (Ed.), Social inequality. New York, NY: Russell Sage: 355–400.

in the global North while those in need take on a far more passive role.[29] Finally, this divide reinforces a skewed relationship with those in need of support being dependent on others mandating over the necessary resources to frame their situation as a crisis worthy of their help, which refers back to data control issues.[30] Considerations of this kind are less contested in cases of natural disasters that are perceived as an immediate emergency and largely out of human control. In contrast, situations of armed conflict are often characterised by complexity, longevity, and multiple narratives of blame and liability, which complicates matters for digital humanitarian actors.

9.3.3 Misuse of Open-Source and Personal Data

Extensive data collection for crisis response "is often justified for 'the greater good'"[31] and privacy and consent fade from the spotlight. This creates a dilemma for humanitarian actors who have to balance quick responses to crises with protection of the personal data of the affected population. During the Haiti earthquake, in an attempt to solve this dilemma, only the most urgent text messages that were sent to the emergency telephone line were published worldwide on the crisis map.[32] Nevertheless, people who sent these messages, primarily to receive help in the aftermath of the disaster, did not explicitly consent to that use of their data. This weighing of the right to privacy and risks of violations against the expected advantages for security and disaster relief raises a number of questions: How can we measure the risks of using this data? Who decides whether benefits outweigh the risks? And ultimately: Which means do people affected by a crisis have to determine how their data is used?[33] The protection of information shared during crisis situations therefore requires more consideration, especially when personal data is stored for later usage.

Open-source satellite imagery used for crisis response further poses a potential threat to vulnerable parts of the population as well as humanitarian workers during armed conflict. Although wide access to these images increases transparency and disaster relief during a crisis, it becomes impossible to control who uses certain data

[29]Burns R (2015) Rethinking Big Data in Digital Humanitarianism: Practices, Epistemologies, and Social Relations. GeoJournal 80: 477–490.

[30]Piccinini C (2014) Communicating Agencies' Humanitarianism in Natural Disaster and Armed Conflicts: Differences, Challenges, and Dilemmas. Journalism and Mass Communication 4(10): 603–623.

[31]Crawford K, Finn M (2015) The limits of crisis data: analytical and ethical challenges of using social and mobile data to understand disaster. GeoJournal 80: 498.

[32]Meier P (2015) Digital Humanitarians. How Big Data is Changing the Face of Humanitarian Response. CRC Press, Boca Raton: 11.

[33]Crawford K, Finn M (2015) The limits of crisis data: analytical and ethical challenges of using social and mobile data to understand disaster. GeoJournal 80: 499.

and for which purpose. As the amount of open-source satellite imagery grows, so does the risk of these images being accessed and potentially misused by hostile groups. To illustrate, making public satellite imagery of internally displaced people in Somalia could put them at risk of terrorist attacks by the militant group Al-Shabaab, which is what they were fleeing from in the first place. This poses a dilemma for humanitarian organisations working in conflict situations. While open-source data is a valuable asset in crisis response, a major principle of humanitarian aid demands that additional harm or threat to individuals must be avoided.[34]

9.4 Conclusion: Geospatial Big Data and the Way Forward

In this contribution it was described how the synergies of commercial Earth observation satellites and Big Data analytics have brought about significant changes to society by fostering new forms of cooperation networks among a broad range of actors. The growing LTS of satellite technology and digital humanitarian action has facilitated fast responses to natural disasters and conflicts by integrating traditional actors and organisations with a growing community of digital humanitarians. We have demonstrated how the commercialisation of satellite imagery has not only brought about new actors but also further increased the volume, velocity and variety of EO data. In addition to government agencies, commercial actors have become important players in the field of satellite imagery production and analysis. Furthermore, the growing availability of satellite imagery coupled with machine-learning algorithms and crowdsourcing techniques have enabled citizens to actively take part in humanitarian actions around the globe.

However, there are shortcomings within the LTS of satellite imagery and humanitarian action. First, if humanitarian efforts are reliant on satellite imagery, the actors involved are often dependent on commercial providers or governments to grant them access to their data, resulting in a data donor dependency. Second, a digital divide persists, in which well-educated males are overrepresented among users of new technologies. Finally, excessive data collection and open-source satellite imagery can, not only violate privacy rights, but also put people in precarious situations at additional risk when this data is used by hostile groups.

The legal and social framework has not kept step with the described techno-political developments so that resulting risks and challenges are currently being addressed by the work, experience and good will of humanitarian practitioners. A general code of conduct for the variety of actors in the field of crisis response is needed as a way to balance opportunities of geospatial Big Data against potential

[34]Raymond N, Card B (2015) Applying Humanitarian Principles to Current Uses of Information Communication Technologies: Gaps in Doctrine and Challenges to Practice. Harvard Humanitarian Initiative. Available via http://hhi.harvard.edu/publications/applying-humanitarian-principles-current-uses-information-communication-technologies Accessed 22 Jan 2016.

risks to the affected population. Here, it will be essential to set up a multi-stakeholder discussion among national and international humanitarian organisations, policy makers, social media, commercial imagery providers, technology developers and also individuals who are willing to share their experiences of being affected by natural disasters such as the Haiti earthquake.

Chapter 10
Some Ethical Constraints on Near-Earth Resource Exploitation

James S.J. Schwartz and Tony Milligan

10.1 Introduction

One of the most common and enduring justifications for space exploration is that resources are limited here but plentiful elsewhere. Exploration has the potential to enable humanity to access the vast store of resources throughout our solar system. Recent discussion (driven to some extent by U.S. legislation and by speculation about the future role of the private sector in space) has tended to focus on what we shall call "near-Earth resources" (NERs), which include those of the Moon and near-Earth asteroids (NEAs). They include: potentially vast stores of water, iron, and platinum-group metals among the NEA population; so-called 'peaks of eternal light', i.e. places of uninterrupted sunlight (for solar energy collection) and crater areas in more or less permanent darkness (able to harbour water ice) on the Moon; Helium-3 (He3) in the lunar regolith; and the relatively banal resource of terrestrial orbital niches. What is often lost in the enthusiasm concerning such NERs is that, in spite of the immensity of space, only a small percentage of the NEA population is profitably accessible in the absence of distant, futuristic technology; only so much of the lunar surface (or elevated areas just above it) is permanently illuminated (or shadowed); the He3 concentration in the lunar regolith is very low, geographically differentiated and reduces with depth (the regolith on asteroids is also likely to be less mature and so He3 concentration levels are likely to be even lower); and,

Based on a presentation of the same title delivered by James S.J. Schwartz at the Ninth ESPI Autumn Conference in Vienna, Austria, 21–22 September 2015. Schwartz thanks those in attendance for discussion and comments.

J.S.J. Schwartz (✉)
Department of Philosophy, Wichita State University, Wichita, KS, USA
e-mail: james.schwartz@wichita.edu

T. Milligan
Department of Theology and Religious Studies, King's College London, London, UK

C. Al-Ekabi et al. (eds.), *Yearbook on Space Policy 2015*, Yearbook on Space Policy, DOI 10.1007/978-3-7091-4860-0_10

as is already well-known, there are only so many available orbital allocations. These practical realities suggest that issues of sustainability will not vanish during at least the early stages of space exploitation, and perhaps at all stages for the foreseeable future. They do not support unrelenting and unregulated utilisation and consumption. In fact, quite the opposite.

Practical scarcity suggests that NER utilisation should be subject to constraints, i.e. behavioural norms or regulative restrictions (or both) in order to limit disputes which that have a negative terrestrial impact, and in order to secure appropriate extraction and use. Doubts about the practicality of securing new international agreements among nations with active space programmes may lead us to endorse only behavioural norms.[1] However, it is arguably too early to tell whether the political climate will remain indefinitely set against agreed regulation. Reluctance to negotiate and compromise in agreement-enabling ways may be eroded by the experienced difficulties of actual resource utilisation—difficulties that have yet to be fully encountered. Given that resource utilisation is going to involve multiple players from different nations, utilisation without agreed regulation may simply turn out to generate too much international disagreement and uncertainty about entitlements, i.e. problems that could impede long-term private investment. Although the need for regulation is widely recognised in the case of orbital allocations (and to a lesser extent in the case of NEA mining[2]), nevertheless little discussion has taken place concerning regulation for other forms of NER exploitation. Given this, it remains difficult to see how deeply the disagreement between behavioural norm-only and regulative approaches actually runs. Especially so given that supporters of regulation also support the development of behavioural norms as part of the processes of building towards regulations and then getting them to work. With this in mind, our aim in this paper is to broach the discussion of NER regulation in broad—form but in a way that is salient to both present and near-future space policy.

Approaches towards regulation fall into various categories, confusion about which sometimes underlies doubts about practicality. Two in particular stand out: ideal theory (an ethico-political approach which examines the best set of arrangements that is consistent with the familiar limits of human nature and therefore, in principle, is attainable) and non-ideal theory (a more pragmatic approach where compromises nonetheless aim to be at least consistent with a move towards a more maximal or ideal state of affairs). Both have a role to play. What follows will be an instance of the former, an attempt to specify the constraints under which an *ideal*

---

[1]The difficulties of securing a workable international agreement are considerable and are an important driver for the recent Spurring Private Enterprise and Competitive Entrepreneurship (SPACE) Act signed into United States law in November 2015. For a norm-based approach see Laura Delgado-Lopez (2015) Beyond the Moon Agreement: norms of responsible behavior for private sector activities on the Moon and celestial bodies. Space Policy. 32: 6–8. For a more norms-plus-regulation approach see Fabio Tronchetti (2014) Private Property Rights on Asteroid Resources: Assessing the Legality of the ASTEROIDS Act. Space Policy. 30: 193–196 and the partner article Fabio Tronchetti (2015) The Space Resource Exploration and Utilization Act: A Move Forward or a Step Back? Space Policy. 34: 6–10.

[2]See, e.g., Pilchman, D (2015) Three Ethical Perspectives on Asteroid Mining. In Galliot, J (ed.) Commercial Space Exploration: Ethics, Policy and Governance, Burlington, VT: Ashgate, pp. 135–147.

NER licensing regime (or other regulatory system) should operate—constraints which, in other terms, will be compromised in practice but which would be the default option under ideal (rarely attainable) conditions. Thus the main concern of the authors *is* with space policy, but not with the whole-cloth construction of an entire NER exploitation framework. Rather, this contribution seeks to identify several adequacy conditions that any NER exploitation regime must meet in order to be ethically sound. The recommendations in brief are that:

R1. Decision-making bodies that form part of the NER exploitation regime must have the authority and expertise to appraise the overall *societal value* of exploitation proposals, and must also have the power to accept or reject proposals accordingly.

R2. The NER exploitation regime must be subject to *independent* scientific and environmental oversight.

These may seem to be relatively non-controversial. However, as is often the case, the devil may be in their supplementation and detail. Following a discussion in Sect. 10.2 of the scarcity of various NERs, we argue in Sect. 10.3 that (R1) and (R2) arise from (a) the ethical dubiousness of wasting scarce resources on trivial and/or frivolous objectives; and (b) the enduring importance of scientific research in space. These considerations are worth emphasizing because they are absent from the existing regimes that might be proposed as models for NER and other space-resource exploitation regimes, such as the International Telecommunication Union's (ITU) orbital allocation regulations and the International Seabed Authority's (ISA) deep sea mining regulations, which we discuss briefly in Sect. 10.4.

10.2 NERs and Scarcity

Though the solar system as a whole contains vast quantities of just about any resource humans could want or need, near-Earth space is relatively resource poor. Even for resources that are in-principle available in significant quantities, such as platinum-group metals from NEAs, there are nevertheless a variety of practical limitations that militate against viewing any NER as abundant:

10.2.1 NEAs

NEAs are asteroids with perihelia below 1.3 Astronomical Units (AU). From a Δv perspective, they are the most attractive mining candidates among the asteroids of the solar system. There is a considerable mass of NEAs that can be accessed at Δv values below that required to retrieve resources from the lunar surface—as much as 10^{14} kg by some estimates.[3] However, the precise quantity of *profitably accessible* NEAs is constrained by the following kinds of factors:

[3] Yárnoz, D, Sanchez, J, McInnes, C (2013) Easily Retrievable Objects Among the NEO Population. Celestial Mechanics and Dynamical Astronomy 116: 367–388.

Fig. 10.1 A visualisation of the volume of *accessible* water from the NEA population (*right*; ~0.8 km^3) as compared to the volume of Lake Nicaragua (*left*; ~108 km^3). The former is based on the following estimates from Sanchez, J, McInnes, C (2013) Available Asteroid Resources in Earth's Neighbourhood, in Badescu, V (ed) Asteroids: Prospective Energy and Material Resources, Springer, p. 439–458: That C-type asteroids comprise approximately 10 % of the 10^{14} kg of accessible NEA mass, and that C-type asteroids contain on average 8 % water by mass. Source: James S.J. Schwartz

1. Whether the NEA is of the appropriate type, e.g., M-type for platinum-metal group extraction; C-type for water extraction.
2. Whether the NEA is of sufficiently small mass to allow for low-energy capture in cislunar space.
3. Whether the NEA is of sufficiently high mass and resource concentration to justify the mission/energy expenses.
4. Whether the NEA is on an orbit that allows for a low-energy capture in cislunar space.
5. Whether the NEA's orbit aligns with Earth's so as to facilitate timely capture.

Even without taking factor (5) into consideration, factors (1)–(4) significantly reduce the number of attractive mining candidates. According to the estimates given by Martin Elvis,[4] there are only on the order of 10 NEAs suitable for platinum-group metals (PGM) mining, and on the order of 9000 NEAs suitable for water extraction (see also Fig. 10.1).[5]

[4]Elvis, M (2014) How Many Ore-bearing Asteroids? Planetary and Space Science 91: 20–26.

[5]Elvis' estimates assume a maximum LEO to NEA Δv of 4.5 km/s. For the number of PGM candidates his calculation assumes (1) that approximately 1/50 of the NEA population are of the appropriate type and concentration; (2) that approximately 2.5 % of this population can be accessed from LEO at $\Delta v \leq 4.5$ km/s; (3) that a minimum diameter of 100 m is necessary for profitable return and extraction; and (4) that there are approximately 20,000 NEAs of diameter

Tempering this picture even further is the cold reality that identification and orbital repositioning can take a great deal of time—NEA intercept and capture missions must generally be planned years in advance. The result is that any resources derived from NEAs will come in a slow, staggered trickle; there is not likely to be a gold rush for NEAs.

10.2.2 Lunar Polar Real Estate

Craters near the lunar north and south poles offer two tantalizing prospects. In rare cases, along the crater rims, there are peaks that are in direct sunlight all or most of the time, or where an elevated surface will be in more or less permanent illumination. As Elvis, Milligan and Krolikowski[6] point out, these "peaks of eternal light" would be ideal locations to position solar energy collectors. Also, some craters contain areas that are permanently shadowed, and that might contain significant quantities of water ice. The combination of the two would constitute a very scarce and valuable resource.

By some estimates there is somewhat greater than 31,000 km^2 of permanently shadowed lunar surface.[7] If the entirety of the shadowed area contains water at the concentration measured by LCROSS, then there should be approximately 2.9×10^{12} kg of water ice in permanently shadowed craters.[8] By volume this amounts to approximately 0.013 % of the water in the Great Lakes, or slightly less than 3 % of the water in Lake Nicaragua (see Fig. 10.2). It is a considerable quantity of water, to be sure, but it does not amount to anything close to the amount of water found in many freshwater lakes on Earth.

10.2.3 Lunar He3

Though the total mass of lunar He3 is unknown, the highest measured concentrations in lunar regolith samples are around 10 parts per *billion*, although collection

---

$\geq$100 m. Thus, $20{,}000 \times 0.025 \times 0.02 = 10$. For the number of water candidates his calculation assumes (1) that approximately 3.1 % of the NEA population are of the appropriate type and concentration; (2) that approximately 3 % of this population can be accessed from LEO at $\Delta v \leq 4.5$ km/s; (3) that a minimum diameter of 18 m is necessary for profitable return and extraction; and (4) that there are approximately 10^7 NEAs of diameter $\geq$18 m. Thus, $10^7 \times 0.03 \times 0.031 = 9300$.

[6]Elvis, M, Milligan, T, Krolikowski, A (forthcoming) The Peaks of Eternal Light: A Near-term Property Issue on the Moon.

[7]McGovern, J, Bussey, D, Greenhaven, B, et al. (2013) Mapping and Characterization of non-Polar Permanent Shadows on the Lunar Surface. Icarus 223: 566–581.

[8]Crawford, I (2015) Lunar Resources: A Review. Progress in Physical Geography 39: 137–167.

Fig. 10.2 A visualisation of the volume of water accessible from permanently shadowed craters (*right*; 2.9 km^3) as compared to the volume of Lake Nicaragua (*left*; 108 km^3). The former is based on the estimate of Crawford, op. cit. Source: James S.J. Schwartz

and handling may have resulted in loss. The highest estimated concentrations are in the neighbourhood of 20 parts per billion (which occur in the Oceanus Procellarum and in the Mare Tranquillitatis near to the lunar equator) with an average concentration of approximately 4 ppb.[9] Given the exceptionally low concentrations of He3, extracting the estimated 2×10^8 kg total of He3[10] would be an especially destructive process, as it would entail the strip mining of a great deal of the lunar surface. The destructiveness of the process, together with the availability of an (admittedly tougher) alternative option of extraction from asteroids, may be a sufficient reason to place a moral injunction on lunar He3 mining.[11]

10.2.4 Orbital Allocations

An apparently more banal space resource, but one that brings us closer to where we are now, in terms of the problems of resource access, is circumterrestrial space. We derive incommensurable value from operations in various Earth orbits, including geostationary orbit (GEO). The limited nature of these resources is already recognised by the relevant governing body, the ITU, whose constitution states in Article 44 that,

---

[9]Ibid., 144–145.

[10]Ibid.

[11]Milligan, T (2013) Scratching the Surface: The Ethics of Helium-3 Extraction. Paper presented at the 8th IAA Symposium on the Future of Space Exploration: Towards the Stars, Torino, Italy, 3–5 July 2013. Crawford (2015) takes into account a variety of additional energy costs in extraction and transportation which puts the annual destructiveness even higher, also making the process economically questionable.

> ...Member States shall bear in mind that radio frequencies and any associated orbits, including the geostationary-satellite orbit, are limited natural resources and that they must be used rationally, efficiently and economically, in conformity with the provisions of the Radio Regulations, so that countries or groups of countries may have equitable access to those orbits and frequencies, taking into account the special needs of the developing countries and the geographical situation of particular countries.

One important feature distinguishing allocations in GEO and other orbits from the above resources is that orbital allocations are, in principle, limited at any given time but are nonetheless *inexhaustible* resources. Assuming that debris mitigation guidelines are adhered to and have their intended effect, the present occupation of an orbital allocation is not a permanent barrier to another's use of the same allocation at a later time.

That regulations exist and are necessary for the sustainable and equitable use of GEO and other orbits can be appropriated as evidence that a governance structure will also be necessary to ensure the sustainable and equitable use of other NERs. Anarchic exploitation of NERs is unlikely to help advance the interests of humanity as a whole. But what kind of constraints would be appropriate for a regime charged with regulating the exploitation of, e.g., NEAs and lunar polar real estate? As will emerge, the constraints we advocate even provide grounds for re-evaluating orbital allocation regulations.

10.3 Rawls in Space

In recent work we have discussed how Rawlsian deliberation might work as an attractive moral compass for space regulations.[12] At least in its near-term form, ideal deliberations would involve states agreeing to regulations in a condition of ignorance about their space capabilities. Such a condition would tend to maximise the benefits of the least-space-capable states, thereby mitigating the ability of the already wealthy and powerful states to stack the NER exploitation game in their own favour. As Schwartz argues,[13] this kind of thought experiment can help provide a more substantive interpretation of notions such as 'fair', 'equitable', and 'rational' in application to issues in space policy, such as orbital allocations, debris mitigation, and planetary protection. One key result of Schwartz's analysis is that, in addition to addressing *distributive* justice, regulation should also address *inter-generational* justice, since most NERs are both scarce and non-renewable.

---

[12]Schwartz, J (2015) Fairness as a Moral Grounding for Space Policy. In Cockell, C (ed.) The Meaning of Liberty Beyond Earth, Heidelberg, London and New York: Springer, pp. 69–89. Milligan, T (2015) Rawlsian Deliberation About Space Settlement. In Cockell, C (ed.) Human Governance Beyond Earth – Implications for Liberty, Heidelberg, London and New York: Springer, pp. 9–22.

[13]Schwartz, op. cit.

Given that our present focus is on the policy implications of NER *scarcity*, what we have to say below concerns inter-generational justice more than distributive justice. We do not mean to suggest that distributive justice is irrelevant or unimportant. Rather, our goal is to shine a light on the increasingly urgent need to address long-term NER management. In this regard, we believe that in ideal deliberations, no state would agree to permit the trivial or frivolous use of limited space resources. Rather, parties would agree that NER exploitation must only take place for sufficiently good reasons. If, as is suggested by Crawford, op. cit., economic factors militate against NER exploitation for *terrestrial* consumption, what we are then forced to consider is what might count as a trivial or frivolous use of NERs by *space-based* consumers. Likely consumer sectors include:

6. Space tourism—NER exploitation could facilitate the construction, maintenance, and supply of orbital and lunar hotels, tourist spacecraft, etc.
7. Space science—NER exploitation could facilitate the construction, maintenance, and supply of scientific outposts.
8. Space settlements—NER exploitation will be a vital source of resources for any permanent settlement in circumterrestrial space.

Item (8) introduces a range of complicating factors. If indeed there are independent, permanent space settlements that depend for their survival on NER exploitation, then it is difficult to resist concluding that they would have first-claim on NERs, or, at least, should be strongly favoured in any resource dispute. As a simplifying assumption our analysis below is not intended to cover situations in which there exist independent space settlements or states. Thus our provisional exploration of this issue will be confined to stakeholders such as (6) and (7).

Ethically speaking, (6) and (7) are not on a par. We know of no ethical duty that space tourism uniquely helps satisfy.[14] However, scientific exploration can be defended on moral grounds, and there is at least some reason to think that the moral justification for space science is stronger than the more common arguments for settlement and exploitation based on long-term species survival concerns.[15] That is not to automatically relegate space tourism to the trivial or frivolous, but only to state that it is *comparatively* less important, from an ethical point of view, than space science. Thus it would be ethically objectionable should an NER exploitation regime permit space-tourism based exploitation to such an extent that important scientific objectives were compromised. We would even go so far as to argue that for the foreseeable future there should be a strong presumption of

[14]It might be objected that space tourism could be defended on capability-building grounds. But so could a wide array of other human missions. There is nothing *uniquely* beneficial about space tourism as far as any duty we might have to extend human life is concerned.

[15]Schwartz, J (2014) Prioritizing Scientific Exploration: A Comparison of the Ethical Justifications for Space Development and for Space Science. Space Policy 30: 202–208.

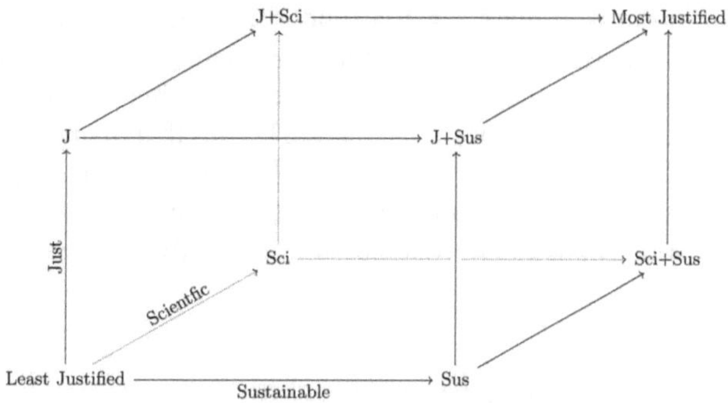

Fig. 10.3 A visualisation of the justifiability/societal value of NER exploitation proposals. Source: James S.J. Schwartz

scientific value in space, since scientific objectives are constantly evolving and it is not in general possible to predict what will be of interest in the future.

Consider NEAs. Though perhaps comparatively less interesting for scientific purposes than planetary bodies, there is still a wealth of scientific research to be conducted on the NEA population. NEAs are especially valuable as clues to the history of the inner solar system. There may even be some reason to invoke planetary protection concerns given the small chance that some NEAs might harbour microbial life or traces of historical microbial life (of which there might no longer be any traces at the point of origin). At the very least, an appraisal of the scientific value of an NEA should take place in advance of any commercial exploitation as a required step in the prospecting/exploration phase, and perhaps it will even be necessary for scientific oversight during exploitation, since the scientific value of an NEA may not be obvious until exploitation begins.

Another consideration that is relevant to adjudicating disputes between (6) and (7) is that, in certain cases, the sheer destructiveness of certain types of NER exploitation should rule against such activity. Because He3 density falls away with depth, any profitable lunar He3 mining regime would require the disruption of surprisingly large tracts of the lunar surface, even to the point of being visible to observers on Earth.[16] In contrast, scientific analyses are almost never destructive to this extent.[17]

Our basic ethical insight, then, is that for any NER exploitation proposal weight must be given to the objectives this exploitation would serve. Scientific objectives should be preferred to non-scientific objectives; sustainable objectives should be preferred to unsustainable objectives; and, all other things being equal, objectives that promote social and economic justice should be preferred to those that do not. This reasoning can be visualised according to Fig. 10.3, in which the axes represent increasing values for science, sustainability, and justice (which, admittedly, lack metrics at present). The idea here is that the overall justifiability of an exploitation

[16]Milligan, T (2013), op. cit.

[17]Counterexamples include the deliberate impacting of spacecraft to perform seismic and other analyses, such as the Saturn V third-stage seismic study impacts (Apollo 13 onward), and the LCROSS lunar impactor.

proposal is at least partly a function of the "volume" of its corresponding cube—ceteris paribus, a decision between competing proposals could be based on which proposal had the greatest "volume." For this insight to make its way into actual policy formation, decision making bodies that operated as part of an ethically sound NER exploitation regime would have to have the requisite authority, power, and expertise to make decisions, and to induce the compliance of contractors. Are any existing regimes likely to count as ethically sound according to these standards?

10.4 Existing Frameworks

In this section we will briefly discuss the extent to which two existing regulatory bodies—the ITU and the ISA—fare in meeting the ethical adequacy conditions raised in the previous section.

10.4.1 ITU Orbital Allocation Regulations

The ITU is best known in space contexts as the regulatory body in charge of orbital allocations in GEO and other Earth orbits. As has already been mentioned, article 44 of the ITU constitution recognises the limited nature of the resource of earth orbits, and even contains language that is sensitive to considerations of distributive justice. To that end the ITU pursues a policy of reserving orbital allocations for all states, regardless of their space capabilities or interests in engaging in operations in earth orbit. Beyond these reservations, however, access to GEO and other orbits is first-come, first-served. Thus beyond a very meagre minimum, the ITU regulations are not designed to address injustices in the distribution of the benefits of orbital allocations.

Our principal concern here, however, is with the ITU's inability to reject applications that would constitute a waste, or at least a very poor use, of an orbital allocation. Beyond ensuring that parties abide by international and space law, the ITU does not concern itself with the purposes for which orbital allocations are actually used.[18] As an extreme example, let us suppose that some state were to grant a launch licence to a corporation that intended to place a commemorative statue of some television personality in GEO.[19] The ITU would not be able to cite the frivolousness of such an exercise as a reason to reject the application. But clearly, if this application were in competition with other, more legitimate objectives (such

[18]We thank Christopher Newman for discussion on this point.

[19]For an interesting discussion of some of the legal issues involved in the related activity of space-based advertising, see O'Brien, Z (2015) Advertising in Space: Sales at the Outer Limits. In Galliot, J (ed) Commercial Space Exploration: Ethics, Policy and Governance, Burlington, VT: Ashgate, pp. 91–106.

as the stationing of a satellite that would make novel scientific observations), the statue application should be rejected in favour of another, more worthy application. And, in practice, we assume that this is exactly what would happen. However, the rejection would be ad hoc and would no doubt raise concerns that other cases, where rejection might be warranted, need not be quite so clear-cut. Suppose, for example, a state were to successfully acquire an orbital allocation in order to secure a minor increase in GPS network capacity, when this allocation could have gone to another proposal that would have collected important earth observation data. The precise extent to which these conflicts will arise is, of course, an empirical matter, but it at least seems likely that sometime in future they will occur with some regularity and with considerable debate about what goals are more significant.

The first-come, first-served nature of the ITU allocations is even less appropriate for other forms of NER exploitation. In most cases there simply isn't enough of each resource to reserve profitable portions for each state. For instance, if Elvis' estimates are at all reasonable, then there are only on the order of 10 profitably accessible PGM NEAs—not even enough for each *space-capable* state! And given the vast tracts of lunar regolith that would have to be processed to extract economic quantities of He3, only a handful of operators could expect to turn a profit extracting He3. Rather than reserving a portion of each resource for each state, it seems a more equitable arrangement would be a system under which each state, or some international body with allocation responsibilities (and a good deal of oversight) receives a portion of the *proceeds* of any exploitation deemed justifiable.[20]

10.4.2 ISA Deep Sea Mining Regulations

A system that does a better job of attending to social justice concerns is the ISA regulations for deep sea mining. For prospecting, exploration, and exploitation, states must partner with developing nations (or the ISA itself), to whom they must provide technical training and to whom they must offer fair pricing on related technology. The partnering state will ultimately be granted access to a portion of the prospected area. It is, however, unclear whether partnering with developing states will be an effective practice for NER exploitation—if, as we think is most likely, the bulk of NER exploitation will take place in support of space activities, rather than for terrestrial consumption, then it is not clear to what extent those not already in space actually are disadvantaged stakeholders in NER exploitation.

We feel that at present the environmental risks associated with deep sea mining are unacceptable, given that very little is known about the deep sea bed, how deep sea ecologies interact with one another, how deep sea mining would impact these

[20]An obvious model here would be the Norwegian Oil Fund, the *Oljefondet*, currently heading towards a total value of 1 trillion euros and arguably the largest stock-holder in Europe. Only a portion of the profits from the fund are disbursed annually.

ecologies, etc. We also agree with Hartley[21] that the ISA's tolerance for environmental damage is too high. However we are willing to grant that these kinds of environmental concerns do not apply in the case of many, if not most, instances of viable NER exploitation.[22] However, the ISA's monitoring of environmental and scientific impacts does raise concerns that are relevant for the case of NER exploitation.

According to the ISA regulations it is the responsibility of *contractors* to perform environmental and scientific impact assessments. Although the ISA requires the use of best practices for these activities, this does not avoid the evident conflict of interest for contractors who (a) presumably have a strong interest in prospecting and exploration (leading to exploitation), and who are also (b) responsible for evaluating the threat these activities might pose to science and to the environment.[23] A more environmentally and scientifically sound approach would require independent, third party oversight on these matters.[24]

Environmental compliance may be easier to come by in space. The threats posed by careless exploitation practices (proliferation of orbital debris; NEAs set on unsafe trajectories; lunar dust harming equipment on the lunar surface; etc.) aren't pure externalities, since they would likely affect exploiters as much as bystanders. However, restraint concerning sources of scientific curiosity may not be so easy. NERs are and will for some time remain legitimate objects of scientific curiosity. It stands to reason that, at present, the scientific value of NERs is greater than their economic value. Thus the protection of objects of scientific curiosity should be a priority for any NER exploitation regime. Thus, any NER exploitation regime should include an independent scientific oversight panel that has the capacity to exert strong influence over exploitation proposals. Perhaps proposals could include fees for the scientific appraisal of the NER in question (a process that would vary as appropriate for the NER in question). Such a process would be tasked with identifying anomalies (phenomena falling outside the remit of scientific understanding) in the site, and would include active oversight during exploitation should any anomalies be discovered in the process of exploitation. Strict anti-bribery laws, possibly including waivers for certain kinds of future personal advancement, should

[21]Hartley, D (2012) Guarding the Final Frontier: The Future Regulations of the International Seabed Authority. Temple International and Comparative Law Journal 26: 335–366.

[22]Which is not to say that no environmental concerns arise in space. In addition to well-known concerns about debris production, NEA exploitation raises concerns about the safety of NEA trajectories and of discarded tailing, and lunar resource exploitation raises concerns about disrupting regolith (which could have deleterious effects on unshielded equipment on the lunar surface).

[23]Nelson, J (2005) The Contemporary Seabed Mining Regime: A Critical Analysis of the Mining Regulations Promulgated by the International Seabed Authority. Colorado Journal of International Environmental Law and Policy 16: 27–75.

[24]For a more optimistic discussion of an ISA-like governance in space, see Pinault, L (2015) Towards a World Space Agency. In Cockell, C (ed.) Human Governance Beyond Earth, Heidelberg, London and New York: Springer, 2015, pp. 173–196.

also perhaps be put in place to minimise corruption (and hence enhance the authority) of the scientific panel. An ever stronger NER protection regime, which treats NERs as scientifically valuable until proven otherwise, may even be attractive initially to industry, since preserving NERs for scientific study would externalise the costs of exploration and prospecting (assuming adequate funding for science).

10.5 Conclusion

Given both the scarcity and importance of NERs to future space operations, it is imperative that any NER exploitation be undertaken for good reasons and, ideally, the best possible reasons. Thus any appropriate NER exploitation regime must have greater power to reject inappropriate exploitation proposals than would be possible under an ITU-like arrangement, and stronger environmental and scientific protections, and a more independent evaluation of risks, than would occur under an ISA-like arrangement. In short, an NER regime should strive to satisfy *more* than mere distributive justice concerns and it should also strive for *more* than a mere token acknowledgment of the potential scientific and environmental consequences of exploitation. Ideally, a mutualism between distributive justice, science, and development should be sought. How far such a system of regulation can be realised will, of course, depend upon contingent political factors. Nonetheless, it should be the default option.

Chapter 11
Governance, Strategies and Approaches in 2030: "European Options"

Jesse Phaler

11.1 Inspired Human Mind

> *"Imagination will often carry us to worlds that never were. But without it we go nowhere."—Carl Sagan*

Establishing the outlook for space activities in 2030 and outlining the public sector's path to arriving there is a challenging task, essentially for two reasons.

Firstly, the principles and the very purpose of various governmental services are increasingly being debated, already with a much shorter time-perspective than 2030. In the context of financial austerity and privatisation, the borderline between the private and public sectors is consequently being redefined in numerous domains.

Public investments and undertakings are increasingly being scrutinised, in terms of both benefits to society at large and the principles and modalities for what to fund and how. This trend has been accentuated recently by concerns about fairness, including with respect to how and where individuals and companies pay taxes and issues concerning tax avoidance and tax evasion. What the role of the public sector is in preparing space 2030, its benefits to society, and what arguments can be used to convince both politicians and the public at large to resource this, must be established.

Secondly, and perhaps as important, such an outlook relies on a process of discovery and innovation, including materials, technologies and ground-breaking solutions, changing technology and lifetime cycles, and specific game changers and paradigm shifts. And, as history has shown us, while an inspired human mind is very, very resourceful, its workings are difficult to predict.

Who would have ever imagined the effects of a story that was written 150 years ago, in 1865, about a merry group of weapons enthusiasts in Baltimore and their

J. Phaler (✉)
European Space Agency, Paris, France
e-mail: Jesse.Phaler@esa.int

C. Al-Ekabi et al. (eds.), *Yearbook on Space Policy 2015*, Yearbook on Space Policy, DOI 10.1007/978-3-7091-4860-0_11

attempts to build an enormous sky-facing space gun in Florida, with the ultimate goal of landing on the Moon?

It is said that on the basis of this imaginative story about "three men, that carried into space, all the resources of art, science, and industry", written by Jules Verne in his book "From the Earth to the Moon", a 14-year-old school student became inspired and sketched his first model rocket in what is today Sibiu, Romania. By the age of 23 he had developed designs for a missile using liquid propellant with a range of 180 miles. Six years later he published a book titled "The Rocket into Interplanetary Spaces".

That book, written by Hermann Oberth, was picked up in the 1920s by another youngster, allegedly also 14 years old, who became famous as the engineer, Wernher von Braun. The next year his mother gave him his first telescope and he soon decided to devote his life to rocketry and the exploration of space. What followed was a not a small chapter in the history of human space activities but impressive results from a series of inspired minds. And this is just one example.

The exercise to define and evaluate European options for prospective public sector strategies for the future should therefore not be based on an attempt to draw up what could be the exact shape, size and features of space activities in 2030. In the spirit of the aviator, the narrator of Little Prince by Saint-Exupery, efforts should not focus on drawing the exact ideal image of the vision of a sheep. Instead, recognising its complexity, one needs to take a step back, leaving room for endless variations and chance, and focus on drawing and characterising the box in which the sheep, or in this case, space in 2030 and its European context, could be framed.

The following sections aim to reflect on these challenges and outline and assess the features of such a box.

11.2 The Role of the Public Domain

The role that the European public domain will play in space 2030 depends on the evolution of its relations to the private sector, its ability to establish political and public support, and its capacity to consolidate a coherent European space programme.

The role of states is continuously evolving and has already come a long way from its purpose at the time when many of the current nation-states were created. Over time, public services have climbed up the stairs of social needs, from provision of basic infrastructure, to provision of a social safety net and security, to building up social cohesion, a key factor in the rise and fall of cities, nations and civilizations, to promoting competing ideologies. The question is, in order to succeed, to what extent can and should states go further in providing the framework and context for society to excel in the future.

It is this very question that underpins debate about the optimum role of the public domain in future space activities. To what extent should states focus on the development of society as a whole, rather than considering their citizens simply as users or customers of public services?

While the questions are being framed in terms of how the public sector can contribute to advancing the wellbeing of its citizens, the answers are being provided through economic policies implemented with the aim of reducing government budget deficits. In recent years there has been a clear trend to reducing public costs in order to stimulate the private sector, while shifting public services to market and customer orientation. Long and costly public investment projects have been pushed out of the limelight, in favour of decisions that deliver quicker returns supported by business models focusing on the private sector.

This trend for public investments and services to move closer to the private sector model has been increasingly accompanied by the expansion of the private sector to classically publicly-held functions, for example through privatisation of infrastructure and services. Interestingly enough, there is also a growing number of private philanthropic initiatives that promote "Global Development, Health, Policy and Advocacy", support leadership in "Education, Immigration and Innovation" and pursue "radical breakthroughs for the benefit of humanity". New types of entrepreneurs are even entering into the domain of educational services, such as home schooling provided for their own employees.

Space industry, infrastructure and services have not been immune to this. In the U.S. and Europe concrete policy steps have been taken to increase private sector involvement and responsibility in launch systems development and the provision of launch services. This aims to capitalise on the private sector's focus on competitiveness, cost efficiency, profitability and customer orientation that *may* result e.g. in lower launch prices.

There is not though enough experience and hard data yet to confirm that the new commercial approaches and practices are better. Preparation of long-run growth, break-through technologies, and new space infrastructures for 2030 must therefore be clearly supported by a proactive public sector, as in the past. What that means in terms of partnering between the public and private sectors, and in terms of autonomy, geopolitics, and dependence, is to be assessed.

11.3 Revisiting Roles

"The failure to recognise the role of the government in driving innovation may well be the greatest threat to rising prosperity."—Martin Wolf[1]

The key argument I would like to put forward for European options in space for 2030, is that the trend to public and private sector convergence must be reversed.

Mariana Mazzucato, Professor of the Economics of Innovation in Sussex University has raised interesting reflections on the need to re-address public sector

[1]Martin Wolf, Financial Times, 5 August 2013, p. 8.

undertakings. In her book "The Entrepreneurial State",[2] she emphases the public sector's role in risk-taking and radical growth that fosters technological change. She describes scenarios where the state has provided the main source of dynamism and innovation in advanced industrial economies, pointing out that the public sector has been the lead player for an economy driven by technological change.

The strategic nature of public investments, including for research that usually will result in profitable activities only over a long period of time, falls beyond the appetite of the market configuration and therefore can only be pursued by public means. Europe has been successful in such investments in the past, and while there are examples that show such efforts have not disappeared, they are under heavy scrutiny.

But are we securing the right future for Europe? Are we aware that without public enterprises, Europe risks jeopardising long term strategic investments, essential for building the framework for activities and services for its citizens, or in other words, securing the box to put the sheep in?

These type of investments have become increasingly challenging, also for the public domain, as investments that greatly exceed political and electoral cycles, that extend even beyond individual interests, and our lifetimes, are laborious and difficult undertakings, in particular when faced with the quick pace of modern day life.

They are difficult because their business cases are not evident. Such investments and projects are often driven by vision, and rallying general support for them is not easy or simple.

To tackle this problem, in some countries long-term investments and innovation have been declared a matter of strategic priority. For some it is a matter of military might, and investments in breakthrough technologies are a matter of national security.

An example is the U.S. Defense Advanced Research Projects Agency (DARPA) whose mission statement is to pursue pivotal investments in breakthrough technologies. The genesis of that mission and of DARPA itself dates back to the launch of Sputnik in 1957, and a commitment by the United States that, from that time forward, it would be the initiator and not the victim of strategic technological surprises.[3] Looking at the U.S. today, it is on its way to defining what for it is that visionary path, and it is resourcing it, and has the organisations to see that it this is efficiently executed.

Looking at the past, state-funded organisations can be nimble and innovative, transforming economies forever. In this regard, let us not forget, as an illustration, that the algorithm behind Google was funded by a U.S. state grant.

---

[2]Entrepreneurial State, Maria Mazzucato, 2011, www.demos.co.uk

[3]http://www.darpa.mil/about-us/mission

11.4 Courage

"We are living in a depressive era in which we no longer have courage. We no longer think governments should have missions. But the market never chooses anything."—Mariana Mazzucato

The main thesis of Maria Mazzucato in the "Entrepreneurial State" is that it is the state, not the private sector, which has kick-started and developed the engine of growth, precisely because of its willingness to take risk in areas where the private sector has not and cannot.

Her argumentations are accompanied by examples. From the development of aviation, nuclear energy, computers, the Internet, the biotechnology revolution, nanotechnology, and now even green technology, Mariana Mazzucato argues that it has been the state that has chosen to put these in motion.

But now, following the convergence with the private sector, the public services have been subject to heavy criticism in terms of comparable performance, and in the spirit of austerity, in particular in terms of costs and return on investments.

11.5 Mind-Set

"Success consists of going from failure to failure without loss of enthusiasm."—Winston Churchill

But what perhaps remains the biggest challenge in the near term to a public or commercial success of Europe is the European mind-set itself. Not only that traditionally, we have not had a big risk appetite, and perhaps been more modest in our successes than our counterparts, Europe's biggest problem is perhaps the way it deals with failures.

It is easy to admire the results of Silicon Valley. But what we should admire is how it treats its failures. It is said that, in Silicon Valley, if you fail three times in a row with your business ideas, you get an invitation to address a Rotary club.

In Europe, on the other hand, if you go bankrupt once, you are easily cast-out, and it is very hard to convince others, and yourself, that there is a future to be built on lessons learned.

This dilemma is accentuated in the public sector where risk avoidance and de-risking has gained a stronghold in convergence with the private sector and respective benchmarking.

11.6 Vision and Heritage

How are we going to achieve these game changers and how are we going to capitalise on them? How are we going to ensure coherence and adequacy in the efforts that are being deployed to structure towards 2030, starting now? How are we going to ensure that there is a landing point in 2030, that there are able and innovative managers that will capitalise on this groundwork?

When seeking to answer these questions we should bear in mind that the managers of these future space activities are today around the age of those 14 years olds referred to at the beginning.

If we don't know where we want to go, we are sure to end up somewhere where we do not want to be. And if we want others to join us where we want to go it must be accompanied by an inspiring vision.

For example, even if they were free, would the reduced cost of launch services also reduce the costs of missions in such a way that would become a game-changer? Launch service represents only a part of mission costs. Other solutions are in the emerging to satisfy infrastructural needs, such as unmanned aerial vehicles, balloons, drones and even gliders.

History has shown that launchers are the foundation of all space activities. In building up the framework that will enable the future of European space activities, we should start being daring in that area and influence the future from the onset. Europe should take benefit from the richness and diversity of its nations and dare to dream the impossible. It must unite under a vision and resources to drive change, focus on basic research and build on the strong industrial base that it has for both launcher and space applications.

The public sector must assume its responsibility in enabling the conditions necessary for exploring the limits of the possible; providing the infrastructure for both innovation and implementation; fully exercising its role for setting up the box in which space 2030 and its European context can be implemented. Obviously this will require collaboration between the public and private sectors, but it is the public sector that must pursue investments that aim at unconditionally providing benefits to society at large, and from which the private sector can also reap benefits.

We need to keep pursuing the space guns of tomorrow, like those characters 150 years ago. We are obliged to keep building the ships that are destined to cross the new oceans.

Such projects are difficult, they are lengthy and they are costly. As they are destined to bring benefits to society at large they cannot rely on individual investments, but must be funded by public means.

We owe this to the generations that brought us this far, and we owe it to the next ones who rely on us to provide them the means to dream and succeed. We must tell them that "Success is not final, failure is not fatal; it is the courage to continue that counts".

Chapter 12
The Economics of Launch Vehicles: Towards a New Business Model

Keith Hayward

12.1 Introduction

For over 50 years, the economics of satellite launchers has reflected the military and strategic interests of government customers. Launcher economics have generally been shaped by explicit or implicit subsidies and a range of market distortions. This business model is now under threat from new entrants offering to undercut established players. While this challenge may be in some degree more illusory than real, there is sufficient novelty in the new approach to disturb established launch service providers. A more disruptive threat is possibly on the far horizon of innovation. There is the prospect—although highly problematic at this stage—of a fully-fledged technological transformation in accessing space. Radically new technologies could revolutionise the economics of space operations—the equivalent of the jet engine's impact on air travel.

12.2 Launch Vehicle Economics: The Traditional Business Model

In contrast to other markets for goods and services, "the market conditions for the space and launch vehicle industries have always been a stepchild of government policy and decision-making. Needs and missions of governments and agencies in the United States and throughout the rest of the world have controlled the access to outer

K. Hayward (✉)
Air and Space Academy of France, Toulouse, France
e-mail: keith.hayward48@btinternet.com

C. Al-Ekabi et al. (eds.), *Yearbook on Space Policy 2015*, Yearbook on Space Policy, DOI 10.1007/978-3-7091-4860-0_12

space. In short, defence, security and politics have trumped market economics".[1] This has been true for many other aspects of aerospace and aviation, but the space sector remains an area where politics pervades decision-making and technological development. Herzfeld goes on to describe the political context of launch vehicle economics: "vehicle R&D to meet military or industry policy objectives; launch site subsidies; market controls, domestic over exports (such as banning domestic satellites from using foreign launchers); technology transfer limitations". These factors severely constrain the operation of market forces, especially international competition. In many cases, the domestic market structure is a monopoly or an oligopoly of suppliers matched by a monopsony—a market situation whereby one buyer interacts with many would-be sellers.

The dominance of political factors was at its height during the years of the Soviet-American space race. This period ran from the late 1950s to the mid-1980s, at which point the duopoly was broken by the entry of the European Ariane. While the European programme was also stimulated by political motives—to ensure an independent access to space—the Ariane rocket was optimised to launch commercial communications satellites into geo-stationary orbit, "the only market large enough for a potentially profitable business venture in space at the time".[2] In this market place, performance was more relevant than efficiency, with a commensurate effect on costs.[3]

From the mid-1980s, a "quasi-commercial" market emerged. In part this reflected the maturing of the Ariane as a player in the rapidly growing market for communications satellite launches. It was also a function of the hiatus in American launch capability caused by the loss of the Space Shuttle in 1986, and the need subsequently to stimulate the American domestic supply of expendable rockets. This led to the emergence of a number of U.S. launch vehicle companies. At the same time, the Soviet Union (and then Russia) entered the world market, followed by China, Japan and then India—although in the case of Russia and China, U.S. technology transfer controls and other sanctions deterred or obstructed some potential customers from using their launch services. Nevertheless, the new entrants were often prepared to offer launches at prices determined by discounting American or European prices rather than a reflection of real costs.

[1]Herzfeld (2005): Henry R. Herzfeld, Ray A. Williamson, Nicolas Peter, *Launch Vehicles: an Economic Perspective*, Space Policy Institute, The George Washington Institute, September 2005: https://www.gwu.edu/~spi/.../NASA%20L.Vehicle%20Study%20V-5.pdf, p. 3.

[2]Herzfeld, *op cit.*, p. 5; this is still largely the case. While scientific payloads are important and there is an emerging market for Earth Resources Satellites, the commercial satellite operators are the dominant customers for commercial launches.

[3]Herzfeld, *op cit.*, p. 7.

12.3 The "Quasi" Commercial Launcher Industry

Since the late 1980s to the early years of the twenty-first century there has been significant progress towards a more market-led launcher industry. However, assessing the true economics of the launcher business remains problematic. The cost of a launch is not necessarily the same as the price paid to a company for a launch; government R&D is still viewed as a sunk cost and not included as a marginal cost; given the relatively few launches per year (even for the U.S. military) industrial and operating efficiencies will have only a limited (though not insignificant) effect on costs (i.e., economies of scale are low). Equally, politically imposed barriers to international competition are still evident, especially in accessing the relatively huge American public market.[4]

12.3.1 Technology Transfer Controls and the Launcher Market

Technology transfer controls continue to create barriers to international competition and a fully transparent commercial market. The U.S. in particular has applied strict regulations under the International Traffic in Arms Regulations (ITAR) system that has effectively prohibited the use of Russian (Soviet) and Chinese rockets for launching American satellites or foreign-built satellites incorporating American components and technology. During the 1990s, these procedures were beginning to be relaxed but revelations that U.S. manufacturers had provided the Chinese with missile design information and skills through the improper transfer of launch failure analysis (ostensibly required to launch U.S.-built satellites on the Long March rocket) led Congress to impose an even stricter regime. Reaction to the Tiananmen Square incident subsequently strengthened American opposition to the use of "non-allied" launch vehicles.

These restrictions created significant hindrances for the U.S. space industry and for customers using most foreign built satellites. The process was highly bureaucratic—one company reported it had 15 full-time staff handling ITAR issues—incurring high overheads and excessive delay. However, the system not only negatively affected the U.S. space industry, the controls had a limited, if any, impact on the pace of the Chinese space programme.[5]

Since 2013, the ITAR system has been liberalised and relaxed to some extent, but U.S.-built satellites still need a State Department licence to use foreign launchers. Similarly, with the exception of close U.S. allies, the presence of DoD

[4] Attempts to force European public customers to use the Ariane have proven nugatory, although there are implicit rules, especially on the part of Ariane's main government sponsors, to use Ariane for military launches and some European funded civilian satellites.

[5] *Space News*, 22 June 2010, "The ITAR Quagmire".

monitors is still required for launches of U.S. hardware on foreign rockets. Chinese launches require congressional approval; the presumption is that this will be denied—Intelsat, for example, has stated that despite the price advantages of a Chinese launcher, it will not even try to seek approval[6] (Space News, 1 July 2013).

12.4 A New Business Model?

Since the beginning of the twenty-first century, the appearance of consciously entrepreneurial players such as SpaceX, Blue Origin and a number of other American-based companies, as well as the space tourism initiative of Virgin Galactic whose Scaled Composite SpaceShipOne system could be developed to boost satellites into transfer orbits, has begun to disrupt the traditional launcher business model.[7] The economic impact these players have on the market would be significant if they could regularly and reliably offer launch costs significantly less than those of conventionally developed vehicles. This is precisely what SpaceX is now offering the market, posing a considerable challenge both to its American-based competitors and Arianespace, the European world leader in the supply of commercial satellite launches.[8]

12.4.1 SpaceX and the U.S. Launcher Market

SpaceX was founded by the American entrepreneur Elon Musk in 2002 and has risen rapidly to challenge both U.S. and European launch service companies. Its emergence as a player in the U.S. market has been felt primarily by United Launch Alliance (ULA). For much of the previous decade, the company's Atlas V and Delta family monopolised the American public launch market. However, the Atlas V uses a Russian-built engine that is now subject to American government sanctions; consequently ULA will be permitted only a limited number of military launch contracts pending the development of an all-American rocket motor. But even without this politically imposed constraint, ULA was under pressure from the SpaceX approach to launcher development. The Atlas V has had 96 successes since the formation of ULA in 2006. But it has failed to hold on to commercial

[6]*Space News,* 1 July 2013, Jeff Foust, "The ITAR Shift".

[7]Blue Origin is funded by Amazon's Jeff Bezos, although recorded an important milestone in April 2015 with the successful launch of its test rocket New Shepard with a prototype BE-4 engine. It is also designed to be re-usable. Its initial purpose is space tourism.

[8]Arianespace is also responsible for marketing and launching the Ariane 5, a "Europeanized" Soyuz and the ESA developed Vega small rocket.

customers who have preferred the cheaper and reliable European Ariane, or the even cheaper Chinese Long March and the Russian Proton.

In May 2015, the Falcon 9 launcher was certified by the USAF and able to compete in the launching of national security payloads. With the U.S. government helping to defray costs, SpaceX will be even better placed to increase its market share of commercial satellite launches. However, as the Falcon 9 cannot lift the U. S.'s heaviest military satellites, and until the Falcon Heavy rocket is ready in 2017, ULA will still retain its monopoly of this business. However, the long-term future of the Atlas V is uncertain without the development a new all-American engine to replace the Russian motor. In order to maintain competition, the U.S. Air Force may have to underwrite a set number of launches for SpaceX and ULA. This would be a *de facto* admission that commercial forces are insufficient to maintain two domestic sources of launch vehicles (Space News, 27 July 2015). It would also maintain the implicit subsidisation of U.S. launchers for security reasons, and further help SpaceX in particular to establish its presence in the world commercial market.

12.4.2 The SpaceX Approach

Using private capital to fund development and innovation (that is to say non-NASA or Department of Defence procurement and management practices) approaches to design and development, SpaceX is offering public and commercial customers substantial savings compared to traditional launch vehicles. This is something of a disingenuous claim: Space X may be funded commercially; it is nevertheless hoping to defray initial costs through accessing a largely closed domestic public market. However, there is no doubting SpaceX's determination to change the business model for launch vehicles.

The programme is also working towards adopting significant levels of reusability, which could cut launch costs still further. SpaceX already offers commercial launches at around 63 million euros against ULA's 173 million euros for U.S. government satellites. Overall, SpaceX intends to undercut existing platforms significantly with kilogram-to-orbit costs 25 % lower than those of Ariane (Table 12.1).

Table 12.1 Payload into low earth orbit: cost per kg in U.S. Dollars

Falcon Heavy	$2200
Falcon 9 v 1.1	$4109
DNEPR	$3784
Ariane 5	$10,476
Delta IV	$13,072
Atlas V	$13,182[a]

[a]Considerable caution should be used in citing these data, which should be regarded as illustrative only: only SpaceX have published payload costs. See http://space.stackexchange.com/questions/1989/what-is-the-current-cost-per-pound-to-send-something-into-leo

In the future, ULA hopes to offer launches using its next generation rocket for below 100 million euros—better than Atlas, but this is still likely to be considerably more than SpaceX.

Along with the highly reliable European Ariane rocket, SpaceX has won the endorsement of space insurance, and with confidence in the Proton severely battered by a number of recent costly failures, the Falcon 9 is now second only to Ariane as a prime commercial launch vehicle. Development of the Falcon Heavy will open up the even more lucrative market for heavyweight telecommunications satellite launches. But this will have to demonstrate its reliability before attracting commercial customers—another benefit of winning future United States Air Force (USAF) heavy lift business.

The focus for future U.S. launcher development is the Evolved Expendable Launch Vehicle (EELV) programme that will feature an all-American rocket engine. This programme is driven primarily for national security reasons and the need to have an all-American launcher. The USAF has recently issued a request for proposals for a new rocket engine to replace the RD-180 by 2019. However, unless the U.S. government relaxes its policy on domestic-only launches for its most sensitive military satellites, it might have to accept a monopoly situation, at least until ULA can develop a new engine for the Atlas 5 or the EELV enters service.

SpaceX has already begun to make a strong impression on the civil market and is emerging as the most likely rocket to match Ariane for market acceptability, especially if the Proton cannot regain its reputation as a cheap, reliable launch vehicle. Further endorsement by the U.S. government will benefit new entrants like SpaceX by covering some of their development costs and, more important, add credibility to their commercial business bringing confidence to a highly risk-averse civil market.

SpaceX hopes further to increase its cost advantages over rival launchers through reusing elements of the booster system. So far, SpaceX has had mixed results with this development, but appears still to be on schedule. More important, both SpaceX and Blue Origin have demonstrated successful re-landing. Proof of reusability is still to be confirmed.

SpaceX's determination to cut launch costs has led to a response from Airbus Defence and Space, makers of the Ariane launcher. It has already launched a 5-year programme to develop a reusable first stage for the next generation Ariane 6, aiming to reuse some 80 % of the first stage's economic value. This work includes a 155 million euros investment in the U.S. to tap advanced technologies. The form of reusability adopted by conventional rockets might save between 20 % and 30 % of launch costs, but to return full value it would need to be used for 30 launches a year—a very optimistic number given the average number of launches generally.[9]

[9] *Space News*, 26 October 2015, Peter de Selding, "With Revenue Looking Up, Arianespace Seeks to Bring Costs Down". This might be feasible if LEO satellite constellations become the norm, with more initial launches and then regular replenishment.

12.5 Ariane and Arianespace

The Ariane family of rockets was created with the explicit political aim of guaranteeing European independent access to space. It has evolved as a premier commercial launcher and now attracts a premium price due to its impressive record for reliability with over 50 % of the accessible market for satellite launches. However, the launch marketing and operational arm of the European launcher industry—Arianespace—has singularly failed to make a profit. Despite posting a record 1.7 billion euros in earnings in 2014, it still required an operating subsidy of over 74 million euros a year. This has entailed an explicit and implicit subsidy on the part of its major shareholders, especially the French government. Ariane 5 manufacturing costs have also been affected by the operation of ESA's *juste retour* industry policy, where shares in hardware development and production balance national contributions to development costs. This has ensured widely spread political support for the system, but at the price of economic inefficiency.

Towards the end of 2015, ESA approved a 4.9 billion euros contract to develop and build a new generation of the Ariane satellite launch rocket, the Ariane 6. This was aimed at underpinning Europe's position as the market leader in the commercial satellite launch business. But it was also influenced by the growing impact, and potential for further disruption, posed by the likes of SpaceX, which has helped to trigger significant changes in the way Ariane rockets are developed and built and in how Arianespace delivers launch services.

The Ariane 6 will be developed under stricter commercial terms than past members of the family, with a large element of development costs covered by a new industry team—Airbus Safran Launchers (Airbus Defence and Space and the Ariane rocket engine manufacturer Safran). The aim is to halve the cost of launching a standard communications satellite. Equally important, subcontractors will be selected on the grounds of efficiency and not governmental contributions to development costs. The company will explore re-usability options further to reduce launch costs. Airbus Safran Launchers also will acquire the French government's 34 % stake in Arianespace; the loss-making satellite launcher services organisation, although the French government will retain an oversight role. Airbus Safran will have to operate without ESA annual price support for launches—worth on average 100 million euros a year. On the other hand, European governments have agreed to guarantee a number of missions a year.[10]

A closer link between construction and operations should enhance control over the complete process of delivering and launching rockets. However, with ESA still paying the bulk of development costs for the Ariane 6 and another 5.5 billion euros on launch-related programmes, there may be some political resistance to ceding so much control to industry. But given the growing commercial threat posed by the American SpaceX launcher, Airbus Safran Launchers will have to cut construction and launch costs to remain competitive.

---

[10] *Space News*, 26 October 2015.

12.6 Transformational Possibilities

SpaceX notwithstanding, the basic economics of accessing space have remained broadly constant for much of the last 20 years. There have been some reductions in costs on the margin, but even SpaceX does not constitute a transformative or truly disruptive approach. Such developments have both military and commercial implications, with the former especially important in the U.S.—a factor that may again reinforce the links between launcher development and the state. However, the long-term impact on the cost of accessing space could be profound.

12.6.1 Hypersonic Vehicles

The emergence of radically new launchers centres on the development of hypersonic engines capable of lifting a vehicle directly into orbit and returning to Earth. In the late 1990s, NASA funded the Space Launch Initiative programme aimed at developing hypersonic demonstrators; the X-33 and X-34 were designed to explore hypersonic technologies. The latest of this series, the X-37 built by Boeing, has already successfully returned from space. The U.S. Defence Department funds most of American hypersonic research and there are no published plans to move rapidly towards development of a launch vehicle based on this technology, but the convergence of civil and military uses points in this direction if the technology is proven. The U.S. Defence Advanced Research Agency (DAPRA) is also funding the XS-1 reusable booster demonstrator, which would have aircraft-like reliability.[11]

In the 1980s, Rolls-Royce and British Aerospace proposed the hypersonic Hotol, but failed to attract government support. However, the design team—Reaction Engines—continued work on the hypersonic propulsion system funded by a mixture of private investment and British and European research grants. In October 2015, BAE SYSTEMS (BAES) agreed to buy a 20 % stake in Reaction Engines for 30 million euros, which will unlock a 90 million euros grant from the British Government.[12] The Reaction Engines project centres on the Sabre design for a reusable hybrid rocket/jet engine that would power a hypersonic vehicle called Skylon. The team aim to have a demonstrator engine ready by 2020.

[11] *Aviation Week*, 3 August 2015, Guy Norris, "Reusable Race".

[12] *Financial Times*, 1 November 2015: Peggy Hollinger and Clive Cookson, "BAE Systems to pay £20.6m for 20% of space engine group".

12.7 The Impact of Low Cost Access to Space

The high cost of reaching orbit means that satellites today are built to maximise their yield per kilogram, which results in high costs to develop, assemble, and test satellites. Reaction Engines aims to reduce the cost per kilogram of payload carried directly to low Earth orbit from the current 22,000 euros per kg to around 1000 euros per kg. Low cost access would provide designers with several options: an unchanged satellite with sharply lower launch costs; heavier but less expensive satellites; bigger, more capable satellites; and smaller, less capable satellites. In general, lower launch costs would allow satellite designers more freedom and could encourage more innovative solutions. Low cost access would strengthen current trends toward distributing functions among many satellites and building picosatellites, nanosatellites (which weigh between 1 kg and 10 kg), and microsatellites, which weigh between 10 and 100 kg. Lower launch costs would also transform the economics of creating and maintaining LEO constellations or clusters of satellites, each specialised but operating together.[13]

By the same token, low cost launchers could encourage orbital refuelling and replenishing of satellites. Low cost access could provide an incentive to develop these and related technologies to rendezvous, dock, transfer liquids, and build larger satellites in Earth orbit. The development of standard structures, components, and modules may finally bring the efficiency of large-scale production to satellites, further reducing satellite costs and thus encouraging more actors to explore and exploit space. Expanding existing services such as communications and remote sensing is an obvious market for low cost launchers. By significantly reducing the barriers to entry posed by high launch costs, low cost access should create obvious new markets such as providing propellants, water, and other bulk supplies to satellites and larger facilities in orbit.

In short, by developing low cost launch vehicles, space would be on the threshold of fully commercial economics, with business models more akin to contemporary airline operation. It would not necessarily reduce the politicisation of space, but it would add a very different dimension to the exploitation of space and the value of space-based services. However, for the foreseeable future—certainly not before 2030—space operations will be dominated by conventional launch vehicles, albeit with increasing degrees of reusability and some downward pressure on launch costs.

[13]The emergence of highly capable micro satellites is itself a potentially transformative event in satellite technology. However, low cost launching would still facilitate more extensive use of constellations, especially in replenishing satellites that have failed or re-entered the Earth's atmosphere.

12.8 Conclusions

The traditional business model that has shaped the launch vehicle industry—largely determined by public support for technological development and closed public markets—is being challenged at least in part by a new wave of entrepreneurs. Some of their claims to be free of state bureaucracy and funding may not escape scrutiny, but the promise of radical reductions in payload cost-to-orbit cannot be dismissed. This does not yet presage an era of airliner like economics in the space sector—that will require another round of technological innovation: but if the space economy is still waiting for its Jumbo jet moment, we are close to the DC-3 era of space economics.

Part III
Facts and Figures

Chapter 13
Chronology: 2015

Cenan Al-Ekabi and Arne Lahcen

13.1 Access to Space[1]

All launch dates are calculated using Greenwich Mean Time (GMT), hence the date at the launch site may differ from the date listed here by 1 day.

Europe	Other countries
Launch log	
January 2015	
	10 Falcon 9 v1.1-R—Dragon CRS-5 (ISS)*, CATS (ISS), and 3 cubesat secondary payloads: Flock-1d 1 (R), Flock-1d 2 (R), AESP-14 (S) 20 Atlas 5 (551)—MUOS 3 (C) 31 Delta II 7320-10C—SMAP (S), and 4 cubesat secondary payloads: ExoCube (D), GRIFEX (D), FIREBIRD-IIA (D), FIREBIRD-IIB (D)
February 2015	
11 Vega—IXV (D), and AVUM VV04 (D)	01 H-IIA 202—IGS Radar Spare (I) 01 Proton M Briz-M (Ph.3)—Inmarsat-5 F2 (C)* 02 Safir 2—Fajr (D) 11 Falcon 9 v1.1-R—DSCOVR (R) 17 Soyuz U—Progress M-26M (ISS) 27 Soyuz 2.1a—Bars M (I)

(continued)

[1]Federal Aviation Administration. The Annual Compendium of Commercial Space Transportation: 2016. Washington, DC: FAA, Jan. 2016: 63.

C. Al-Ekabi (✉) • A. Lahcen
European Space Policy Institute, Vienna, Austria
e-mail: cenan.al-ekabi@espi.or.at

C. Al-Ekabi et al. (eds.), *Yearbook on Space Policy 2015*, Yearbook on Space Policy, DOI 10.1007/978-3-7091-4860-0_13

Europe	Other countries
March 2015	
27 Soyuz STB Fregat-MT (Soyuz 2.1b)—Galileo FOC-3 (N), Galileo FOC-4 (N)	01 Falcon 9 v1.1—ABS 3A (C)*, Eutelsat 115 West B (Satmex 7) (C)* 12 Atlas 5 (421)—MMS 1 (S), MMS 2 (S), MMS 3 (S), MMS 4 (S) 19 Proton-M Briz-M (Ph.3)—Ekspress-AM7 (C) 25 Delta 4 Medium+ (4,2)—GPS-2F 09 (N) 25 Dnepr—KOMPSat 3A (R) 26 H-IIA 202—IGS Optical-5 (I) 27 Soyuz FG—Soyuz TMA-16M (MF/ISS) 28 PSLV XL—IRNSS 1D (N) 30 Long March 3C—BDS I1-S (N) 31 Rokot KM—Gonets-M21 (C), Gonets-M22 (C), Gonets-M23 (C), Cosmos 2504 (D)
April 2015	
26 Ariane 5 ECA—Thor 7 (C)*, SICRAL 2 (Syracuse 3C) (C)	14 Falcon 9 v1.1-R—Dragon CRS-6 (ISS)*, and 16 cubesat secondary payloads: Flock-1e 1 (D), Flock-1e 2 (D), Flock-1e 3 (D), Flock-1e 4 (D), Flock-1e 5 (D), Flock-1e 6 (D), Flock-1e 7 (D), Flock-1e 8 (D), Flock-1e 9 (D), Flock-1e 10 (D), Flock-1e 11 (D), Flock-1e 12 (D), Flock-1e 13 (D), Flock-1e 14 (D), Arkyd-3-Reflight (D), Centennial 1 (D) 27 Falcon 9 v1.1—TürkmenÄlem 52E/MonacoSat-1 (C)* 28 Soyuz 2.1a—Progress M-27M (ISS)* (launch failure: failure due to loss of telemetry prior to separation from the upper stage)
May 2015	
27 Ariane 5 ECA—DirecTV 15 (C)*, Sky-Mexico 1 (C)*	16 Proton M Briz-M (Ph.3)—MexSat 1 (C)* (launch failure: failure of the launcher's third stage engine) 20 Atlas 5 (501)—X-37B OTV-4, and 11 cubesat secondary payloads: GEARRS 2 (D), LightSail A (D), OptiCube 1 (D), OptiCube 2 (D). OptiCube 3 (D), USS Langley (D), AeroCube 8A (D), AeroCube 8B (D), BRICSat-P (D), ParkinsonSat (D), ULTRASat (D)
June 2015	
22 Vega—Sentinel 2A (R)	05 Soyuz 2.1a—Cosmos 2505 (Kobalt-M #10) (I) 23 Soyuz 2.1b—Cosmos 2506 (Persona #3) (EW) 27 Long March 4B—Gaofen 8 (R) 28 Falcon 9 v1.1-R—Dragon CRS-7 (ISS)* and 9 cubesat secondary payloads: IDA 1 (D), Flock-1f 1 (D), Flock-1f 2 (D), Flock-1f 3 (D), Flock-1f 4 (D), Flock-1f 5 (D), Flock-1f 6 (D), Flock-1f 7 (D), Flock-1f 8 (D) (launch failure: failure due to too much pressure in the Falcon 9's upper stage liquid oxygen tank about 139 seconds into the flight)

(continued)

Europe	Other countries
July 2015	
15 Ariane 5 ECA—Star One C4 (C)*, MSG 4 (M)	03 Soyuz U—Progress-M 28M (ISS) 10 PSLV XL—DMC-3 A (R)*, DMC-3 B (R)*, DMC-3 C (R)*, CBNT 1 (D), DeOrbitSail (D) 15 Atlas 5 (401)—Navstar GPS 2F-10 (N) 22 Soyuz FG—Soyuz TMA-17M (MF/ISS) 24 Delta 4 Medium+ (5,4)—WGS 7 (C) 25 Long March 3B—BDS M1-S (N), BD-3 M2-S (N)
August 2015	
20 Ariane 5 ECA—Eutelsat 8 West B (C)*, Intelsat 34 (C)*	19 H-IIB 304—HTV 5 (Kounotori 5) (ISS) and 18 cubesat secondary payloads: Flock-2b 1 (D), Flock-2b 2 (D), Flock-2b 3 (D), Flock-2b 4 (D), Flock-2b 5 (D), Flock-2b 6 (D), Flock-2b 7 (D), Flock-2b 8 (D), Flock-2b 9 (D), Flock-2b 10 (D), Flock-2b 11 (D), Flock-2b 12 (D), Flock-2b 13 (D), Flock-2b 14 (D), SERPENS (D), S-CUBE (D), GOMX 3 (D), AAUSAT 5 (D) 27 Long March 4C—Yaogan 27 (I) 27 GSLV Mk.2—GSAT 6 (C) 28 Proton-M Briz-M (Ph.3)—Inmarsat-5 F3 (C)*
September 2015	
10 Soyuz 2.1b—Galileo FOC-5 (N), Galileo FOC-6 (N) 30 Ariane 5 ECA—NBN Co. 1A (C)*, Arsat-2 (C)*	02 Soyuz FG—Soyuz TMA-18M (MF/ISS) 02 Atlas 5 (551)—MUOS 4 (C) 12 Long March 3B—TJSSW 1 (EW) 14 Proton M Blok DM 03—Exspress-AM8 (C) 14 Long March 2D—Gaofen 9 (R) 19 Long March 6—XY 2 and 19 cubesat secondary payloads: DCBB (D), ZDPS 2A (D), ZDPS 2B (D), Xiwang 2A (C), Xiwang 2B (C), Xiwang 2C (C), Xiwang 2D (C), Xiwang 2E (C), Xiwang 2F (C), LilacSat 2 (C), NS 2 (D), ZJ 1 (D), ZJ 2 (D), TianTuo 3 (D), NUDT-Phone-Sat (D), Xingchen 1 (D), Xingchen 2 (D), Xingchen 3 (D), Xingchen 4 (D) 24 Rokot KM—Strela-3M #13 (C), Strela-3M #14 (C), Strela-3M #15 (C) 25 Long March 11—Pujiang 1 (D), Tianwang 1A (D), Tianwang 1B (D), Tianwang 1C (D), 28 PSLV XL—AstroSat (S), LAPAN A2 (R), ExactView 9 (C), Lemur-2 1 (R), Lemur-2 2 (R), Lemur-2 3 (R), Lemur-2 4 (R) 29 Long March 3B—BDS I2-S (N)
October 2015	
	01 Soyuz U—Progress M-29M (ISS) 02 Atlas 5 (421)—MEXSAT 3 (C)* 07 Long March 2D—Jilin-1A (R), Jilin-1B (R), Jilin-1C (R), Jilin-1D (R) 08 Atlas 5 (401)—USA 264 (NOSS) (EW), and 13 cubesat secondary payloads: SNaP-3A (D),

(continued)

Europe	Other countries
	SNaP-3B (D), SNaP-3C (D), LMRST-Sat (D), SINOD-D 1 (D), SINOD-D 2 (D), AeroCube 5C (D), AeroCube 7 (D), ARC 1 (D), BisonSat (R), AMSAT Fox-1 (C), PropCube 1 (D), PropCube 2 (D) 16 Long March 3B—APStar 9 (C) 16 Proton-M Briz-M (Ph.3)—Türksat 4B (C)* 26 Long March 2D—Tianhui 1C (R) 31 Atlas 5 (401)—Navstar GPS 2F-11 (N)
November 2015	
10 Ariane 5 ECA—Arabsat 6B (C)*, GSAT 15 (C)	03 Long March 3B—ChinaSat 2C (C) 04 Super Strypi—HiakaSat (D) and 12 cubesat secondary payloads: Supernova-Beta (D), STACEM (D), Argus (D), EDSN 1 (D), EDSN 2 (D), EDSN 3 (D), EDSN 4 (D), EDSN 5 (D), EDSN 6 (D), EDSN 7 (D), EDSN 8 (D), PrintSat (D) (launch failure: failure due to malfunction in its first stage) 08 Long March 4B—Yaogan 28 (I) 17 Soyuz 2.1b Fregat-M—Cosmos 2510 (EW) 20 Long March 3B—LaoSat 1 (C) 24 H-IIA-204—Telstar 12V (C)* 26 Long March 4C—Yaogan 29 (I)
December 2015	
03 Vega—LISA Pathfinder (S) 17 Soyuz-STB Fregat-MT—Galileo FOC 8 (N), Galileo FOC 9 (N)	05 Soyuz 2.1v Volga—Kanopus-ST (I), and 1 cubesat secondary payload: KYuA 1 (D) 06 Atlas 5 (401)—Cygnus CRS-4 (ISS)*, and 20 cubesat secondary payloads: SIMPL (D), AggieSat 4 (D), Bevo 2 (D), MinXSS 1 (S), CADRE (D), STMSat 1 (D), Nodes 1 (D), Nodes 2 (D), Flock-2e 1 (D), Flock-2e 2 (D), Flock-2e 3 (D), Flock-2e 4 (D), Flock-2e 5 (D), Flock-2e 6 (D), Flock-2e 7 (D), Flock-2e 8 (D), Flock-2e 9 (D), Flock-2e 10 (D), Flock-2e 11 (D), Flock-2e 12 (D) 09 Long March 3B—ChinaSat 1C (C) 11 Zenit 3F—Elektro-L No. 2 (M) 13 Proton-M Briz-M (Ph.1 mod. 2)—Garpun No. 12L (C) 15 Soyuz-FG—Soyuz TMA-19M (MF/ISS) 16 PSLV CA—TeLEOS 1 (R)*, VELOX C1 (S), Kent Ridge 1 (R), and 3 cubesat secondary payloads: VELOX 2 (D), Athenoxat 1 (R), Galassia (S) 17 Long March 2D—DAMPE (S) 21 Soyuz 2.1a—Progress MS-1 (ISS) 22 Falcon 9FT—ORBCOMM-2 F2 (C)*, ORBCOMM-2 F5 (C)*, ORBCOMM-2 F8 (C)*, ORBCOMM-2 F10 (C)*, ORBCOMM-2 F12 (C)*, ORBCOMM-2 F13 (C)*, ORBCOMM-2 F14 (C)*,

(continued)

Europe	Other countries
	ORBCOMM-2 F15 (C)[a], ORBCOMM-2 F16 (C)[a], ORBCOMM-2 F17 (C)[a], ORBCOMM-2 F18 (C)[a] 24 Proton-M Briz-M (Ph.3)—Ekspress-AMU 1 (C) 28 Long March 3B—Gaofen 4 (R)

C communications, *D* development, *I* intelligence, *ISS* International Space Station, *M* meteorological, *MF* manned flight, *N* navigation, *R* remote sensing, *S* scientific, *EW* early warning system

*Commercial

13.2 Space Science and Exploration

Europe	Other countries
Earth sciences	
03 December—Launch of LISA Pathfinder (ESA, Europe) proof-of-concept mission testing two masses in a near-perfect gravitational free-fall and control to measure their motion with unprecedented accuracy.[a]	*10 January*—Launch of AESP-14 scientific cubesat (ITA, INPE, Brazil).[a] *31 January*—Launch of SMAP (NASA, USA) science mission.[a] *12 March*—Launch of MMS 1, MMS 2, MMS 3, MMS 4 (NASA, USA) magnetosphere research mission.[a] *28 September*—Launch of AstroSat satellite (ISRO, India) to provide optical, ultraviolet, and X-ray images of black holes and other related phenomena.[b] *16 December*—Launch of VELOX C1 (Nanyang Technological University, Singapore), and Galassia (National University of Singapore, Singapore) scientific satellites.
Astronomy	
21 August—Gaia (ESA), completed the first year of its main scientific observations for its 5-year scientific mission to measure the positions and motions of a billion stars of the roughly 100 billion stars in our galaxy to create the most accurate map yet of the Milky Way.[c]	*28 September*—Successful launch of Astrosat mission (ISRO, India) to provide optical, ultraviolet, and X-ray images of black holes and other related phenomena.[d] *17 December*—launch of the Dark Matter Particle Explorer (DAMPE) mission (China) to conduct a 3-year mission to shed new light on the nature of dark matter.[e]
Exploration	
13 August—Rosetta spacecraft (ESA, Europe) moved to a safe distance until mid-September as it observed 67P/Churyumov-Gerasimenko's closest approach to the Sun.[f]	*12 January*—Chang'e 5-TI (CAST, China) returned to lunar orbit as part of a test-run for an eventual lunar sample return mission.[g] *15 January*—images taken by the Mars Reconnaissance Orbiter (MRO) (NASA, USA) helped to reveal the location of the UK-led Beagle-2 Mars lander which lost communications on Mars' surface on

(continued)

Europe	Other countries
	25 December 2003.[h] *10–18 February*—Mars Atmosphere and Volatile EvolutioN (MAVEN) (NASA, USA) began a series of five deep-dip manoeuvres designed to gather measurements closer to the lower end of the Martian upper atmosphere.[i] *04 May*—Lunar Reconnaissance Orbiter (LRO) (NASA, USA) completed two station keeping burns to lower its orbit to within 20 km of the Moon's South Pole and 165 km from its North Pole, to enable improved measurements of the LRO's Lunar Orbiter Laser Altimeter (LOLA) and its high resolution Diviner instrument over the South Pole.[j] *16 July*—Arkyd 3 Reflight (A3R) spacecraft (Planetary Resources, USA) launched from the ISS to validate several core technologies that will be incorporated into a future fleet of Arkyd spacecraft that will be launched into the solar system.[k] *12 August*—the MRO (NASA, USA) reached its tenth year, completing over 40,000 orbits of Mars, while in its fourth extended mission.[l] *07 October*—SpaceIL (Israel) was the first team to be confirmed to have made launch arrangements prior to the end of 2015 to meet the Google Lunar X Prize requirement to extend the competition to the end of 2017.[m] *01 November*—Chang'e 5-TI (CAST, China) began moving to the Earth-Moon Lagrange (L2) point; the service module then completed three circles around that point prior to returning to lunar orbit.[n] *11 November*—ISRO announces the first science results of the Mars Orbiter Mission (MOM) (ISRO, India) to observe Mars' surface, atmosphere and exosphere for a detailed understanding of the planet's evolution.[o] *08 December*—Moon Express (USA) was the second team to be confirmed to have made launch arrangements for the Google Lunar X Prize competition.[p]
Manned spaceflight & cargo transfers	
24 November (2014)–11 June—Samantha Cristoforetti (Italy) participated in ISS Expeditions 42 & 43 as a flight engineer for ESA's Futura mission.[q] *15 February*—ATV-5, George Lemaître, ISS resupply vehicle (European Space Agency, Europe), ended its mission on the ISS.[a]	*10 January*—launch of Dragon CRS-5* (SpaceX, USA) commercial resupply services mission to ISS.[a] *17 February, 28 April (failed launch), 03 July, 01 October, 21 December*—launches of 5 Progress (Roscosmos, Russia) cargo transfer vehicles to the ISS.[a]

(continued)

Europe	Other countries
04 September–12 September—Andreas Mogensen (Denmark) participated in a short-duration visit to the ISS for ESA's IrISS mission.[r] *15 December*—Tim Peake (Italy) began a 6-month mission on ISS Expedition 46 & 47 as a flight engineer for ESA's Principia mission.[s]	*27 March*—launch of Soyuz TMA-16M (Roscosmos, Russia) with three ISS Expedition 43/44 crew.[a] *14 April*—launch of Dragon CRS-6* (SpaceX, USA) commercial resupply services mission to ISS.[a] *28 June*—launch of Dragon CRS-7* (SpaceX, USA) commercial resupply services mission to ISS (*failed launch*).[a] *22 July*—launch of Soyuz TMA-17M (Roscosmos, Russia) with three ISS Expedition 44/45 crew.[a] *19 August*—launch of HTV 5 (SpaceX, USA) commercial resupply services mission to ISS.[a] *02 September*—launch of Soyuz TMA-18M (Roscosmos, Russia) with three ISS Expedition 45/46 crew.[a] *06 December*—launch of Cygnus Orb-4* (Orbital Sciences, USA) commercial resupply services mission to ISS.[a] *15 December*—launch of Soyuz TMA-19M (Roscosmos, Russia) with three ISS Expedition 46/47 crew.[a]

[a]Federal Aviation Administration. The Annual Compendium of Commercial Space Transportation: 2016. Washington, DC: FAA, Jan. 2016: 63
[b]De Selding, Peter B. "PSLV Rocket Launches India's 1st Astronomy Satellite, 4 Spire Cubesats." 28 Sept. 2015. SpaceNews 5 Jan. 2016 <http://spacenews.com/pslv-rocket-launches-indias-1st-astronomy-satellite/>
[c]"Gaia's first year of scientific observations." 25 Aug. 2015. ESA 22 May 2016 <http://m.esa.int/Our_Activities/Space_Science/Gaia/Gaia_s_first_year_of_scientific_observations>
[d]De Selding, Peter B. "PSLV Rocket Launches India's 1st Astronomy Satellite, 4 Spire Cubesats." 28 Sept. 2015. SpaceNews 5 Jan. 2016 <http://spacenews.com/pslv-rocket-launches-indias-1st-astronomy-satellite/>
[e]Yue, Huang. "China Launches Country's First Dark Matter Satellite." 17 Dec. 2015. CRI English News 22 Jan. 2016 <http://english.cri.cn/12394/2015/12/17/4201s908753.htm>
[f]Crew, Bec. "Our last chance to contact the Philae comet lander probably failed." 13 Jan. 2016. Science alert 20 May 2016 <http://www.sciencealert.com/our-last-chance-to-contact-the-philae-comet-lander-probably-failed>
[g]David, Leonard. "Chinese Spacecraft Enters Orbit around the Moon." 20 Jan. 2015. SpaceNews 8 Jan. 2016 <http://spacenews.com/chinese-spacecraft-enters-orbit-around-the-moon/>
[h]"Beagle-2 lander found on Mars." 16 Jan. 2015. ESA 25 Apr. 2016 <http://www.esa.int/Our_Activities/Space_Science/Mars_Express/Beagle-2_lander_found_on_Mars>
[i]"NASA's MAVEN Spacecraft Completes First Deep Dip Campaign." 19 Feb. 2015. NASA Mars Exploration 17 May 2016 <http://mars.nasa.gov/news/whatsnew/index.cfm?FuseAction=ShowNews&NewsID=1780>
[j]"NASA's LRO Moves Closer to the Lunar Surface." 5 May 2015. NASA 24 Apr. 2016 <http://www.nasa.gov/feature/goddard/nasas-lro-moves-closer-to-the-lunar-surface>
[k]"Planetary Resources' First Spacecraft Successfully Deployed, Testing Asteroid Prospecting Technology in Orbit." 16 July 2015. Planetary Resources 27 Nov. 2015 <http://www.planetaryresources.com/2015/07/planetary-resources-first-spacecraft-deployed/>

(continued)

Europe	Other countries

[l]"One Decade after Launch, Mars Orbiter Still Going Strong." 10 Aug. 2015. NASA Jet Propulsion Laboratory 16 May 2016 <http://mars.nasa.gov/mro/news/whatsnew/index.cfm?FuseAction=ShowNews&NewsID=1849>
[m]Foust, Jeff. "Israeli X Prize Team Announces Launch Contract for Lunar Mission." 7 Oct. 2015. SpaceNews 5 Jan. 2016 <http://spacenews.com/israeli-x-prize-team-announces-launch-contract-for-lunar-mission/>
[n]"Service Module of Chinese Probe Successfully Enters Lunar Orbit." 11 Jan. 2015. Sputnik News 22 Jan. 2016 <http://sputniknews.com/science/20150111/1016776437.html>
[o]"India's Mars Orbiter Mission delivers first Science Results, looks at future Challenges." 11 Nov. 2015. Spaceflight 101 17 May 2016 <http://spaceflight101.com/mom/indias-mars-orbiter-mission-delivers-first-science-results-looks-at-future-challenges/>
[p]Foust, Jeff. "Moon Express Buys Rocket Lab Launches for Lunar Missions." 1 Oct. 2015. SpaceNews 5 Jan. 2016 <http://spacenews.com/moon-express-buys-rocket-lab-launches-for-lunar-missions/>
[q]"Samantha's longer stay on Space Station." 12 May 2015. ESA 27 May 2016 <http://m.esa.int/Our_Activities/Human_Spaceflight/Futura/Samantha_s_longer_stay_on_Space_Station>
[r]"ESA astronaut Andreas Mogensen begins busy International Space Station tour." 4 Sept. 2015. ESA 27 May 2016 <http://www.esa.int/Our_Activities/Human_Spaceflight/iriss/ESA_astronaut_Andreas_Mogensen_begins_busy_International_Space_Station_tour>
[s]"ESA astronaut Tim Peake begins six-month stay on Space Station." 15 Dec. 2015. ESA 27 May 2016 <http://m.esa.int/Our_Activities/Human_Spaceflight/Principia/ESA_astronaut_Tim_Peake_begins_six-month_stay_on_Space_Station>
[*]Commercial

13.3 Applications

Europe	Other countries
Earth observation	
22 June—Launch of Sentinel 2A (ESA, Europe) optical imaging satellite contributing to the EU Copernicus Earth Observation flagship programme.[a] *15 July*—Launch of MSG 4 (Eumetsat, Europe) meteorology satellite.[a]	*11 February*—Launch of DSCOVR (NOAA, USA).[a] *25 March*—Launch of KOMPSat 3A (KARI, South Korea).[a] *27 June*—Launch of Gaofen 8 (CNSA, China).[a] *10 July*—Launch of DMC-3 A, B, and C (DMCii, UK).[a] *14 September*—Launch of Gaofen 9 (CNSA, China).[a] *28 September*—Launch of LAPAN A2 (LAPAN, Indonesia) and Lemur-2 1, 2, 3, and 4 (Spire Global, USA) remote sensing cubesats.[a] *07 October*—Launch of Jilin-1A, -1B, -1C, and -1D (CIOMP, China).[a] *08 October*—Launch of BisonSat (Salish Kootenai College, USA).[a] *26 October*—Launch of Tianhui 1C (People's Liberation Army, China).[a]

(continued)

Europe	Other countries
	11 December—Launch of Elektro-L No.2 (Roscosmos, Russia) meteorology satellite.[a] *16 December*—Launch of TeLEOS 1 (AgilSpace, Singapore), and the Kent Ridge 1 (National University of Singapore, Singapore) and Athenoxat 1 (Nanyang Technological University, Singapore) remote sensing cubesats.[a] *28 December*—Launch of Gaofen 4 (CNSA, China).[a]
Intelligence and early warning	
	01 February—Launch of IGS Radar Spare (Japan Self-Defense Forces, Japan) radar reconnaissance satellite.[a] *27 February*—Launch of Bars M (Russian Ministry of Defense, Russia) optical reconnaissance satellite.[a] *26 March*—Launch of IGS Optical-5 (Japan Self-Defense Forces, Japan) optical reconnaissance satellite.[a] *05 June*—Launch of Cosmos 2505 (Kobalt-M #10) reconnaissance satellite.[a] *23 June*—Launch of Cosmos 2506 (Persona #3) (Russian Ministry of Defense, Russia) reconnaissance satellite.[a] *27 August*—Launch of Yaogan 27 (People's Liberation Army, China) remote sensing reconnaissance satellite.[a] *12 September*—Launch of TJSSW 1 (People's Liberation Army, China) early warning satellite.[a] *08 October*—Launch of USA 264 (NOSS) (NRO, USA) reconnaissance satellite.[a] *08 November*—Launch of Yaogan 28 (People's Liberation Army, China) reconnaissance satellite.[a] *17 November*—Launch of Cosmos 2510 (Russian Ministry of Defense, Russia) early warning satellite.[a] *29 November*—Launch of Yaogan 29 (People's Liberation Army, China) reconnaissance satellite.[a] *05 December*—Launch of Kanopus-ST (Russian Ministry of Defense, Russia) reconnaissance satellite.[a]

(continued)

Europe	Other countries
Navigation	
27 March—Launch of Galileo FOC 3 and FOC 4 (ESA, Europe) satellites. *10 September*—Launch of Galileo FOC 5 and FOC 6 (ESA, Europe) satellites. *17 December*—Launch of Galileo FOC 8 and FOC 9 (ESA, Europe) satellites.	*25 March*—Launch of the Navstar GPS 2F-09 (USAF, USA) space-based radio positioning, navigation, and time distribution system.[a] *28 March*—Launch of IRNSS 1D (ISRO, India) satellite.[a] *30 March*—Launch of BDS I1-S (CNSA, China) satellite.[a] *15 July*—Launch of the Navstar GPS 2F-10 (USAF, USA) satellite.[a] *25 July*—Launch of BDS M1-S and BD-3 M2-S (CNSA, China) satellites.[a] *29 September*—Launch of BDS I2-S (CNSA, China) satellite.[a] *31 October*—Launch of the Navstar GPS 2F-11 (USAF, USA) satellite.[a]
Telecommunications/broadcasting	
01 February—Launch of Inmarsat-5 F2* (Inmarsat, UK).[a] *26 April*—Launch of Thor 7* (Telenor, Norway) and SICRAL 2 (Syracuse 3C) (Italian Ministry of Defence/French Defence Procurement Agency, Italy/France).[a] *20 August*—Launch of Eutelsat 8 West B* (Eutelsat, France) and Intelsat 34* (Intelsat, Luxembourg).[a] *28 August*—Launch of Inmarsat-5 F3* (Inmarsat, UK).[a]	*20 January*—Launch of MUOS 3 (U.S. Navy, USA).[a] *01 March*—Launch of ABS 3A* (Asia Broadcast Satellite, China) and Eutelsat 115 West B* (Satmex 7) (Satellites Mexicanos S.A. de C.V., Mexico).[a] *19 March*—Launch of Ekspress-AM7 (RSCC, Russia).[a] *31 March*—Launch of Gonets-M21, -M22, and -M23 (Gonets Satcom, Russia).[a] *27 April*—Launch of TürkmenÄlem 52E* (Turkmenistan National Space Agency, Turkmenistan).[a] *16 May*—Launch of MexSat 1* (Secretariat of Communications and Transportation, Mexico).[a] *27 May*—Launch of DirecTV 15* (DirecTV, USA) and Sky-Mexico 1* (SKY Mexico, Mexico).[a] *15 July*—Launch of Star One C4* (Star One, Brazil).[a] *24 July*—Launch of WGS 7 (USAF, USA).[a] *27 August*—Launch of GSAT 6 (ISRO, India).[a] *02 September*—Launch of MUOS 4 (U.S. Navy, USA).[a] *14 September*—Launch of Exspress-AM8 (RSCC, Russia).[a] *19 September*—Launch of Xiwang 2A, 2B, 2C, 2E, and 2F (CAMSAT, China) communications cubesats. *24 September*—Launch of Cosmos 2507, Cosmos 2508, and Cosmos 2509 (Russian Ministry of Defense, Russia).[a]

(continued)

Europe	Other countries
	28 September—Launch of ExactView 9 (exactEarth, Canada) communications cubesat.[a] *30 September*—Launch of NBN Co. 1A* (NBN Co Limited, Australia) and Arsat-2* (AR-SAT SA, Argentina).[a] *02 October*—Launch of MEXSAT 3* (Ministry of Communications and Transportation, Mexico).[a] *8 October*—Launch of AMSAT Fox-1 (AMSAT, USA) communications cubesat.[a] *16 October*—Launch of the APStar 9 (APT Satellite Holdings, China).[a] *16 October*—Launch of Türksat 4B* (Turksat, Turkey).[a] *03 November*—Launch of ChinaSat 2C (People's Liberation Army, China).[a] *10 November*—Launch of Arabsat 6B* (Arabsat, Saudi Arabia) and GSAT 15 (ISRO, India).[a] *20 November*—Launch of LaoSat 1 (Laos National Authority for Science and Technology, Laos).[a] *24 November*—Launch of Telstar 12V (Telesat, USA).[a] *09 December*—Launch of Meridian 7 (People's Liberation Army, China).[a] *13 December*—Launch of Garpun No. 12L (Russian Ministry of Defense, Russia).[a] *22 December*—Launch of ORBCOMM-2 F2, F5, F8, F10, F12, F13, F14, F15, F16, F17, and F18 (ORBCOMM, USA).[a] *24 December*—Ekspress-AMU 1 (RSCC, Russia).[a]
Technology development	
11 February—Launch of IXV (ESA, Europe) spacecraft and AVUM VV04 (Arianespace, Europe) Vega fourth stage module.[a] *10 July*—Launch of CBNT 1 (SSTL, UK) microsat and DeOrbitSail (Surrey Space Centre, UK) development cubesat.[a] 19 August—GOMX 3 (GOMSpace, Denmark) and AAUSAT 5 (Aalborg University Cubesat, Denmark) development cubesats.[a]	*10 January*—Launch of Flock-1d 1-to-2 (Planet Labs, USA) development cubesats.[a] *31 January*—Launch of ExoCube (California Polytechnic State University, USA), GRIFEX (NASA, USA), FIREBIRD-IIA and IIB (Montona State University, USA) development cubesats.[a] *02 February*—Launch of Fajr (Iranian Department for Science and Technology, Iran) development microsat.[a] *31 March*—Launch of Cosmos 2504 (14F153 #3) (Russian Ministry of Defense, Russia) development microsat.[a] *14 April*—Launch of Flock-1e 1-to-14 (Planet Labs, USA), Arkyd-3-Reflight (Planetary Resources, USA), Centennial 1 (Booz Allen

(continued)

Europe	Other countries
	Hamilton, Air Force Research Laboratory (AFRL), USA) development cubesats.[a] *20 May*—Launch of X-37B OTV-4 (USAF, USA) spacecraft, ULTRASat (NASA, USA) small satellite, and GEARRS 2 (Taylor University, USA), LightSail A (The Planetary Society, USA), OptiCube 1-to-3 (California Polytechnic State University, USA), USS Langley (U.S. Naval Academy, USA), AeroCube 8A and 8B (The Aerospace Corporation, USA), BRICSat-P (U.S. Naval Academy, USA), ParkinsonSat (U.S. Naval Academy, USA) development cubesats.[a] *28 June*—Launch of Flock-1f 1-to-8 (Planet Labs, USA) development cubesats (*failed launch*).[a] *19 August*—Launch of Flock-2b 1-to-14 (Planet Labs, USA), SERPENS (SERPENS, Brazil), and S-CUBE (PERC/Chitech, Tohoku University, Japan) development cubesats. *19 September*—XY 2 (CASC, China) smallsat, and DCBB (CAMSAT, China), ZDPS 2A and 2B (Zhejiang University, China), LilacSat 2 (Harbin Institute of Technology, China), NS 2 (Tsinghua University, China), ZJ 1 and 2 (Tsinghua University, China), TianTuo 3 (NUDT, China), NUDT-Phone-Sat (NUDT, China), and Xingchen 1-to-4 (NUDT, China) development cubesats.[a] *25 September*—Launch of Pujiang 1 (SAST, China) smallsat and Tianwang 1A, 1B, and 1C development cubesats.[a] *08 October*—Launch of SNaP-3A, 3B, and 3C (U.S. Army, USA), LMRST-Sat (NASA/JPL, USA), SINOD-D 1 and 2 (SRI International, USA), AeroCube 5C and 7 (The Aerospace Corporation, USA), ARC 1 (University of Alaska Fairbanks, USA), and PropCube 1 and 2 (Tyvak Nano-Satellite Systems, USA) development cubesats.[a] *04 November*—Launch of HiakaSat (Operationally Responsive Space (ORS) Office, USA) microsat, and Supernova-Beta (Pumpkin Inc., USA), STACEM (Utah State University, USA), Argus (St. Louis University/ Vanderbilt University, USA), EDSN 1-to-8 (NASA, USA), and PrintSat (Montana State University, USA) development cubesats (*failed launch*).[a]

(continued)

Europe	Other countries
	05 December—Launch of KYuA 1 (Russian Ministry of Defense, Russia) development cubesat.[a] *06 December*—Launch of SIMPL (NovaWurks, USA) and AggieSat 4 (Texas A&M University, USA) microsats, and Bevo 2 (University of Texas at Austin, USA), MinXSS 1 (University of Colorado at Bolder, USA), CADRE (University of Michigan, USA), STMSat 1 (St. Thomas More Cathedral School, USA), Nodes 1 and 2 (NASA Ames Research Center, USA), Flock-2e 1-to-12 (Planet Labs, USA) development cubesats.[a] *16 December*—Launch of VELOX 2 (Nanyang Technological University, Singapore) development cubesat.[a]
Business	
05 June—Airbus Defence & Space revealed its own design for a reusable first-stage engine and avionics package intended for a next generation Ariane rocket.[b] *16 July*—Airbus D&S was selected by ESA for a 350.8 million euros ($389 million) contract to build the Juice Orbiter.[c] *12 August*—ESA signs contract with Airbus Safran Launcher's (ASL) to develop the Ariane 6 launcher in a contract valued at around 2.4 billion euros.[d] *12 August*—ESA signs contract with ELV and ASI to develop the upgraded Vega-C rocket in a contract valued at 395 million euros.[d] *5 October*—Eumetsat and ESA signed a cooperation agreement that secures the development of the six-satellite Metop-SG system.[e]	*22 January*—Russian President Vladimir Putin approves the merger of its Federal Space Agency (Roscosmos) and the newly-founded state-owned holding company United Rocket and Space Corporation (URSC).[f] *16 April*—NewSat Ltd. of Australia filed for bankruptcy protection with the U.S. Bankruptcy Court of Delaware.[g] *21 May*—India's office of the Prime Minister announced that it had approved an order to build 15 additional PSLV launchers to be used between 2017 and 2020, at a rate of 4–5 launches per year; raising ISRO's docket of PSLV launchers to 50 in total.[h] *1 July*—the U.S. Ex-Im Bank's authority lapses, closing its doors to new satellite and other export business, but continuing all pre-existing loans, guarantees, and insurance policies into their maturity.[i] *August*—Boeing, Lockheed Martin, and Raytheon each won study contracts worth approximately $9.7 million to develop MOKV concepts which are expected to be completed by May 2016.[j] *17 August*—Thaicom satellite operator and Thailand's Ministry of Defense signed a memorandum of understanding for collaboration of satellite communications and the possibility in co-developing a satellite project between the two parties.[k] *14 September*—International Criminal Court (ICC) ruling directs ISRO's commercial arm

(continued)

Europe	Other countries
	Antrix Corp. to pay $672 million in damages to Devas Multimedia (an Indian company backed by German and U.S. investors); the Indian government has announced that it will challenge the ruling and apply for remedy.[l] *4 December*—U.S. President Obama signs legislation reauthorizing the Ex-Im Bank, extending its life through September 2019.[m]

[a]Federal Aviation Administration. The Annual Compendium of Commercial Space Transportation: 2016. Washington, DC: FAA, Jan. 2016: 63
[b]De Selding, Peter B. "Meet Adeline, Airbus' Answer To SpaceX Reusability." 5 June 2015. SpaceNews 5 Jan. 2016 <http://spacenews.com/meet-adeline-airbus-response-to-reusable-spacex-rocket/>
[c]De Selding, Peter B. "ESA, Airbus Formalize Jupiter Icy Moons Contract." 11 Dec. 2015. SpaceNews 11 Jan. 2016 <http://spacenews.com/esa-airbus-formalize-jupiter-icy-moons-contract/>
[d]De Selding, Peter B. "ESA Inks $3.8 Billion in Contracts for Ariane 6, Vega-C and Spaceport Upgrades." 12 Aug. 2015. SpaceNews 13 Jan. 2016 <http://spacenews.com/esa-inks-3-8-billion-in-contracts-for-ariane-6-vega-c-and-spaceport-upgrades/>
[e]De Selding, Peter B. "ESA, Eumetsat Finalize Contract for Next-gen Weather Satellites." 6 Oct. 2015. SpaceNews 13 Jan. 2016 <http://spacenews.com/esa-eumetsat-finalize-contract-for-next-gen-weather-satellites/>
[f]Bodner, Matthew. "Putin Approves Roscosmos Merger with Conglomerate." 23 Jan. 2015. SpaceNews 12 Jan. 2016 <http://spacenews.com/putin-approves-roscosmos-merger-with-conglomerate/>
[g]De Selding, Peter B. "NewSat Seeks Bankruptcy Protection." 17 Apr. 2015. SpaceNews 13 Jan. 2016 <http://spacenews.com/newsat-seeks-bankruptcy-protection/>
[h]De Selding, Peter B. "India OKs Budget for Building, Launching 15 PSLV Rockets by 2020." 22 May 2015. SpaceNews 5 Jan. 2016 <http://spacenews.com/india-to-build-and-launch-15-pslv-rockets-by-2020/>
[i]De Selding, Peter B. "Ex-Im is Closed to New Business; Will the U.S. Space Industry Suffer?" 2 July 2015. SpaceNews 13 Jan. 2016 <http://spacenews.com/ex-im-is-closed-to-new-business-will-the-u-s-space-industry-suffer/>
[j]"Russia involves 10 reconnaissance satellites in Syria operation—General Staff." 17 Nov. 2015. TASS 20 Mar. 2016 <http://tass.ru/en/science/837273>
[k]"The Next Phase 2016—Annual Report 2015." 3 Mar. 2016. Thaicom Public Company Limited 15 June 2016: 051° E–053° E <http://thcom.listedcompany.com/ar.html>
[l]Jayaraman, K. S. "Antrix To Appeal $672M Judgment for Nixed Satellite Contract." 30 Sept. 2015. SpaceNews 13 Jan. 2016 <http://spacenews.com/antrix-to-appeal-672m-judgment-for-nixed-satellite-contract/>
[m]Calmes, Jackie. "Ex-Im Bank Is Reopened, but Big Loans Are Stalled." 7 Dec. 2015. The New York Times 11 Mar. 2016 <http://www.nytimes.com/2015/12/08/business/ex-im-bank-is-reopened-but-big-loans-are-stalled.html>
[*]Commercial

13.4 Policy and International Cooperation

Europe	Other countries
General policy	
03 March—British government, its UK Space Agency, and other agencies issued a joint public statement on the outcome of a consultation prepared by the UK's Civil Aviation Authority (CAA) on criteria for a commercial spaceport.[a] *27–31 July*—The meeting on "Multilateral Negotiations on an international code of conduct for outer space activities", convened at the initiative of the European Union, with the assistance of the United Nations Office for Disarmament Affairs, resulted with the placement of the draft International Space Code of Conduct for Outer Space Activities (ICoC) within the framework of the UN through a mandate of the General Assembly.[b]	*09 January*—Japan's Space Policy Commission published a third iteration of its Basic Plan for Space Policy focusing on security and commerce from its earlier emphasis on the peaceful use of outer space.[c] *26 May*—China's Ministry of National Defense (MOD) released a white paper on China's Military Strategy, in response to what it views as provocative actions of the U.S. and other offshore neighbours, outlining China's strategic guideline of active defence, along with modernizing its military to adapt to new changes in its maritime security environment.[d] *26 October*—China releases its Middle and Long Term Development Plan for State Civil Space Infrastructure (2015–2025) to produce comprehensive application demonstrations in 12 fields (including territory, mapping, energy, communications, and environmental protection) to provide core business with timely, accurate, and stable space information service.[e] *29 October*—China's Communist Party preliminarily approves its 13th Five Year Plan (2016–2020) which is expected to address the following four economic thematic areas: (1) secure a high growth target for 2020; (2) place more emphasis on domestic consumption with less focus on investment and exports; (3) further embrace competitive market mechanisms as part of its structural reform; and (4) enhance individual well-being through social welfare and healthcare reforms.[f] *11 November*—Japan releases revised version of its Basic Plan for Space Policy recommending that the IGS system be expanded to eight satellites (plus two relay satellites to support the constellation).[g] *25 November*—the U.S. Commercial Space Launch Competitiveness Act (CSLCA) signed into law CSLCA grants commercial U.S. prospectors an entitlement to any asteroid resource or space resource obtained.[h] *16 December*—U.S. Congress approved the National Defense Authorization Act for 2016 (NDAA-16) which lifted the ban on the Russian-built RD-180 engine that was enacted in the previous year's NDAA.[i]

(continued)

Europe	Other countries
General cooperation	
05 January—the French space agency (CNES) announced that it had begun a small technology research programme with Germany and other ESA governments to develop the technology needed for a reusable rocket stage.[j]	*02–13 February*—UN COPUOS 52nd Scientific and Technical Subcommittee meeting took place in Vienna, Austria *13–24 April*—UN COPUOS 54th Legal Subcommittee meeting took place in Vienna, Austria *10–19 June*—UN COPUOS 58th Plenary Meeting took place in Vienna, Austria *28 September*—U.S. State Department and China agree to consult with one another as part of a new U.S.-China Civil Space Cooperation Dialogue for the purpose of satellite collision avoidance and the long-term sustainability of space.[k] *12–16 October*—66th International Astronautical Congress (IAC) held in Jerusalem, Israel. *08 December*—the U.S. Missile Defense Agency and Japan's Ministry of Defence had conducted a second live-fire test of the new Standard Missile-3 (SM-3) Block IIA missile, following another test conducted on 6 June 2015, under the Aegis Ballistic Missile Defense program.[l]

[a]Foust, Jeff. "British Government Narrows List of Potential Spaceport Sites." 3 Mar. 2015. SpaceNews 14 Dec. 2015 <http://spacenews.com/british-government-narrows-list-of-potential-spaceport-sites/>

[b]"Multilateral Negotiations on an International Code of Conduct for Outer Space Activities, New York 27–31 July 2015 | Chair's Summary." 31 July 2015. Ministry of Foreign Affairs of Japan 29 June 2016 <http://www.mofa.go.jp/mofaj/files/000098436.pdf>

[c]Kallender-Umezu, Paul. "Japan Boosts Space Spending In Support of Security Focus." 2 Feb. 2015. SpaceNews 10 Dec. 2015 <http://spacenews.com/japan-boosts-space-spending-in-support-of-security-focus/>

[d]"III. Strategic Guideline of Active Defense." 26 May 2015 Ministry of National Defense—The People's Republic of China 3 Feb. 2016 <http://eng.mod.gov.cn/Database/WhitePapers/2015-05/26/content_4586711.htm>

[e]"Middle and Long Term Development Plan for Civil Space Infrastructure Issued." 13 Jan. 2016. LexisNexis 2 Feb. 2016 <https://hk.lexiscn.com/latest_message.php?access=show_detail&id=184408>

[f]Reeves, Martin, and David He. "What China's 13th Five-Year Plan Means for Business." 7 Dec. 2015. Harvard Business Review 1 Feb. 2016 <https://hbr.org/2015/12/what-chinas-13th-five-year-plan-means-for-business>

[g]Kallender-Umezu, Paul. "What's Behind Japan's Sudden Thirst for More Spy Satellites." 13 Nov. 2015. SpaceNews 7 Jan. 2016 <http://spacenews.com/whats-behind-japan-sudden-thirst-for-more-spy-satellites/>

[h]"President Obama Signs Bill Recognizing Asteroid Resource Property Rights into Law." 25 Nov. 2015. Planetary Resources 26 Nov. 2015 <http://www.planetaryresources.com/2015/11/president-obama-signs-bill-recognizing-asteroid-resource-property-rights-into-law/>

[i]Section 1608. Carl Leven and Howard P. "Buck" McKeon National Defense Authorization Act for Fiscal Year 2015, Pub. L. no 113–291 (2014) <http://www.gpo.gov/fdsys/pkg/CPRT-113HPRT92738/pdf/CPRT-113HPRT92738.pdf>

(continued)

Europe	Other countries

[j]De Selding, Peter B. "With Eye on SpaceX, CNES Begins Work on Reusable Rocket Stage." 5 Jan. 2015. SpaceNews 30 Dec. 2015 <http://spacenews.com/with-eye-on-spacex-cnes-begins-work-on-reusable-rocket-stage/>
[k]Smith, Marcia S. "U.S., China Agree to Bilateral Civil Space Cooperation Dialogue." 26 June 2015. SpacePolicyOnline.com 22 Jan. 2016 <http://www.spacepolicyonline.com/news/u-s-china-agree-to-bilateral-civil-space-cooperation-dialogue>
[l]"US, Japan Test Advanced SM-3 Missile for Aegis System—US Navy." 10 Dec. 2015. Sputnik News 21 Jan. 2016 <http://sputniknews.com/military/20151210/1031509097/us-japan-aegis.html>

13.5 Country Profiles 2015

Austria

Population[a]	8.576 million
GDP[b]	339.90 billion euros
Responsibility[c]	The Austrian Space Programme is financed by the Federal Ministry for Transport, Innovation and Technology (BMVIT) and managed by the Aeronautics and Space Agency (ALR), which is integrated into the Austrian Research Promotion Agency (FFG), the central organisation for fostering science and technology activities in Austria, and serves as a docking station to the international aerospace world for Austrian business and science.
Activities[d]	ALR of FFG implements national aerospace policy and represents Austria in numerous European and international aerospace institutions–from ESA and EUMETSAT through EU committees and other international and national organisations responsible for aerospace policy. Austrian representation in the European Space Agency is of special strategic importance in this context. In addition to ESA programmes, the Austrian Space Applications Programme ASAP and the Space Programmes under the Seventh EU Framework Programme are of special relevance for space research projects.
Budget[e]	68.8 million euros (ESA contribution: 51.5 million euros; National Civil Space Budget: 10 million euros; and EUMETSAT: 7.3 million euros)
Staff[f]	12 (ALR)
Direct employment in space manufacturing industry[g]	387

[a]All Population figures are from Eurostat: http://ec.europa.eu/eurostat/data/database. Accessed 25 July 2016
[b]All GDP figures are from Eurostat: http://ec.europa.eu/eurostat/data/database. Accessed 25 July 2016

(continued)

[c]All information on the ESA Member States "Responsibilities" was obtained from direct email exchange with the different agencies and organisations responsible for space in each country
[d]All information on the ESA Member States "Activities" was obtained from direct email exchange with the different agencies and organisations responsible for space in each country
[e]All Contributions to 2015 ESA Budget figures are from ESA: http://www.esa.int/For_Media/Highlights/ESA_budget_2015. Accessed 25 July 2016. All Contributions to 2015 EUMETSAT Budget figures are from EUMETSAT: http://www.eumetsat.int/website/wcm/idc/idcplg?IdcService=GET_FILE&dDocName=PDF_AR2015_EN&RevisionSelectionMethod=LatestReleased&Rendition=Web. Accessed 30 July 2016. All budget figures on National expenditures, both civil and military, originates from EUROCONSULT
[f]All information on the ESA Member States "Staff" in their national public organisations and agencies for space was obtained from direct email exchange with the different agencies and organisations responsible for space in each country
[g]All "Direct Employment in Space Manufacturing Industry" figures are taken from ASD-Eurospace. "Facts and Figures—The European Space Industry in 2015"

Belgium

Population	11.21 million
GDP	409.41 billion euros
Responsibility	Two-level competence system: – Federal: space in the frame of international agreements; – Regions and communities: basic and applied research. The Belgian Federal Science Policy Office (BELSPO) with its Department of Space Research and Applications manages the scientific aspects of the space policy. The Belgian High Representation for Space Policy (BHRS), acting under the auspices of the federal minister in charge of Science Policy, defends Belgian interests in the international institutions dealing with space (ESA and EU levels). Other federal ministries involved with space are the Ministries of Defence and Foreign Affairs. On the regional level, the Ministries for Economic Development, Transport and Environment have stakes in space applications and industrial development.
Activities	Strategy and operational goals: • to create and/or strengthen expertise and industrial capabilities in high-growth areas of the space sector; • to increase the return on investment of the public money invested; • in space at the ESA level, in the frame of the EU as well as in bilateral cooperation projects; • to support industry to be able to tackle new markets, positive impact on economy; • to give to the public authority the space means necessary to define and to implement its policies; • to represent Belgium's interests in the definition and implementation of the European space policy. • Cooperation with ESA, EUMETSAT, EC and other aerospace related agencies.

(continued)

	Within ESA, Belgium traditionally has major interests in Launchers, Telecommunications, Integrated Applications and Technology Support programmes, Earth Observation.
Budget	220.8 million euros (ESA contribution: 189.5 million euros; National Civil Space Budget: 22 million euros; and EUMETSAT: 9.3 million euros)
Staff	Space Research and Applications directorate: 15
Direct employment in space manufacturing industry	1652

Czech Republic

Ministry of Transport

Population	10.538 million
GDP	166.96 billion euros
Responsibility	The Ministry of Transport of the Czech Republic is the coordinator of all space activities in the Czech Republic. For the coordination purposes it established the Coordination Council for Space Activities under its leadership, which includes all relevant ministries and institutions. The Ministry of Transport is also directly responsible for the major space activities in the Czech Republic, including the membership in ESA, EU space policy or international cooperation in space.
Activities	The Czech Republic is subscribed in ESA Optional Programmes such as: MTG, MetOp-SG, EOEP, GSC, ARTES, EGEP, GSTP, MREP, ELIPS, SSA, FLPP and PRODEX.
Budget	20.7 million euros (ESA contribution: 14.2 million euros; National Civil Space Budget: 3.1 million euros; and EUMETSAT: 3.5 million euros)
Direct employment in space manufacturing industry	176

Denmark

DTU Space
Institut for Rumforskning og -teknologi

Population	5.66 million
GDP	266.18 billion euros
Responsibility	NSI (National Space Institute) is the only research organisation in Denmark with the basic funding and mandate to carry out space research. Although space research is also conducted at other universities and organisations, these activities primarily rely on funding from research agencies and other external funding sources. Through national and international cooperation, NSI leverages its research capabilities to create new knowledge and technology of value for Danish research infrastructure,

(continued)

	industry, and society at large. NSI's scientific competences further give it the ability to provide qualified advice to national and international agencies and organisations, and also forms the basis for educating researchers and engineers in both basic and specialised disciplines. The common denominator in NSI's research activities is the application of satellites for studying Earth and space physics and the development of systems, methods and instruments which support these activities. The department is characterised by a close-knit interplay between engineering disciplines and natural science. The scientific challenges require development of new technologies, and the technological research opens up new scientific opportunities. The combination of the technical and scientific competencies enables the NSI to deliver unique end-to-end solutions within selected areas, i.e. to both develop the necessary instrumentation and process the collected data to scientific results.
Activities	Within the ESA programmes, the Danish areas of strength encompass Earth observation and climate monitoring, launchers, the international space station, telecommunications (ARTES) and technology development (GSTP). In addition, Denmark has decided to contribute to the development of new meteorological satellites in cooperation with the European meteorological organisation EUMETSAT and in the ESA's new programme, Space Situational Awareness that ensures research and warnings of soar flares' effect on the security of supply infrastructure.
Budget	34 million euros (ESA contribution: 26.8 million euros; National Defence Space Budget: 1 million euros; and EUMETSAT: 6.2 million euros)
Staff	29 (DTU)
Direct employment in space manufacturing industry	222

Estonia

EAS
Enterprise Estonia

Population	1.313 million
GDP	20.25 billion euros
Responsibility	Space activities in Estonia are coordinated by the Space Affairs Council (SAC), established in 2010 by Ministry of Economic Affairs and Communications. The main task of SAC is initiation and governance space related programs and activities on national and international level and coordination of the utilisation of resources. SAC is supported by secretariat, containing representatives of Enterprise Estonia and Estonian Research Council. Since 2006 Estonian public business development and support agency Enterprise Estonia (www.eas.ee) is assigned as implementing body (ESA

(continued)

	delegation) for the ESA-Estonian agreements. Enterprise Estonia also carries out the tasks of Estonian Space Office, being intermediary between Estonian companies/institutions and international space community. Main tasks concluded mapping and auditing space related competences and capabilities, organizing trainings for entrepreneurs and public sector officials, taking companies to expositions and study trips, encouraging the technology transfer etc. Enterprise Estonia is a member of EURISY since 2008 and member of IAF since 2009. Space science activities in Estonia are carried out by Tartu Observatory (www.to.ee), having long tradition in astronomy, cosmology and Earth observation. Nowadays Tartu Observatory serves as Estonian space technology development and testing centre.
Activities	Estonian space focus of space activities is on earth observation, navigation, technology, especially related to ICT and science. Estonian space activities are mainly related to Horizon 2020 projects and cooperation with ESA and PECS programme, where Estonia has 27 projects. In 2015 Estonia will be 21st ESA member state. Estonia is actively preparing for Galileo and Copernicus programme implementation in public and private sector.
Budget	Estonian contribution to ESA as ECS is ca 1.2 million euros per year, public funding of space science research activities is ca 1 million euros per year. Estonia joined ESA in September 2015 so no figure is included for that year.
Staff	10 (related to space in Enterprise Estonia) 80 (at Tartu Observatory)
Direct employment in space manufacturing industry	39

Finland

Tekes

Population	5.472 million
GDP	209.15 billion euros
Responsibility	Finland's public sector space activities are funded by the Ministry of Employment and Economy (ESA membership), Tekes (ESA and national programmes), Academy of Finland (EISCAT, ESO, NOT) under the Ministry of Education and Culture, and Finnish Meteorological Institute (EUMETSAT) under the Ministry of Transport and Communications. National coordination is provided by the Finnish Space Committee under Ministry of Economy.
Activities	In addition to cooperation with ESA and EC (e.g. Space Programme under the Horizon 2020 EU Framework Programme), Tekes funds space technologies and applications nationally. The Finnish focus of space activities is on Earth Observation, telecommunications, technology development and science.

(continued)

Budget	46.3 million euros (ESA contribution: 19.6 million euros; National Civil Space Budget: 22 million euros; and EUMETSAT: 4.7 million euros)
Staff	6 (Tekes)
Direct employment in space manufacturing industry	178

France

Population	66.415 million
GDP	2181.06 billion euros
Responsibility	Space activities in France are under parent Ministries of Higher Education and Research and of Defence. Among different actors, France relies on its national space agency CNES (Centre National d'Etudes Spatiales) for space policy proposal and application.
Activities	France inscribes its space activities in the frame of European space activities and intent to play there a leading role. Activities are conducted in strong interface with EC, ESA, EUMETSAT and others Members States or at international level. National orientations include five acting domains, namely Ariane (autonomous European access to space), Observation, Sciences, Telecommunications and Defence.
Budget	2212.1 million euros (ESA contribution: 718.2 million euros; National Civil Space Budget: 1003 million euros; National Defence Space Budget: 440 million euros; and EUMETSAT: 50.9 million euros)
Staff	2400 (CNES)
Direct employment in space manufacturing industry	13,393

Germany

Population	81.197 million
GDP	3032.82 billion euros
Responsibility	Acting on behalf of the Federal Government, the Space Administration designs and implements Germany's Space Programme, which integrates all German space activities on the national and European plane. These activities include Germany's national Space Program, DLR's 'Space' research and development programme and Germany's contributions to the European Space Administration (ESA) as well as the European Organisation for the Exploitation of Meteorological Satellites (EUMETSAT). In addition, the Administration designs and supervises space and security research projects under the sixth and seventh EU research framework program.

(continued)

	The Space Administration's principal client is the Federal Ministry for Economics and Technology. However, the Space Administration also works for other ministries, mainly in application related fields such as Earth observation, navigation and satellite communication. The Federal Government's Space Programme provides both business and science with a reliable political framework for independent planning and action, thus ensuring that public funds are used efficiently. The national Programme is implemented by the Space Administration. Its major function is to promote and support Germany's strategic goals in the European programs of ESA and the EU by purposefully equipping Germany's industry and scientific institutions for competition within the EU as well as for their tasks within the ESA framework.
Activities	Germany has a national civil programme, which includes bi- or trilateral co-operations, a participation in the ESA programmes, and a defence programme (e.g. SARLupe, MUSIS). The major areas of interest for Germany's domestic programmes are Earth observation, navigation, telecommunications, space transportation systems, space infrastructures, microgravity research, space science, space technologies and space exploration.
Budget	1738.4 million euros (ESA contribution: 797.4 million euros; National Civil Space Budget: 810 million euros; National Defence Space Budget: 65 million euros; and EUMETSAT: 66 million euros)
Staff	8000 (DLR)
Direct employment in space manufacturing industry	7293

Greece

Population	10.858 million
GDP	176.02 billion euros
Responsibility	The General Secretariat for Research and Technology (GSRT), under the responsibility of the Ministry of Research and Innovation, is responsible for Greek space activities.
Activities	Space R&D is a priority area of the Greek Space Strategy, mostly pursued through ESA's technological activities' EOEP, GSTP, ARTES-1, ARTES-3-4 and ARTES-5, ELIPS, and GNSS Evolution Programme. Greece and Cyprus have created the HellasSat consortium to launch the Hellasat telecommunication satellites. Due to the economic crisis Hellasat was sold to Arabsat in 2013. Greece is a member of EUMETSAT and Eutelsat.
Budget	17.2 million euros (ESA contribution: 12.1 million euros; and EUMETSAT: 5.1 million euros)
Direct employment in space manufacturing industry	34 (GSRT)

Hungary

Population	9.856 million
GDP	109.67 billion euros
Responsibility	The Hungarian Space Office (HSO), integrated into the body of the Ministry of National Development, manages, coordinates and represents Hungarian space activities. Hungary's space activity is supervised by the Minister, who makes his decisions based on the advices of the Scientific Council on Space Research in scientific issues and the Hungarian Space Board in strategic questions.
Activities	Participation in microgravity, Earth observation, life and material sciences and GTSP programmes of ESA.
Budget	The Hungarian contribution for ESA ECS programme is 2.0 million euros per year.
Direct employment in space manufacturing Industry	57

Ireland

Population	4.629 million
GDP	255.82 billion euros
Responsibility	Enterprise Ireland (EI) is the Irish Government's business development agency. EI manages industrial and technology development programmes relating to space activity on behalf of the Department of Jobs, Enterprise & Innovation (DJEI). EI promotes ESA and EU space programmes in Ireland, primarily to the industrial sector, as well as to the academic community and end-user organisations.
Activities	ESA programmes (launchers, telecoms, satnav, Earth observation, science); Horizon 2020 Space; EUMETSAT
Budget	21.3 million euros (ESA contribution: 18 million euros; and EUMETSAT: 3.3 million euros)
Staff	740 (Enterprise Ireland)
Direct employment in space manufacturing industry	61

Italy

Population	60.796 million
GDP	1642.44 billion euros
Responsibility	The Italian Space Agency, Agenzia Spaziale Italiana (ASI), defines, coordinates and manages national space programs

(continued)

	and the Italian participation to European and international space projects, under the supervision of the Ministry of Education, University and Research and in coordination with the Ministry of Foreign Affairs and International Cooperation. ASI's main headquarter is located in Rome and the agency has also three additional centres in Matera (Space Geodesy Centre); in Malindi, Kenya (Luigi Broglio Space Centre); and Rome (ASI Science Data Centre).
Activities	Italy is the third country in terms of contribution to the European Space Agency's (ESA) programs. Italy participates in the European Union (EU) programs as well such as Galileo and GMES/Copernicus. ASI has a long tradition in space scientific missions at ESA level and in cooperation with NASA. Indeed, ASI has significantly contributed to space exploration embarking many scientific payloads aboard NASA and ESA satellites for discovering the secrets of Mars, Jupiter, Saturn and Venus and for the study of cosmology. Other scientific payloads have been carried out for the astrophysical study of high energy and black matter such as AMS on-board the ISS. Italy is the first European country in terms of investment in the ISS through the participation in ESA and through the MoU with NASA. ASI is deeply engaged in international bilateral cooperation with many partners in the five continents. At multilateral governmental level Italy is member of the UNCOPUOS and GEO. ASI is also member of CEOS, IADC, ICG, ESPI, EURISY and IAF.
Budget	896.9 million euros (ESA contribution: 329.9 million euros; National Civil Space Budget: 281 million euros; National Defence Space Budget: 247 million euros; and EUMETSAT: 39 million euros)
Staff	250 (ASI)
Direct employment in space manufacturing industry	5019

Luxembourg

Luxembourg Space Cluster

Population	562.96 thousand
GDP	51.22 billion euros
Responsibility	The Ministry of the Economy, Directorate of ICT & Space Affairs, is in charge of space affairs in Luxembourg. Luxinnovation, the National Agency for Innovation and Research, also acts on behalf of the Ministry, representing Luxembourg interests in the ESA and EU committees.
Activities	The Ministry of the Economy defines the National Space Policy, manages relationship with ESA, the EU and space agencies worldwide. The Ministry is responsible for financial, political and programmatic aspects related to ESA

(continued)

	activities. Luxinnovation supports the Ministry for the implementation of the National Action Plan. Moreover, Luxinnovation is the national point of contact for space matters. Luxinnovation provides companies and research organisations with information and advices on ESA and EC programmes. Luxembourg is active in telecommunications, Earth observation, navigation, security and space technologies.
Budget	27.71 million euros (ESA contribution: 23.0 million euros; National Civil Space Budget: 2 million euros; National Defence Space Budget: 2 million euros; and EUMETSAT 710,000 euros)
Staff	5 (3 Ministry of the Economy + 2 Luxinnovation)
Direct employment in space manufacturing industry	26

Netherlands

Population	16.901 million
GDP	676.53 billion euros
Responsibility	The Netherlands Space Office acts as the Dutch agency for space affairs. The Netherlands Space Office (NSO) was established by the Dutch government in order to develop the Netherlands' space programme and to bring that programme to action. The NSO is the face of the Dutch space community for international space organisations like ESA, NASA and JAXA as well as the central point of contact for the space community within the Netherlands. The NSO also works to innovatively and openly bring the story of spaceflight science, usage and exploration to teachers, students and the general public. The Ministry of Economic Affairs, Ministry of Education, Culture and Science, Ministry of Transport, Public Works and Water Management and the Netherlands Organisation for Scientific Research (NWO) signed an agreement in October 2008 for the establishment of the Netherlands Space Office.
Activities	In the ESA context, the Netherlands participate in all major programmes with a special interest in earth observation and environmental measurements and in robotics. The Netherlands Institute for Space Research (SRON) develops and uses innovative technology for ground-breaking research in space, focusing on astrophysical research, Earth science and planetary research. In addition to this, SRON has a line of research into new and more sensitive sensors for X-rays and infrared radiation. Space policy in the Netherlands is primarily focused on international cooperation in European contexts within the European Space Agency, the European Organisation for the Exploitation of Meteorological

(continued)

	Satellites (EUMETSAT), and the European Union. The Netherlands also has a national programme guided by the ministries of Economic Affairs, of Education, Culture and Science, and of Traffic, Public Works and Water Management.
Budget	140.6 million euros (ESA contribution: 74.7 million euros; National Civil Space Budget: 40 million euros; National Defence Space Budget: 11 million euros; and EUMETSAT: 14.9 million euros)
Staff	26 (NSO) Approx. 200 (SRON)
Direct employment in space manufacturing industry	968

Norway

Norsk Romsenter
NORWEGIAN SPACE CENTRE

Population	5.166 million
GDP	348.33 billion euros
Responsibility	The Norwegian Space Centre (NSC), under the Ministry of Trade, Industry, and Fisheries, is responsible for organising Norwegian space activities, particularly with respect to ESA and the EU, and for coordinating national space activities.
Activities	In addition to ESA programmes (in particular Earth observation, telecommunications and launchers), Norway has national support programmes and commercial activities (Telenor). Moreover, Norway operates the Andøya rocket range and the Svalbard and Antarctica ground stations. Norway has also a bilateral agreement with Canada on the use of Radarsat 2 data.
Budget	90.6 million euros (ESA contribution: 59.8 million euros; National Civil Space Budget: 12.3 million euros; National Defence Space Budget: 9.6 million euros; and EUMETSAT: 8.9 million euros)
Staff	NSC: 40
Direct employment in space manufacturing industry	322

Poland

POLISH SPACE AGENCY

Population	38.001 million
GDP	427.74 billion euros
Responsibility	The recently established Polish Space Agency (POLSA) is the implementing entity for Polish space activities under the responsibility of the Ministry of Economy, Department of Industry.

(continued)

Activities	The five main SCR's research areas are: physics of the Sun, study of planets and small solar system bodies, interplanetary space physics and astrophysics, plasma physics, and planetary geodesy and geodynamics. The SRC is the body cooperating with ESA. Poland has participated or is participating in several scientific missions of ESA, such as: Ulysses, ISO, Soho, XMM, Cluster, DoubleStar, Huygens, Mars Express, Herschel, Planck, XEUS, Integral, Rosetta, ExoMars and BepiColombo.
Budget	44.6 million euros (ESA contribution: 30.0 million euros; National Civil Space Budget: 4.9 million euros; and EUMETSAT: 9.7 million euros)
Direct employment in space manufacturing industry	60

Portugal

ANI AGÊNCIA NACIONAL DE INOVAÇÃO

Population	10.375 million
GDP	179.54 billion euros
Responsibility	Portuguese Space institutional activities are under responsibility of the Ministry of Education and Science. The main tool to develop national space capabilities is ESA. The National Innovation Agency (ANI) coordinates the national participation in ESA, with the executive support of the Portuguese Space Office, and in close collaboration with the Foundation for Science and Technology (FCT). Most of Portuguese ESA budget is funded by FCT, and in part by the telecom regulator (Anacom), mainly for the ARTES programmes. Other Ministries and entities are also involved with ESA funding and activities according to their specific needs and priorities. ANI is also responsible for promoting the Portuguese participation in H2020 Space programme, through the Office for the promotion of EU H2020 research programme (GPPQ).
Activities	Portuguese space activities are implemented mainly through ESA. The EU space programmes, Galileo and Copernicus, alongside H20 20 Space, are also of relevance for the space community. Additionally Portugal is a member state of ESO and EUMETSAT.
Budget	20.8 million euros (ESA contribution: 16.7 million euros; and EUMETSAT: 4.1 million euros)
Staff	Approx. 5
Direct employment in space manufacturing industry	173

Romania

Population	19.871 million
GDP	160.35 billion euros
Responsibility	The Romanian Space Agency (ROSA) is the coordinator of Romania's national and international space activities. The Romanian Space Agency (ROSA) is a public institution entirely self-funded, operating under the decisions of the Ministry of Education and Research—National Authority for Scientific Research and Innovation (A.N.C.S.I.). As a coordinator of national space research and applications programme, ROSA designs and coordinates the implementation of the National Space Programme. Following its objectives, the Agency is authorised to establish research and development centres.
Activities	Main areas of interest are space exploration, space applications, technology, and security. On behalf of the Government, ROSA is the national representative in the cooperative agreements with international organisations, such as European Space Agency (ESA) and Committee on Space Research (COSPAR), as well as bilateral governmental agreements. Together with the Ministry of Foreign Affairs, ROSA is representing Romania in the sessions of the United Nations Committee on the Peaceful Use of Outer Space (COPUOS) and its Subcommittees.
Budget	40.3 million Euro (ESA contribution: 25.4 million euros; National Civil Space Budget: 11.7 million euros; and EUMETSAT: 3.2 million euros)
Direct employment in space manufacturing industry	64

Spain

Population	46.45 million
GDP	1075.64 billion euros
Responsibility	The Centre for the Development of Industrial Technology (CDTI), under the Ministry of Economy and Competitiveness, channels the funding and support applications for national and international R&D&i projects of Spanish companies, including the Spanish space activities in coordination with the Ministry of Industry, Energy and Tourism. The National Institute of Aerospace Technology (INTA) is the other important Spanish actor in the space field. INTA is an independent body related to the Ministry of Defence. INTA is responsible for organising the Spanish participation in HELIOS and is responsible for approval and certification, research, technical assistance and services in several areas related to space.

(continued)

Activities	In addition to ESA and EUMETSAT programmes, Spain has several national space programmes in the field of Earth observation (SEOSAT/INGENIO), communication satellites (Hispasat, Amazonas), defence space systems (SPAINSAT, XTAR-EUR, HELIOS, SECOMSAT, Pleiades and SEOSAR/PAZ), small satellites, space exploration (MSL-REMS) and ground control stations.
Budget	227.3 million euros (ESA contribution: 131.7 million euros; National Civil Space Budget: 41 million euros; National Defence Space Budget: 29 million euros; and EUMETSAT: 25.6 million euros)
Staff	approx. 300 (CDTI)
Direct employment in space manufacturing industry	3143

Sweden

RYMDSTYRELSEN
Swedish National Space Board

Population	9.747 million
GDP	446.95 billion euros
Responsibility	The Swedish National Space Board (SNSB), a central governmental agency under the Ministry of Higher Education and Research, is responsible for national and international activities relating to space and remote sensing, primarily research and development. The Swedish space programme is carried out by means of extensive international cooperation, in particular through Sweden's membership of the European Space Agency, ESA. SNSB's responsibility for international activities includes the Swedish involvement in ESA as well as bilateral cooperation within space. Most of the activities funded by the Swedish National Space Board are carried out in cooperation with other countries.
Activities	The Swedish space programme is mostly carried out through international cooperation. Sweden participates in almost all the optional ESA programmes, in addition to the mandatory basic and scientific programmes. Optional programmes of special interest are the remote sensing and launcher technology programmes. Bilateral co-operation particularly concerns the co-operation with France on SPOT, Vulcain MK2 and Proteus. Sweden also has a participation in the French Pleiades programme. Other Memoranda of Understanding have been concluded as basis for co-operation with SNSB's counterparts in Austria, Brazil, China, India, Italy, Japan, Kazakhstan, Norway and Thailand. On a governmental level Sweden has framework agreement with USA and Russia for cooperative activities in the exploration and use of outer space. Sweden is also engaged in other bilateral co-operative projects carried out on ad hoc basis. Sweden is

(continued)

	a member of the UN Committee for Peaceful Use of Outer Space, COSPAR (Committee on Space Research), Intelsat, Eutelsat, Inmarsat and EUMETSAT.
Budget	119.8 million euros (ESA contribution: 80.3 million euros; National Civil Space Budget: 29.8 million euros; and EUMETSAT: 9.7 million euros)
Staff	17
Direct employment in space manufacturing industry	822

Switzerland

Schweizerische Eidgenossenschaft
Confédération suisse
Confederazione Svizzera
Confederaziun svizra

Population	8.238 million
GDP	604.51 billion euros
Responsibility	The Swiss Space Office (SSO) under the authority of the State Secretariat for Education, Research and Innovation (SERI) and the Federal Department of Economic Affairs, Education and Research (EAER) is the administrative body in charge of planning and implementing the Swiss space policy. The Space Office serves as the secretariat for the Federal Commission for Space Affairs (CFAS), advises the Federal Council on matters relating to a coherent and forward-looking space policy by taking into consideration the European and global development and the national interests (amongst others of political, application-oriented, scientific, technological and industrial nature); • issues recommendations to the Federal Council regarding the implementation of this policy; • advises the Federal Council in terms of the evaluation of proposals for projects with important scientific or technological relevance. The SERI's Swiss Space Office serves as the secretariat for the CFAS. The Federal Council has given the Interdepartmental Coordination Committee for Space Affairs (IKAR) a mandate to prepare official Swiss position papers on space on the basis of CFAS recommendations. IKAR also coordinates the activities of various federal agencies involved in space affairs. IKAR is chaired by the SERI's Swiss Space Office, which also serves as the secretariat for IKAR. Space Office, which also serves as the secretariat for IKAR.
Activities	Most of the Swiss activities are undertaken within ESA programmes: space science, Earth observation, microgravity, human spaceflight, small satellites, telecommunications, navigation, and technology.

(continued)

Budget	153.6 million euros (ESA contribution: 134.9 million euros; National Civil Space Budget: 6.8 million euros; and EUMETSAT: 11.9 million euros)
Direct employment in space manufacturing industry	863

United Kingdom

UK SPACE AGENCY

Population	64.767 million
GDP	2577.28 billion euros
Responsibility	The UK Space Agency, an executive agency of the Department for Business, Innovation and Skills (BIS), and reporting to the Minister of State for Universities and Science, is responsible for the strategic decisions on all UK space activities. The UK Space Agency also interfaces with other departments on security and military programmes. Although it does not manage these programmes directly, the UK Space Agency is kept informed and involved in decisions relating to the programmes as necessary.
Activities	UK space activities include bilateral cooperation with JAXA, India and the U.S. and within ESA in Earth observation and space exploration (Cassini-Huygens, James Webb Space Telescope, Herschel, and Planck missions).
Budget	641.9 million euros (ESA contribution: 322.3 million euros; National Civil Space Budget: 87 million euros; National Defence Space Budget: 187.2 million euros; and EUMETSAT: 45.4 million euros)
Staff	50 (UKSA)
Direct employment in space manufacturing industry	3535

Chapter 14
Bibliography of Space Policy Publications 2015

Blandina Baranes

14.1 Monographs

Al-Ekabi C (ed) (2015) European autonomy in space. Springer, Wien

Aliberti M (2015) When China goes to the Moon..., Studies in space policy. Springer, Wien

Bell J (2015) The interstellar age. The story of the NASA men and women who flew the forty-year Voyager Mission. Dutton, New York, NY

Benhaddou D, Al-Fuqaha A (eds) (2015) Wireless sensor and mobile ad-hoc networks. Vehicular and space applications. Springer, Cham

Bittencourt Neto O (2015) Defining the limits of outer space for regulatory purposes, SpringerBriefs in space development. Springer, New York, NY

Blair-Smith H (2015) Left brains for the right stuff. Computers, space, and history. SDP Publishing, South Shore

Bordeneuve-Guibé J, Drouin A, Roos C (eds) (2015) Advances in aerospace guidance, navigation and control. Selected papers of the third CEAS Specialist Conference on Guidance, Navigation and Control held in Toulouse. Springer, Cham

Brown T (2015) The digital galactic complex. Springer, New York, NY

Burgess C (2015) Friendship 7. The epic orbital flight of John H. Glenn Jr, Springer Praxis Books. Springer, Cham

Burgess C, Vis B (2015) Interkosmos. The Eastern Bloc's early space program, Springer Praxis Books. Springer, Cham

B. Baranes (✉)
ESPI, Schwarzenbergplatz 6, 1030 Vienna, Austria
e-mail: blandina.baranes@espi.or.at

C. Al-Ekabi et al. (eds.), *Yearbook on Space Policy 2015*, Yearbook on Space Policy, DOI 10.1007/978-3-7091-4860-0_14

Chai Y, Li S (2015) Intelligent testing, control and decision-making for space launch. Wiley, Singapore

Cockell C (ed) (2015a) The meaning of liberty beyond earth, Space and society. Springer, Cham

Cockell CS (ed) (2015b) Human governance beyond earth, Space and society. Springer, Cham

Convay EM (2015) Exploration and engineering. The jet propulsion laboratory and the quest for Mars. Johns Hopkins University Press, Baltimore, MD

Cooper S (ed) (2015) Handbook of navigation and observation, vol 1–2. Clanrye International, New York, NY

Cruzen C, Schmidhuber M, Dubon L (eds) (2015) Space operations. Innovations, inventions, and discoveries. The American Institute of Aeronautics and Astronautics (AIAA), Reston, VA

David DE (2015) Spies and shuttles. NASA's secret relationships with the DoD and CIA. University of Florida Press, Gainesville, FL

Driggers GW (2015) Mars close to home. The case for a Mars simulation in earth orbit. Create Space Independent Publishing Platform, North Charleston, SC

Esposito F (2015) Fascism, aviation and mythical modernity. Palgrave Macmillan, London

Feron E (ed) (2015) Advances in control system technology for aerospace applications. Springer, Cham

Friedman L (2015) Human spaceflight: from Mars to the stars. The University of Arizona Press, Tucson, AZ

Garan R (2015) The orbital perspective. Lessons in seeing the big picture from a journey of 71 million miles. Berrett-Koehler Publishers, Oakland

Garan Col R (2015) The orbital perspective: an astronaut's view. John Blake Publishing, London

Gerovitch S (2015) Soviet space mythologies. Pubic images, private memories, and the making of cultural identity. University of Pittsburgh Press, Pittsburgh, PA

Gibson DC (2015) Terrestrial and extraterrestrial space dangers. Outer space perils, rocket risks and the health consequences of the space environment. Bentham Science Publishers, Sharjah

Gilliland A (2015) How to build a universe. From the Big Bang to the end of the universe. Octopus Publishing, London

Gruntman M (2015) Intercept 1961. The birth of Soviet missile defense. The American Institute of Aeronautics and Astronautics (AIAA), Reston, VA

Grzymisch J (2015) Bearing-only guidance and navigation for in-orbit rendezvous. Shaker Verlag, Aachen

Harris D, Li W-C (eds) (2015) Decision making in aviation. Ashgate Publishing Limited, Burlington, VT

Heer M et al (2015) Nutrition physiology and metabolism in spaceflight and analog studies, SpringerBriefs in life sciences. Springer, New York, NY

Hobe S, Schmidt-Tedd B, Schrogl K-U (eds) (2015) Cologne commentary on space law (CoCoSL), vol III. Carl Heymanns Verlag, Köln

Hoerber T, Stephenson P (eds) (2015) European space policy. European integration and the final frontier, Routledge advances in European politics. Routledge, London

Hu W (2015) Fundamental spacecraft dynamics and control. Wiley, Singapore

Impey C (2015) Beyond. Our future in space. W.W. Norton & Company, New York, NY

Kanas N (2015) Humans in space: the psychological hurdles, Springer Praxis Books. Springer, Cham

Kim Y et al (2015) Acquisition of space systems. Past problems and future challenges, vol 7. Rand Corporation, Santa Monica, CA

Laney M (2015) German rocketeers in the heart of dixie. Making sense of the Nazi past during the civil rights era. Yale University Press, New Haven, CT

Larry R (ed) (2015) China's space programs. Progress and military implications. Nova Science Publisher, Hauppauge

Leve FA, Hamilton BJ, Peck MA (2015) Spacecraft momentum control systems. Springer, Cham

Lewis JS (2015) Asteroid mining 101. Wealth for the new space economy. Deep Space Industries, Moffett Field

Logsdon J (2015) After Apollo? Richard Nixon and the American space program. Palgrave Macmillan, London

Madry S (2015) Global navigation satellite systems and their applications, SpringerBriefs in space development. Springer, New York, NY

Magini L (2015) Stars, myths and rituals in Etruscan Rome, Space and society. Springer, Cham

McConnell B, Trolley A (2015) A design for a reusable water-based spacecraft known as the Spacecoach, SpringerBriefs in space development. Springer, New York, NY

Meuser P (2015) Galina Balashova. Architect of the Soviet space programme. DOM Publishers, Berlin

Millard D (ed) (2015) Cosmonauts: birth of a space age. Scala Arts & Heritage Publishers, New York, NY

Montomoli F (2015) Uncertainty quantification in computational fluid dynamics and aircraft engines, SpringerBriefs in applied sciences and technology. Springer, New York, NY

Moro-Aguilar R, Blount PJ, Masson-Zwaan T (eds) (2015) Proceedings of the international institute of space law 2014. Eleven International Publishing, The Hague

Moss S (2015) The International Mars Research Station. An exciting new plan to create a permanent human presence on Mars. Create Space Independent Publishing Platform, North Charleston, SC

Mushta T (ed) (2015) New frontiers in space propulsion. Nova Science Publisher, Hauppauge

Norberg C (ed) (2015) Human spaceflight and exploration, Springer Praxis Books. Springer, Cham

Oleary BL, Capelotti PJ (eds) (2015) Archeology and heritage of the human movement into space, Space and society. Springer, Cham

Paul R, Moss S (2015) We could not fail. The first African Americans in the space program. Texas University Press, Austin, TX

Pelton JN (2015) New solutions for the space debris problem, SpringerBriefs in space development. Springer, New York, NY

Rapp D (2015) Human missions to Mars: enabling technologies for exploring the Red Planet, Springer Praxis Books. Springer, Cham

Rinner A et al (2015) Space law essentials, vol 2, Casebook. Linde Verlag, Wien

Rustamov RB et al (2015) Space technology contribution in engineering services. Nova Science Publisher, Hauppauge

Sambaluk NM (2015) The other space race: Eisenhower and the quest for aerospace security. Naval Institute Press, Annapolis, MD

Sasiadek J (2015) Aerospace robotics II. Selection of papers from CARO'13 (Conference on Aerospace Robotics). Springer, Cham

Schrogl K-U et al (eds) (2015) Handbook of space security. Policies, applications and programs. Springer, New York, NY

Seedhause E (2015) Survival and sacrifice in Mars exploration. What we know from polar expeditions, Springer Praxis Books. Springer, Cham

Seedhouse E (2015a) Virgin Galactic, Springer Praxis Books. Springer, Cham

Seedhouse E (2015b) Microgravity and vision impairments in astronauts, SpringerBriefs in space development. Springer, New York, NY

Seedhouse E (2015c) Mars via the Moon. The next giant leap, Springer Praxis Books. Springer, Cham

Sgobba T, Rongier I (eds) (2015) Space safety is no accident. The 7th IAASS Conference. Springer, Cham

Soucek A (2015) Space law essentials, vol 1, Textbook. Linde Verlag, Wien

Sun J et al (eds) (2015) China Satellite Navigation Conference (CSNC) 2015 proceedings, vol 1–3. Springer, Berlin

Suresh BN, Sivan K (2015) Integrated design for space transportation system. Springer, Cham

Tabor J (ed) (2015) Deciphering spacecraft systems and orbital calculations. Clanrye International, New York, NY

Traphagan J (2015) Extraterrestrial intelligence and human imagination. SETI at the intersection of science, religion, and culture, Space and society. Springer, Cham

Vance A (2015) Elon Musk: Tesla, SpaceX, and the quest for a fantastic future. HarperCollins, New York, NY

Viviani A, Pezzella G (2015) Aerodynamic and aerothermodynamic analysis of space mission vehicles. Springer, Cham

von der Dunk F, Tronchetti F (eds) (2015) Handbook of space law. Edward Elgar, Cheltenham

Ward JH (2015a) Countdown to a moon launch. Preparing Apollo for its historic journey, Springer Praxis Books. Springer, Cham

Ward JH (2015b) Rocket ranch, Springer Praxis Books. Springer, Cham

Young A (2015) The twenty-first century commercial space imperative, SpringerBriefs in space development. Springer, New York, NY

Ziarnick B (2015) Developing national power in space. A theoretical model. McFarland Publishers, Jefferson

14.2 Articles

Accomazzo A et al (2015) Rosetta operations at the comet. Acta Astronaut 115:434–441

Adriaensen M et al (2015) Priorities in national space strategies and governance of the member states of the European Space Agency. Acta Astronaut 117:356–367

Aganaba-Jeanty T (2015) Common benefit from a perspective of "Non-traditional Partners": a proposed agenda to address the status quo in Global Space Governance. Acta Astronaut 117:172–183

Ambrosius JD (2015) Separation of church and space: religious influences on public support for U.S. space exploration policy. Space Policy 32:17–31

Anderson AP, Newman DJ (2015) Pressure sensing for in-situ measurement of space suited biomechanics. Acta Astronaut 115:218–225

Andrews DG et al (2015) Defining successful commercial asteroid mining program. Acta Astronaut 108:106–118

Assad A, de Weck OL (2015) Model of medical supply and astronaut health for long-duration human space flight. Acta Astronaut 106:47–62

Avakyan SV et al (2015) The role of a space patrol of solar X-ray radiation in the provisioning of the safety of orbital and interplanetary manned space flights. Acta Astronaut 109:194–202

Babuscia A et al (2015) Development of cooperative communication techniques for a network of small satellites and CubeSats in deep space: the SOLARA/SARA test case. Acta Astronaut 115:349–355

Bagrov AV et al (2015) Single-satellite global positioning system. Acta Astronaut 117:332–337

Baiocco P et al (2015) System driven technology selection for future European launch systems. Acta Astronaut 107:301–316

Balint TS, Hall A (2015) Humanly space objects—perception and connection with the observer. Acta Astronaut 110:129–144

Barker DC (2015) The Mars imperative: species survival and inspiring a globalized culture. Acta Astronaut 107:50–69

Bauer W, Romberg O, Putzar R (2015) Experimental verification of an innovative debris detector. Acta Astronaut 117:49–54

Benkö M, Zickler A, Röhn G (2015) Space tourism: facts and figures. German J Air Space Law 64:50–54

Betelin VB, Smirnov NN, Nikiti VF (2015) Supercomputer predictive modeling for ensuring space flight safety. Acta Astronaut 109:269–277

Billings L (2015) Words matter: a call for responsible communication about asteroid impact hazards and plans for planetary defense. Space Policy 33:8–12

Boden RC, Hein AM, Kawaguchi J (2015) Target selection and mass estimation for manned NEO exploration using a baseline mission design. Acta Astronaut 111:198–221

Brosnan IG (2015) An analysis of the relative success of applicants to the new STEM Presidential Management Fellowship and the NASA context. Space Policy 34:47–49

Budianu A, Meijerink A, Bentum MJ (2015) Swarm-to-Earth communication in OLFAR. Acta Astronaut 107:14–19

Burov AA et al (2015) Uniform rotations of tethered system connected to a moon surface. Acta Astronaut 116:349–354

Caldwell BS (2015) Spaceflight-relevant STEM education and outreach: social goals and priorities. Acta Astronaut 112:174–181

Capuano V et al (2016) Feasibility study of GNSS as navigation system to reach the Moon. Acta Astronaut 116:186–201

Carta R et al (2015) Sample canister capture mechanism for Mars sample return: functional and environmental test of the elegant breadboard model. Acta Astronaut 117:99–115

Castiglioni AG et al (2015) Spaceship earth. Space-driven technologies and systems for sustainability on ground. Acta Astronaut 115:195–205

Chang Y (2015) The first decade of commercial space tourism. Acta Astronaut 108:79–91

Chapman B (2015) Waste and duplication in NASA programs: the need to enhance U.S. space program efficiency. Space Policy 31:13–20

Cheng AF et al (2015) Asteroid impact and deflection assessment mission. Acta Astronaut 115:262–269

Christophe B et al (2015) A new generation of ultra-sensitive electrostatic accelerometers for GRACE follow-on and towards the next generation gravity missions. Acta Astronaut 117:1–7

Cohen S, Misra AK (2015) Static deformation of space elevator tether due to climber. Acta Astronaut 111:317–322

Corpino S et al (2015) Thermal design and analysis of a nanosatellite in low earth orbit. Acta Astronaut 115:247–261

Correia Mendoca H, Cocco M, Macedo Scavuzzi dos Santos J (2015) International laws regulating satellite communications and their international disruption in times of peace and conflict. Annals of air and space law, vol XL

Davison P et al (2015) Communications satellites: time expanded graph exploration of a tradespace of architectures. Acta Astronaut 115:442–451

de Freitas Dewes M, Dalmarco G, Domingos Padula A (2015) Innovation policies in Brazilian and Dutch aerospace industries: how sectors driven by national procurement are influenced by its S&T environment. Space Policy 34:32–38

Delgado-López L (2015) Beyond the Moon agreement: norms of responsible behaviour for private sector activities on the Moon and celestial bodies. Space Policy 32:6–8

Deloo JJ, Mooij E (2015) Active debris removal: aspects of trajectories, communication and illumination during final approach. Acta Astronaut 117:277–295

Dicaire I et al (2015) Space options for tropical cyclone hazard mitigation. Acta Astronaut 107:208–217

Duchek M et al (2015) Solar electric propulsion orbital debris ferry, vehicle concept and reference mission. Acta Astronaut 116:175–185

Ebisuzaki T et al (2015) Demonstration designs for the remediation of space debris from the International Space Station. Acta Astronaut 112:102–113

Farzana IK et al (2015) Lunar laser communication demonstration operations architecture. Acta Astronaut 111:77–83

Ferrazzani M, Soucek A (2015) Experiences from and prospects for the ESA Convention. German J Air Space Law 64:288–309

Francesconi A et al (2015) An engineering model to describe fragments clouds propagating inside spacecraft in consequence of space debris impact on sandwich panel structures. Acta Astronaut 116:222–228

Froehlich A (2015) Das Verhältnis ESA-EU: von Neapel bis Luxemburg. German J Air Space Law 64:54–65

Gates M et al (2015) The asteroid redirect mission and sustainable human exploration. Acta Astronaut 111:29–36

Ghosh P, Conway BA (2015) Spatial statistical point prediction guidance for heating-rate-limited aeroassisted orbital transfer. Acta Astronaut 111:257–269

Golkar A, Lluch i Cruz I (2015) The Federated Satellite Systems paradigm: concept and business case evaluation. Acta Astronaut 111:230–248

Gosh A, Coverstone V (2015) Optimal cooperative CubeSat maneuvers obtained through parallel computing. Acta Astronaut 107:130–149

Graham AR, Kingston J (2015) Assessment of commercial viability of selected options for on-orbit servicing (OOS). Acta Astronaut 117:38–48

Griffin J (2015) Reflections on the authorship of space technology. Space Policy 33:4–7

Gustetic JL et al (2015) Outcome-driven open innovation at NASA. Space Policy 34:11–17

Hansen R (2015) The role of air-space boundary in regulating military use of outer space. Annals of air and space law, vol XL

Hansen R, Wouters J, de Man P (2015) Why the military use of outer space matters today. (Editorial perspective). Annals of air and space law, vol XL

Harris HE, Russo P (2015) The influence of social movements on space astronomy policy: the cases of "Hubble Huggers", JWST's "Science Warriors" and the ISEE-3 "Reboot Team". Space Policy 31:1–4

Harris R, Baumann I (2015) Open data policies and satellite Earth observation. Space Policy 32:44–53

Hesar SG et al (2015) Lunar far side surface navigation using Linked Autonomous Interplanetary Satellite Orbit Navgation (LiAISON). Acta Astronaut 117:116–129

Hobe S (2015) The meaning of "peaceful purposes" in Article IV of the outer space treaty. Annals of air and space law, vol XL

Horneck G et al (2015) European landscape in astrobiology, results of the AstRoMap consultation. Acta Astronaut 110:145–154

Hornig A, Homeister M (2015) TYCHO: demonstrator and operational satellite mission to Earth-Moon-Liberation point EML-4 for communication relay provision as a service. Acta Astronaut 108:156–170

Howell E, Whalen D (2015) Caution light: the Anik E2 solar disruption and its effects on Telesat Canada. Space Policy 32:32–36

Ikenaga T et al (2015) Interplanetary parking method and its applications. Acta Astronaut 116:271–281

Inamori T et al (2015a) Electromagnetic panel development and retraction in satellite missions. Acta Astronaut 109:14–24

Inamori T et al (2015b) Magnetic plasma deorbit system for nano- and micro-satellites using magnetic torquer interference with space plasma in low Earth orbit. Acta Astronaut 112:192–199

Ivanov D et al (2015) Analytical study of microsatellite attitude determination algorithms. Acta Astronaut 116:339–348

Jakhu R, Chen K-W (2015) How much is a picture worth? Selected legal issues of sourcing military intelligence from commercial satellites. Annals of air and space law, vol XL

Jaramillo C (2015) The multifaceted nature of space security challenges. Space Policy 33:63–66

Johannsson M et al (2015) Space and open innovation: potential, limitations and conditions of success. Acta Astronaut 115:173–184

Jones M (2015) Reconsidering macro artefacts in SETI searches. Acta Astronaut 116:161–165

Kading B, Straub J (2015) Utilizing in-situ resources and 3D printing structures for a manned Mars mission. Acta Astronaut 107:317–326

Kaiser SA (2015) Legal and policy aspects of space situational awareness. Space Policy 31:5–12

Kaiser SA, Mejia-Kaiser M (2015) Cyber security in air and space law. German J Air Space Law 64:396–410

Kezerashvili RY (2015) Space exploration with a solar sail coated by materials that undergo thermal desorption. Acta Astronaut 117:231–237

Kilibarda P (2015) Militarization of outer space and the liability convention. Air Space Law 40:271–290

Klas M et al (2015) Biomining and methanogenesis for resource extraction from asteroids. Space Policy 34:18–22

Kumar A (2015) Appropriation of space? Apollo Lunar Landing Legacy Bill as a trigger for colonization of space? German J Air Space Law 64:653–665

Lambright WH (2015) Launching commercial space: NASA, cargo, and policy innovation. Space Policy 34:23–31

Langston S, Pell SJ (2015) What is in a name? Perceived identity, classification, philosophy, and implied duty of the "astronaut". Acta Astronaut 115:185–194

Lara M (2015) On inclination resonances in Artificial Satellite Theory. Acta Astronaut 110:239–246

Larsen PB (2015) Berlin space protocol: update. German J Air Space Law 64:361–395

Lawrence DJ et al (2015) High resolution mapping of lunar polar hydrogen with a low-resource orbital mission. Acta Astronaut 115:452–462

Lee DY et al (2015) Maximizing photovoltaic power generation of a space-dart configured satellite. Acta Astronaut 111:283–299

Lee S (2015) Space at your fingertips: assessing the public's interest in space activities. Space Policy 34:39–46

Leigh AM, Black JT (2015) Navigation solution to maneuver a spacecraft relative to multiple satellite and ground locations. Acta Astronaut 109:1–13

Lindenroth L et al (2015) A novel approach to activate deep spinal muscles in space – results of a biomechanical model. Acta Astronaut 116:202–210

Loeb HW et al (2015) A realistic concept of manned Mars mission with nuclear-electric propulsion. Acta Astronaut 116:299–306

Logsdon JM (2015) Why did the United States retreat from the Moon? Space Policy 32:1–5

Lundquist CA (2015) Rocketry mentors, rocketry practitioners and unique space pioneers. Acta Astronaut 115:241–246

Macauley MK (2015) The economics of space debris: estimating the costs and benefits of debris mitigation. Acta Astronaut 115:160–164

Maccone C (2015) Statistical Drake – Seager Equation for exoplanet and SETI searches. Acta Astronaut 115:277–285

Machay M, Hajko V (2015) Transatlantic space cooperation: an empirical evidence. Space Policy 32:37–42

MacLeish MM et al (2015) Global space workforce development: a model for partnership building and knowledge transfer to developing space-faring societies. Acta Astronaut 116:106–116

Marchisio S (2015) Security in space: issues at stake. Space Policy 33:67–69

Marmet FX et al (2015) GPS/Galileo navigation in GTO/GEO orbit. Acta Astronaut 117:263–276

Martinez P (2015) The development of space law in South Africa. German J Air Space Law 64:353–360

Mayorova VI et al (2015) Importance of joint efforts for balanced process of designing and education. Acta Astronaut 111:249–256

Mazanek DD et al (2015) Asteroid Redirect Mission concept: a bold approach for utilizing space resources. Acta Astronaut 117:163–171

Messerschmitt DG (2015) Design for minimum energy in interstellar communication. Acta Astronaut 107:20–39

Moltz JC (2015) Brazil's space program: dreaming with its feet on the ground. Space Policy 33:13–19

Nagendras NP et al (2015) Indian space transportation programme: near term outlook and issues for commercialisation. Acta Astronaut 110:161–167

Nazarenko AI (2015) How can we increase the accuracy of determination of spacecraft's lifetime? Acta Astronaut 116:229–236

Newman CJ (2015) Seeking tranquility: embedding sustainability in lunar exploration policy. Space Policy 33:29–37

Ning X et al (2015) An adaptive filter method for spacecraft using gravity assist. Acta Astronaut 109:103–111

Okninski A et al (2015) Development of the Polish Sounding Rocket Program. Acta Astronaut 108:46–56

Pace S (2015) Security in space. Space Policy 33:51–55

Palermo G, Golkar A, Gaudenzi P (2015) Earth orbiting support systems for commercial low Earth orbit data relay: assessing architectures through tradespace exploration. Acta Astronaut 111:48–60

Panov DV et al (2015) Large-scale shielding structures in low Earth orbits. Acta Astronaut 109:153–161

Paton BE, Lobanov LM, Volkov VS (2015) Metal transformable-volume structures for space engineering. Acta Astronaut 110:50–57

Paul J et al (2015) INVERITAS: a facility for hardware-in-the-loop long distance movement simulation for rendezvous and capture of satellites and other autonomous objects. Acta Astronaut 116:1–24

Plattard S (2015) Security in space: should space traffic management also concern payloads management? Space Policy 33:56–62

Quero JM et al (2015) CEPHEUS, a multi-project satellite for technology qualification. Acta Astronaut 117:238–242

Robinson GS (2015) The devolution of space law positivisms and a reassessment of space law philosophy: natural law theory roots of space jurisprudence. Annals of air and space law, vol XL

Rogers BY et al (2015) Establishing cycler trajectories between Earth and Mars. Acta Astronaut 112:114–125

Sandalinas J (2015) Satellite imagery and its use as evidence in proceedings of the International Criminal Court. German J Air Space Law 64:666–675

Schmitz M, Fasoulas S, Utzmann J (2015) Performance model for space-based laser debris sweepers. Acta Astronaut 115:376–383

Scholz A, Juang J (2015) Toward open source CubeSat design. Acta Astronaut 115:384–392

Schreiner SS et al (2015) An overnight habitat for expanding lunar surface exploration. Acta Astronaut 112:158–170

Schrogl K-U (2015) The 2014 and 2015 sessions of the UNCOPUOS Legal Subcommittee – a personal assessment. German J Air Space Law 64:481–488

Scott RL, Ellery A (2015) An approach to ground based space surveillance of geostationary on-orbit servicing operations. Acta Astronaut 112:56–68

Shepherd GG (2015) Development of wind measurement systems for future space missions. Acta Astronaut 115:206–217

Smirnov NN et al (2015) Space traffic hazards from orbital debris mitigation strategies. Acta Astronaut 109:144–152

Spudis PD (2015) The Moon as an enabling asset for space flight. Space Policy 32:9–10

Srinivasan DK, Adams N, Wallis R (2015) In-flight performance of the Van Allen Probes RF telecommunications system. Acta Astronaut 116:211–221

Stephens D, Steer C (2015) Conflicts in space: international humanitarian law and its application to space warfare. Annals of air and space law, vol XL

Straub J (2015) Does the use of space solar power for in-space activities really make sense: an updated economic assessment. Space Policy 31:21–26

Su J (2015) Measures proposed for planetary defence: obstacles in existing international law and implications for space arms control. Space Policy 34:1–5

Suedfeld P et al (2015) Coping strategies during and after spaceflight: data from retired cosmonauts. Acta Astronaut 110:43–49

Summerer L et al (2015) The International Safety Framework for nuclear power source applications in outer space – useful and substantial guidance. Acta Astronaut 111:89–101

Sun R et al (2015) Algorithms and applications for detecting faint space debris in GEO. Acta Astron 110:9–17

Supancanda IBR (2015) How the progressive development of outer space law affects the formulation of national space legislation: the experience of Indonesia. Air Space Law 40:93–106

Tabassi L, Werzi R (2015) The prohibition of nuclear testing in outer space. Annals of air and space law, vol XL

Tafforin C et al (2015) Correlation of etho-social and psycho-social data from "Mars-500" interplanetary simulation. Acta Astronaut 111:19–28

Tanaka T, Kawamura Y, Tanaka T (2015) Development and operations of nano satellite FITSAT-1 (NIWAKA). Acta Astronaut 107:112–129

Tardivel S et al (2015) Human exploration of near Earth asteroids: architecture of proximity operations. Acta Astronaut 110:18–28

Tronchetti F (2015) The Space Resource Exploration and Utilization Act: a move forward or a step back? Space Policy 34:6–10

Tronchetti F, Hao L (2015) The 2014 updated draft PPWT: hitting the spot or missing the market? Space Policy 33:38–49

Ulamec S et al (2015) Rosetta lander Philae: landing preparations. Acta Astronaut 107:79–86

van der Ha JC (2015) Lessons learned from the dynamical behaviour of orbiting satellites. Acta Astronaut 115:121–137

Visagie L, Lappas V, Erb S (2015) Drag sails for space debris mitigation. Acta Astronaut 109:65–75

Vrolijk A, Szajnfarber Z (2015) When policy structures technology: balancing upfront decomposition and in-process coordination in Europe's decentralized space technology ecosystem. Acta Astronaut 106:33–46

Whitman Cobb WN (2015) Trending now: using big data to examine public opinion of space policy. Space Policy 32:11–16

Wille E, Bavdaz M (2015) European x-ray optics for next generation space observatories. Acta Astronaut 116:50–55

Wu X (2015) China and space security: how to bridge the gap between its stated and perceived intentions. Space Policy 33:20–28

Wu Y et al (2015) Low complexity code tracking loop with multipath diversity for GNSS over multipath fading channels. Acta Astronaut 115:367–375

Yau AW et al (2015) The Canadian CASSIOPE small satellite mission: the enhanced polar outflow probe and Cascade technology demonstration payloads. Acta Astronaut 110:155–160

Yee K (2015) America COMPETES Act's effect on NASA's Education and Public Outreach Programs. Space Policy 31:27–30

Yousefian P, Salarieh H (2015) Anti-sway control of tethered satellite systems using attitude control of the main satellite. Acta Astronaut 111:300–307

Yu Z, Zhu S, Cui P (2015) Orbit optimization of Mars orbiters for entry navigation: from an observability point of view. Acta Astronaut 111:136–145

Zannoni D (2015) International law issues concerning the interruption and the degradation of the radio-navigation signal. German J Air Space Law 64:489–510

Zhang F et al (2015) On-line estimation of inertia parameters of space debris for its tether-assisted removal. Acta Astronaut 107:150–162

Biographies

About the Editors

Cenan Al-Ekabi joined the European Space Policy Institute in Vienna in 2011, functioning as a project manager from 2012, and a Resident Fellow from 2013. Prior to that, he obtained two Advanced L.L.M. degrees in Air & Space Law, and European & International Business Law from Leiden University in the Netherlands. He also holds a U.S. JD with concentration in studies in international law from the Thomas M. Cooley law school, and a bachelor's degree in Political Science from McMaster University in Canada.

Blandina Baranes joined the European Space Policy Institute (ESPI) in Vienna in February 2005 and currently holds the position of ESPI Communications Manager. Prior to this, she was the chief librarian of the Jewish Studies Department of the University of Vienna. During the past years she has also worked as a documentalist and librarian for different institutions, such as the Austrian Broadcasting Corporation (ORF), the Der Spiegel magazine and others. She conducted her studies and research in Austria and Israel and graduated with a Master's Degree from Vienna University, Faculty of Philosophy, Department of Social and Cultural Anthropology. She is co-editor of the past three volumes of ESPI's "Yearbook on Space Policy", published several articles and participated as a speaker in a number of space related events on the topic of "Space and Interdisciplinarity".

Peter Hulsroj pursued his legal studies at the University of Copenhagen from 1974 to 1979. After military service, Peter Hulsroj did post-graduate studies at Harvard Law School in the USA, where he earned a Master's Degree in 1981. Returning from the USA, Mr Hulsroj joined a large commercial law firm in Copenhagen, where he became a full member of the Danish Bar. After 3½ years of private practice in Copenhagen, Mr Hulsroj became a Contracts Officer in ESTEC, and thereafter served for almost 14 years as the Head of Contracts and Legal Affairs at Eumetsat in Darmstadt, Germany. Before rejoining ESA, Mr Hulsroj was the Legal Adviser of the Preparatory Commission for the Comprehensive Nuclear-Test-Ban Treaty Organization in Vienna. Peter Hulsroj was appointed ESA Director of Legal Affairs and External Relations (D/LEX) in April 2008, comprising four specialist areas: the Legal Department, the International Relations Department, the Communication Department and the Education and Knowledge Management Office at ESA from 2008 to 2011. Since 2011 Peter Hulsroj is Director of ESPI.

Arne Lahcen is Resident Fellow at the European Space Policy Institute in Vienna, Austria. His research interests include Earth observation governance, international cooperation schemes,

C. Al-Ekabi et al. (eds.), *Yearbook on Space Policy 2015*, Yearbook on Space Policy, DOI 10.1007/978-3-7091-4860-0

innovation management, and the relationship between space and society. Prior to joining ESPI in 2011, he obtained an Advanced Master's degree in Space Studies with a specialisation in space law, policy, business and management at the Faculty of Science of the Catholic University of Leuven, Belgium. He also holds a Bachelor's and Master's degree in Social-Economic Sciences pursued at the Faculty of Applied Economics, University of Antwerp, Belgium. In addition to ESPI research activities, he is editor-in-chief of the ESPI Perspectives series and co-responsible for the follow-up of the ESPI Autumn Conference and the editing of the ESPI Yearbook.

About the Authors

Jean-Marc Astorg is currently Director of Launchers in the French Space Agency (Centre national d'études spatiales—CNES). Mr Astorg was graduated from the Ecole Centrale des Arts et Manufactures in 1985 and joined the CNES Launch Vehicles Directorate the same year. Successively, he held various managerial positions in the field of European launchers, responsible alternately for programmes that were preparing the future or for developing launchers. From 1991 to 1998 he supervised the studies that were to lead to programme decisions for the new version ARIANE 5ECA, and the small Vega launcher. In 1998, he became project leader for development of the new ESC-A upper stage of ARIANE 5 and led its development right through to qualification. This new stage enabled ARIANE 5 to increase its throw-weight capacity for geostationary transfer orbits by more than 3 tons. ARIANE 5 has successfully completed every mission since 2005 and will therefore be used until ARIANE 6 becomes available. He also directed the 'Soyuz in French Guiana' project, from its inception in 2003 up to the successful first launch in October 2011. In November 2011, he was appointed CNES Director of Procurement, Sales and Legal Affairs. He was also CNES Director in charge of Intellectual Property. As such, he negotiated and implemented agreements with other space-related manufacturers regarding contracts and intellectual property. He was also the CNES Mediator for relations with SMEs. On 7 April 2015, he was appointed Head of the CNES Launch Vehicles Directorate. Jean-Marc ASTORG is a Chevalier de l'Ordre National du Mérite.

Cristina T. Chaplain currently serves as a Director, Acquisition and Sourcing Management, at the U.S. Government Accountability Office. She has responsibility for GAO assessments of military space acquisitions, NASA, and the Missile Defense Agency. Among other topics, she has led reviews on the international space station, the Space Launch System, acquisition progress for major NASA projects, the James Webb telescope, commercial cargo and crew, NASA contract management, contract terminations and the evolved expendable launch vehicle. In addition to her work on space and missile system development, Ms. Chaplain has led a variety of DOD-wide contracting-related and best practice evaluations for the GAO. Before her current position, Ms. Chaplain worked with GAO's financial management and information technology teams. Ms. Chaplain has been with the GAO for 23 years. She received a bachelor's degree, magna cum laude, in International Relations from Boston University and a Master's Degree in Journalism from Columbia University.

Richard DalBello is Vice President of Business Development and Government Affairs at Virgin Galactic. In this position he is responsible for managing the company's interactions with the U.S. government. Prior to joining Virgin, Richard served as Director, Space and Aeronautics, in the White House Office of Science and Technology Policy. In this role, he served as the principal advisor on space and aeronautics matters to Dr. John Holdren, the Science Advisor to the President. In this position, Richard played a key role in the development and implementation of the Administration's domestic and international space policy and program priorities.

Prior to joining OSTP, Richard served as the Vice President of Government Affairs for Intelsat General. He served previously as president of the Satellite Broadcasting and Communications Association, and for more than 3 years as the president of the Satellite Industry Association. Earlier, Richard was General Counsel for Spotcast Communications Inc., and Vice President of

Government Affairs, North America, for ICO Global Communications, a provider of mobile satellite communications services. He also served for 4 years as the Assistant Director for Space and Aeronautics in the Clinton White House's Office of Science and Technology Policy. Richard earned a bachelor's degree in political science from the University of Illinois, a master's degree in law from McGill University, and a doctorate in jurisprudence from the University of San Francisco.

Keith Hayward is a consultant and writer on aerospace and aviation issues. He was formally Professor of International Relations at Staffordshire University, Head of Economic and Political Affairs at the UK aerospace trade association and until January 2015, Head of Research at the Royal Aeronautical Society in London. He has been a consultant or advisor to several company and government departments, the latter including the UK Ministry of Defence and the Department of Business, Innovation and Science. He has also acted as an advisor to the UK House of Commons Trade and Industry Committee and the U.S. Congress Office of Technology Assessment. He has taken part in two recent collaborative studies of the space industry on behalf of the Commission of the European Union and the European Space Agency. He is the author of several books and over 100 articles and chapters on aerospace and aviation issues. He is a Fellow of the Royal Aeronautical Society and a corresponding member of the French Air and Space Academy.

Laura Hook currently serves as a Senior Analyst specializing in defense space acquisitions at the U.S. Government Accountability Office. In almost 10 years at the agency, she has led and worked on numerous reviews of U.S. space programs. Most notably, she has reviewed many facets of the U.S. Air Force's Evolved Expendable Launch Vehicle programme, including its transition to competitive acquisitions. In addition, she has worked on reviews of other topics and programmes such as the Global Positioning System, satellite operations and control, commercial space topics including launch and remote sensing, and broader analyses of the U.S. space community and government space responsibilities. Ms. Hook earned a bachelor's degree in international political science and Chinese from Trinity University, and a master's degree in international security from the University of Denver.

Stefaan de Mey holds a PhD in Engineering, Master's degrees in Biomedical Engineering and Civil Engineering, and a Bachelor degree in Law. Dr. de Mey started his career at the Institute of Biomedical Technology (IBITECH, Ghent University, Belgium) conducting research in the field of cardiovascular fluid dynamics and the non-invasive diagnosis of heart failure using ultrasound. In 2001 he joined the European Space Agency's (ESA) Space Research and Technology Centre in Noordwijk, The Netherlands. During this period he worked on various assignments including the development of medical instrumentation for the International Space Station (ISS), the end-to-end system engineering management of the EXPERT re-entry vehicle, the phase B procurement of an observatory to be placed on the exterior of the ISS aiming at measuring high altitude lightning, and providing the integration management function of the payloads and experiments selected for ISS. In 2011, on a secondment from the ESA Director General cabinet, Dr. De Mey has been heading as Secretary General Eurisy, an international association of governmental space offices, space agencies and research institutions with offices in Paris, France. Eurisy's mission is to facilitate the access of professional public and private communities to satellite information and services thus helping to ensure that society fully benefits from European investments in space. Since October 2016, Dr. De Mey is appointed Senior Strategy Officer at ESA's Directorate of Human Spaceflight and Robotic Exploration Programmes (D/HRE) in Noordwijk, the Netherlands, to coordinate the further development and provision of the D/HRE strategy area for human spaceflight and robotic exploration.

Tony Milligan is the co-editor of The Ethics of Space Exploration (2016) and the author of Nobody Owns the Moon: The Ethics of Space Exploitation (2015), The Next Democracy? (2016), Animal Ethics: The Basics (2015), Civil Disobedience: Protest, Justification and the Law (2013), Love (2011), Beyond Animal Rights (2010), as well as the author and co-author of numerous articles on the complex relations between the human and the non-human. He is currently a Teaching Fellow in Ethics and the Philosophy of Religion with the Department of Theology and Religious Studies at King's College London, and is currently working on a co-edited volume on Robots, Relationships and Power.

Philipp Olbrich is a PhD student in the Department of International Relations and International Organization at the University of Groningen. Prior to that he was a researcher at the Austrian Institute for International Affairs in Vienna. His doctoral dissertation focuses on the use and implications of commercial satellite imagery by non-state actors in the context of international security. His broader research interests concern the nexus of international security and technology, governance of space-related activities, new materialist approaches in International Relations and the conflict on the Korean peninsula.

Jesse Phaler has more than 18 years of experience at corporate management in intergovernmental agencies. He is currently the Head of Industrial Return Management Office at the European Space Agency. Since joining the European Space Agency in 2001 he has held several positions in the corporate domain and participated to the strategic planning and formulation and contributed to several corporate reforms aiming at adapting the Agency to its changing environment. Jesse Phaler has a Master degree in Advanced European and International Studies and a Bachelor of Science degree in Business Administration.

Isabella Rebasso has been an intern at the Austrian Institute for International Affairs. She recently graduated in Political Science from University of Vienna. During her studies she has focused on intersections of International Relations and Science and Technology.

James S. J. Schwartz completed his PhD in philosophy at Wayne State University and currently teaches philosophy at Wichita State University. His primary area of research is the philosophy and ethics of space exploration; he also has research interests in environmental ethics, philosophy of mathematics, and metaphyics. Schwartz is lead editor (with Tony Milligan) of The Ethics of Space Exploration (Springer, 2016). His publications have appeared in Philosophia Mathematica, Environmental Ethics, Ethics & the Environment, Space Policy, Advances in Space Research, and in several volumes of Springer's Space and Society series.

Leopold Summerer is currently head of ESA's Future Preparation and Strategic Studies office, and of its Advanced Concepts Team, (ESTEC). Holding a PhD from the University of Vienna, master degrees in theoretical physics and nuclear physics and space studies, he directs and conducts research on advanced space systems and concepts, including artificial intelligence, space power systems, brain machine interfaces, biomimetics, computer science, advanced propulsion, fundamental physics and advanced space propulsion. He conducted European and international space experiments, and has been involved in space strategy developments. Author of close to 100 publications and technical officer of over 100 ESA studies.

Nina Witjes is a Research Associate (postdoc) at the Chair of Innovation Research, Munich Center for Technology in Society (MCTS), Technical University Munich. Before that she was a researcher at the Austrian Institute for International Affairs (oiip) in Vienna since 2013 and head of the research group on Science, Technology and Foreign Policy. Nina holds an MA in Development Studies from Vienna University. From 2009 to 2012, Nina has been a researcher the Austrian Research Foundation for International Development. Her PhD thesis in Science and Technology Studies (STS) thesis is on the role of science, technology and innovation in global politics. Her main areas of interest are satellite imagery analysis on issues of international security, the use of geospatial big data for crisis response, European space policy and geopolitical infrastructures.

Shuang-Nan Zhang is currently distinguished Professor and Director of Key Laboratory of Particle Astrophysics and Division for Particle Astrophysics, Institute of High Energy Physics, Chinese Academy of Sciences, Beijing, China. Professor's Zhang past education and work experience includes: 1984, Bachelor Degree, Tsinghua University; 1989, PhD Degree, University of Southampton, U.K.; 1989–1992, Postdoctoral research associate, University of Pennsylvania, USA; 1992–1998, Senior scientist, Universities Space Research Association and NASA Marshall Space Flight Center, Huntsville, Alabama, USA; 1998–2014, Tenure-track assistant professor, Research associate professor, Research full professor, University of Alabama in Huntsville, USA; 2002–2009, Distinguished professor, Physics Department, Tsinghua University.

MIX
Papier aus verantwortungsvollen Quellen
Paper from responsible sources
FSC® C105338

If you have any concerns about our products,
you can contact us on
ProductSafety@springernature.com

In case Publisher is established outside the EU,
the EU authorized representative is:
Springer Nature Customer Service Center GmbH
Europaplatz 3, 69115 Heidelberg, Germany

Printed by Libri Plureos GmbH
in Hamburg, Germany